Agriscience & Technology

2nd Edition

Delmar Publishers
is proud to support
FFA Activities

Agriscience & Technology

2nd Edition

L. DeVere Burton

Delmar Publishers

an International Thomson Publishing company 

Albany • Bonn • Boston • Cincinnati • Detroit • London • Madrid
Melbourne • Mexico City • New York • Pacific Grove • Paris • San Francisco
Singapore • Tokyo • Toronto • Washington

NOTICE TO THE READER

Publisher does not warrant or guarantee any of the products described herein or perform any independent analysis in connection with any of the product information contained herein. Publisher does not assume, and expressly disclaims, any obligation to obtain and include information other than that provided to it by the manufacturer.

The reader is expressly warned to consider and adopt all safety precautions that might be indicated by the activities herein and to avoid all potential hazards. By following the instructions contained herein, the reader willingly assumes all risks in connection with such instructions.

The publisher makes no representation or warranties of any kind, including but limited to, the warranties of fitness for particular purpose or merchantability, nor are any such representations implied with respect to the material set forth herein, and the publisher takes no responsibility with respect to such material. The publisher shall not be liable for any special, consequential, or exemplary damages resulting, in whole or part, from the readers' use of, or reliance upon, this material.

Cover art courtesy: Michael Dzaman

Delmar Staff
Publisher: Tim O'Leary
Acquisitions Editor: Cathy L. Esperti
Senior Project Editor: Andrea Edwards Myers
Production Manager: Wendy A. Troeger
Marketing Manager: Maura Theriault

Printed in the United States of America

For more information, contact:

Delmar Publishers
3 Columbia Circle, Box 15015
Albany, New York 12212-5015

International Thomson Publishing Europe
Berkshire House 168-173
High Holburn
London WC1V 7AA
England

Thomas Nelson Australia
102 Dodds Street
South Melbourne, 3205
Victoria, Australia

Nelson Canada
1120 Birchmount Road
Scarborough, Ontario
Canada M1K 5G4

International Thomson Editores
Campos Eliseos 385, Piso 7
Col Polanco
11560 Mexico D F Mexico

International Thomson Publishing Gmbh
Konigswinterer Strasse 418
53227 Bonn
Germany

International Thomson Publishing Asia
221 Henderson Road #05-10
Henderson Building
Singapore 1315

International Thompson Publishing—Japan
Hirakawacho Kyowa Building, 3F
2-2-1 Hirakawacho
Chiyoda-ku, 102 Tokyo
Japan

1 2 3 4 5 6 7 8 9 10 XXX 02 01 00 99 98 97

Library of Congress Cataloging-in-Publication Data
Burton, L. DeVere.
Agriscience & technology / L. DeVere Burton.—2nd ed. p. cm.
Includes index.
ISBN 0-8273-6747-3
1. Agriculture. 2. Agricultural biotechnology. 3. Agriculture and energy. 4. Agricultural processing. 5. Technology. I. Title.
S495.B97 1997
630-dc20 96-17752
CIP

Table of Contents

Preface

The greatest accomplishment of Americans in this century has been the building of an agricultural dynasty that is the envy of the world. It is the greatest effort in efficiency in the history of the planet. Never before have so few farmers produced so much food and fiber for so many people.

The movement of our population from the farms to the cities has been painful to some, but it has allowed our people to imagine and accomplish other great feats. Humans would not have walked on the surface of the moon or sent space probes into the vast blackness of the solar system had they been worried about their next meals.

Because of the abundance of inexpensive food, much of the national wealth and energy has been available to invest in business and industry. Because we are free from hunger and oppression, we are free to dream, free to achieve any noble cause. What is "right with America" is that the "right to imagine" has been nurtured and preserved.

Three of the most significant contributors to the success of our great agricultural industry are these:

- scientific research
- agricultural technologies
- agricultural education

Scientific research has led to a greater understanding of our most basic agricultural resources: soils, water, plants, animals, and people. As we have learned more about each of these important resources, we have learned more about the ways that they interact with each other. We have also become more conscious of the needs of the environment in which we live.

Agricultural technologies have grown out of basic research. They include the application of scientific principles to solve agricultural problems. Some of the greatest inventions the world has ever known have solved simple but significant problems such as tilling the soil, harnessing energy sources to perform work, and separating seeds from plant materials.

Agricultural education has succeeded in delivering new knowledge and technologies from the research centers and universities to the farms and ranches. Many of the people who produce agricultural goods today gained experience while they were young through 4-H and FFA programs. Many have been educated through college and university agriculture programs.

The story of the evolution of agriculture is truly exciting. This textbook is about all of these things and more. It focuses on basic science, agricultural research, and the technologies that have been developed to meet agricultural needs. Each of these subjects is exciting, but when they are considered together, a symbiotic relationship is discovered.

The national trend to modernize the secondary agriculture curriculum to emphasize instruction in science and technology has contributed to a need for a textbook that focuses on those subjects. This textbook has been written with that theme in mind.

Chapter 1 is historical in nature. It includes discussion of major technological advances in agriculture from the past to the present. It also focuses on modern agricultural research that will impact the agricultural industry in the immediate future.

The book is divided into six sections:

Section I—Introduction
Section II—Biotechnology
Section III—Technology: Food and Fiber
Section IV—Energy and Power Technology
Section V—Computer Aided Management
Section VI—Environmental Technology

Included in each chapter are features entitled "Objectives," "Terms For Understanding," "Career Options," "Chapter Summary," "Chapter Review," and "Learning Activities." In addition, each chapter contains photographs and illustrations to aid students in understanding the concepts that are discussed. This text will expose students to scientific concepts and technologies that are the foundation of the modern industry of agriculture.

About the Author

L. DeVere Burton, author of *Agriscience & Technology 2E,* is State Supervisor, Agricultural Science & Technology in the Idaho State Division of Vocational Education. He is a past-president of the National Association of Supervisors of Agriculture Education.

The author was a high school agriculture teacher for 15 years, and he has been involved as a professional educator in agricultural education since 1967. He has experienced teaching assignments in large and small schools, and in single and multiple teacher departments. He has taught at four different high schools, and at a major land grant university. Since 1987, he has been involved in program supervision since 1987. All of these experiences have contributed to his philosophy that "education must be fun and exciting for those who learn and for those who teach."

His first experience with agricultural education occurred when his high school agriculture teacher recruited him to a ninth grade agriculture class at Star Valley High School in Afton, Wyoming. By the end of his second year as an FFA member, he had discovered his career in agricultural education. He has never looked back, or regretted his choice.

A wide range of agricultural experiences has prepared the author for his career as an educator in agriculture. He was raised on a farm in Western Wyoming that produced hay, grain, milk, eggs, and a few hogs and sheep. His high school jobs included testing milk for butterfat content, and caring for the hogs on a combination beef, swine, and trout ranch. During college he worked as a maintenance/warehouse worker in a feed mill, manager of a dairy farm, sawmill worker, logger, finish carpenter, and animal research assistant. He has worked in the food processing, metal fabrication, and concrete industries, and he owned and managed a purebred sheep and row crop farm for several years.

Dr. Burton earned his B.S. degree in Agricultural education from Utah State University in 1967. He was awarded his M.S. degree in Animal Science from Brigham Young University in 1972. His Ph.D. degree was earned at Iowa State University in 1987 where he was also an instructor in the Agricultural Engineering Department.

This is the first of two textbooks that Dr. Burton has authored. His second text is entitled *Ecology of Fish and Wildlife.* Both books were written in a serious attempt to strengthen the science content, and to expand the breadth of the curriculum in agricultural education.

Acknowledgments

This textbook is dedicated to serious agriculture students in the hope that they will become as excited and intrigued with learning the science and technology of agriculture as the author has been during the preparation, writing, and revision of this book. The concepts contained herein have been gleaned from reading numerous books and journals, and from discussions with agricultural educators and other agricultural professionals over a period of many years.

Grateful acknowledgment is extended to colleagues, family members, and personal friends who have contributed patience and encouragement to the completion of this work. The reviewers who read the manuscript and provided written comments performed a valuable service to the author and the publisher.

A hearty thank you is due to all who have contributed ideas, materials, technical expertise, and encouragement during the writing and production of this book:

Gaylen Smyer, Agriculture Teacher, Burley, Idaho
Weldon Slight, Ph.D., Utah Agricultural Experiment Station, Logan, Utah
Doyle Matthews, Ph.D., Utah State University, Logan, Utah
Gary Newenswander, Utah Agricultural Experiment Station, Logan, Utah
Utah State University Historical Farm, College Ward, Utah
Utah Agricultural Experiment Station, Logan, Utah
KAID Television Station, Boise, Idaho
Wayne Rush, University of Idaho, Moscow, Idaho
Ronald L. Crawford, Ph.D., University of Idaho, Moscow, Idaho
Agricultural & Extension Education Dept., University of Idaho, Moscow, Idaho
Philip Berger, University of Idaho, Moscow, Idaho
Robert S. Zemetra, Ph.D., University of Idaho, Moscow, Idaho
James E. Butler, Ph.D., University of Idaho, Moscow, Idaho
University of Idaho Dairy Farm, Moscow, Idaho
Washington State University, Pullman, Washington
J. R. Simplot Co., Boise, Idaho
Kelly L. Olson, Boise, Idaho
William Grange, Boise, Idaho
Gregory Nelson, D.V.M., Idaho Department of Agriculture
United States Department of Agriculture
Idaho Department of Fish and Game
1986 Yearbook of Agriculture: Research For Tomorrow
FFA Chapter, Culdesac, Idaho
FFA Chapter, Burley, Idaho
Idaho Cooperative Extension System
Darrell Boltz, Canyon County Extension Service, Caldwell, Idaho
John Throne, Boise, Idaho
The City of Boise, Idaho
Angie Richard, Melba, Idaho
Daniel E. Jones, Boise, Idaho
Bruce Durrant, Kuna, Idaho
The Epcot Center, Orlando, Florida
Monsanto Company, St. Louis, Missouri

Double D Service Center, Meridian, Idaho
USDA, Soil Conservation Service
Webster's New World Dictionary, Second Edition
DuPont Company Biotechnology Systems, Wilmington, Delaware
Verde Technologies, Inc., Watsonville, California
CESCO, Boise, Idaho
Amalgamated Sugar Company, Nampa, Idaho
Sanders Fruit Ranch, Emmett, Idaho
Lewis & Clark Terminal, Lewiston, Idaho

Thanks to the following authors of Delmar texts, from which art was obtained for use in this text:

From *Introductory Horticulture,* Fifth Edition, by H. Edward Reiley and Carroll L. Shry, copyright 1997 by Delmar Publishers.

From *Technology In Your World,* Second Edition, by Michael Hacker and Robert Barden, copyright 1992 by Delmar Publishers.

From *Agricultural Mechanics: Fundamentals and Applications,* Third Edition, by Elmer L. Cooper, copyright 1997 by Delmar Publishers.

From *Managing Our Natural Resources,* Second Edition, by William G. Camp and Thomas B. Daugherty, copyright 1997 by Delmar Publishers.

From *Agriscience: Fundamentals & Applications,* Second Edition, by Elmer L. Cooper, copyright 1997 by Delmar Publishers.

From *Advances in Agriculture,* by Daniel Burns and Patti Thomsen, copyright 1990 by Burrus Research Associates, Inc.

The services of the following reviewers of this text are gratefully acknowledged:

Dale Layfield
Branford, Florida

Celenn Linder
Pecatonica Area Schools
Blanchardville, Wisconsin

Douglas Plant
Runge, Texas

Jim Wells
Cherokee High School
Rogersville, Tennessee

Curtis Barclay
Shenandoah High School
Shenandoah, Iowa

E.C. Conner
Park View Senior High School
South Hill, Virginia

Daniel Lantis
Baker High School
Baker, Montana

Gerald McDonald
Memorial Senior High School
Houston, Texas

Danny Shipley
Allen County Scottsville High School
Scottsville, Kentucky

SECTION I

Introduction

- **Technology in an Agricultural Setting**

CHAPTER 1

Technology in an Agricultural Setting

This chapter introduces students to the study of technology as it relates to agriculture. It contains a brief history of the development of agricultural technology from the beginning of human civilization to the present time.

OBJECTIVES

After completing this chapter, you should be able to:

- identify several agricultural technologies that have had an impact upon the production of agricultural goods.
- describe the importance of agricultural technology in providing an adequate supply of food and clothing.
- relate the adoption of agricultural technologies to the increase in the availability of careers outside the field of agriculture.
- forecast possible technologies to provide solutions to current agricultural problems.
- appraise the effects of genetic engineering on environmental safety.

TERMS FOR UNDERSTANDING

The following vocabulary terms should be studied carefully as you read:

agricultural research
agricultural technology
fiber
draft animals
tillage tools
mechanical power
industrial revolution
textile industry
cotton gin
Bessemer process
land grant university
Agricultural extension educator

- internal combustion engine
- synthetic fertilizer
- hybrid
- hybrid vigor
- heterosis
- biotechnology
- clone
- gene
- chromosomes
- heredity
- genetic engineering
- infrared photography
- radiant heat
- bacteria
- enzymes

Agricultural technology is applied science. It uses the knowledge obtained from scientific research to create machines, processes, and new varieties of plants and animals. These technologies are used to improve production methods on the farms, and to improve methods of processing, transporting, and distributing agricultural goods.

EMERGENCE OF AGRICULTURAL TECHNOLOGY

Agriculture has experienced major changes during the twentieth century. During this time, the world in which we live has advanced from the era of the horse-drawn plow to the age of mechanized farming, Figures 1–1 and 1–2. Powerful tractors have replaced the horses, mules, and oxen upon which farmers once depended for power. Specialized machines have replaced the horse-drawn implements of earlier years, Figure 1–3. Computers are gaining importance on farms. They are used to gather information, keep records, analyze and calculate livestock rations, and to assist in making management and marketing decisions. Agricultural technologies include all of the tools and

FIGURE 1–1 Horses and other draft animals have long been used to provide power for humans' machines. (*Photo courtesy of Utah Agricultural Experiment Station*)

FIGURE 1–2 Manual labor was once required in almost every agricultural job. (*Photo courtesy of Utah Agricultural Experiment Station*)

FIGURE 1–3 Modern horsepower provided by tractors is vastly superior to the power of horses. (*Photo courtesy of Utah Agricultural Experiment Station*)

FIGURE 1–4 Research is the key that opened the door to modern agriculture. (*Photo courtesy of Utah Agricultural Experiment Station*)

processes that are used to produce farm products and prepare them for use by consumers.

Research is the partner of technology, Figure 1–4. Through scientific research, we study the problems of agriculture. **Agricultural research** is the study of why and how plants and animals respond to different stimuli. It is also used to investigate scientific principles and to determine relationships with living things, Figure 1–5. Evidence of science and technology in agriculture can be seen everywhere: on farms, on the highways, in the factories, and in the laboratories. The most convincing evidence of agricultural advancement through science and technology is found by observing the abundance of agricultural products found on the shelves of our food and clothing stores.

FIGURE 1–5 Animal agricultural practices have been improved by performing research on many species of farm animals. Bull testing helps to identify superior sires. (*Photo courtesy of David H. Burton*)

The most basic role of agriculture is the production of food and fiber for use by humans. Food includes a vast number of products that are eaten by people in all parts of the world. **Fiber** includes products such as cotton, wool, linen, and silk that can be made into textiles and cloth. Other common agricultural products include medicines, fuels, cosmetics, soaps, and paper products. The influence of technology in agriculture is not limited to production of food and clothing.

New farming methods have made it possible for one farmer to produce enough food and fiber for over one hundred people. Humans are no longer closely tied to the land. Because of new agricultural technologies, many people have left the farms to pursue other interests. The standard of living and leisure time activities (sports, entertainment, etc.) enjoyed by modern society would be impossible without agricultural technology, Figure 1–6.

The emergence of agricultural technology has contributed to the current trend in which farming units have become much larger in size, but are managed by fewer farmers. While this trend has resulted in social and economic adjustments in rural areas where farmers have been displaced from the land, it has also contributed to greater efficiency in the production of farm commodities. In those countries where agricultural technologies have been widely adopted, food prices are relatively low in comparison with the amount of income that is available to use for human needs.

Agricultural technologies began to emerge when early humans learned to shape tools to control the environment. Some cultures tamed wild animals, thereby providing a dependable supply of food and skins for clothing. Other early cultures learned to gather seeds and later to cultivate seed plants. Early tools

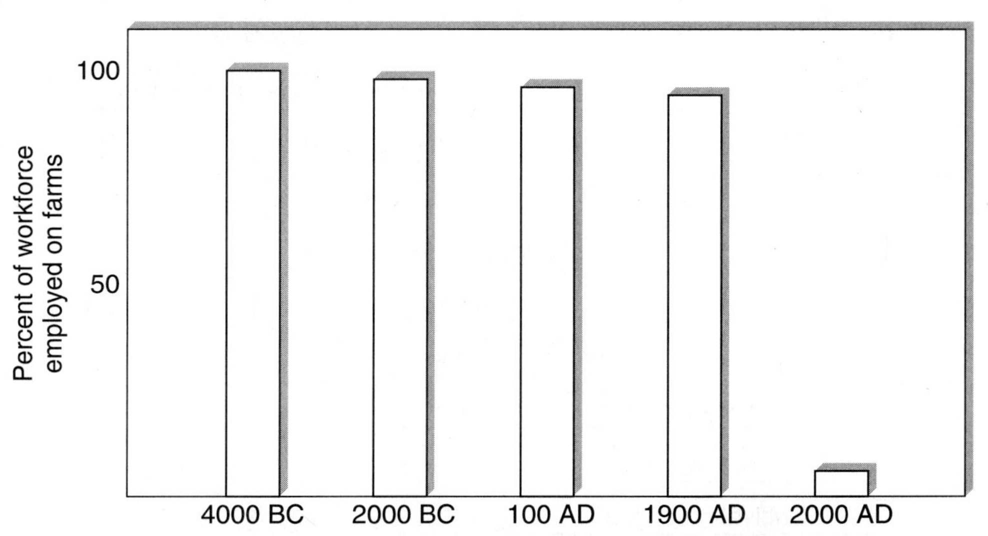

FIGURE 1–6 Modern technology in agriculture has released much of the population from farm labor, allowing them to pursue other careers and interests.

used to obtain food and clothing often consisted of little more than a stick used for planting seeds and stone tools used for cutting and grinding. Eventually bronze tools replaced stone tools. Centuries later, iron tools were developed to replace those made of bronze. Some of the most important technologies to affect human life are outlined in Figure 1–7.

Use of animals to transport people and materials led to trading between cultures. **Draft animals** that pulled simple **tillage tools** for soil preparation brought about major advancements in agricultural technology. They made it possible to cultivate much more land than had been possible using only human labor. One of the most significant developments in agriculture was the iron plow. It replaced the wooden plow, which had been in use since 3000 B.C. in Mesopotamia and Egypt.

The invention of the wheel and its use in transportation was a major technological milestone. With the wheel came carts and wagons that could be used to transport crops and trade wares.

Water wheels provided one of the earliest sources of **mechanical power** by applying the force of moving water to shafts and gears to drive machinery. Mechanical power supplemented the labor that had been previously supplied by humans and animals. Windmills added another dimension to this technology.

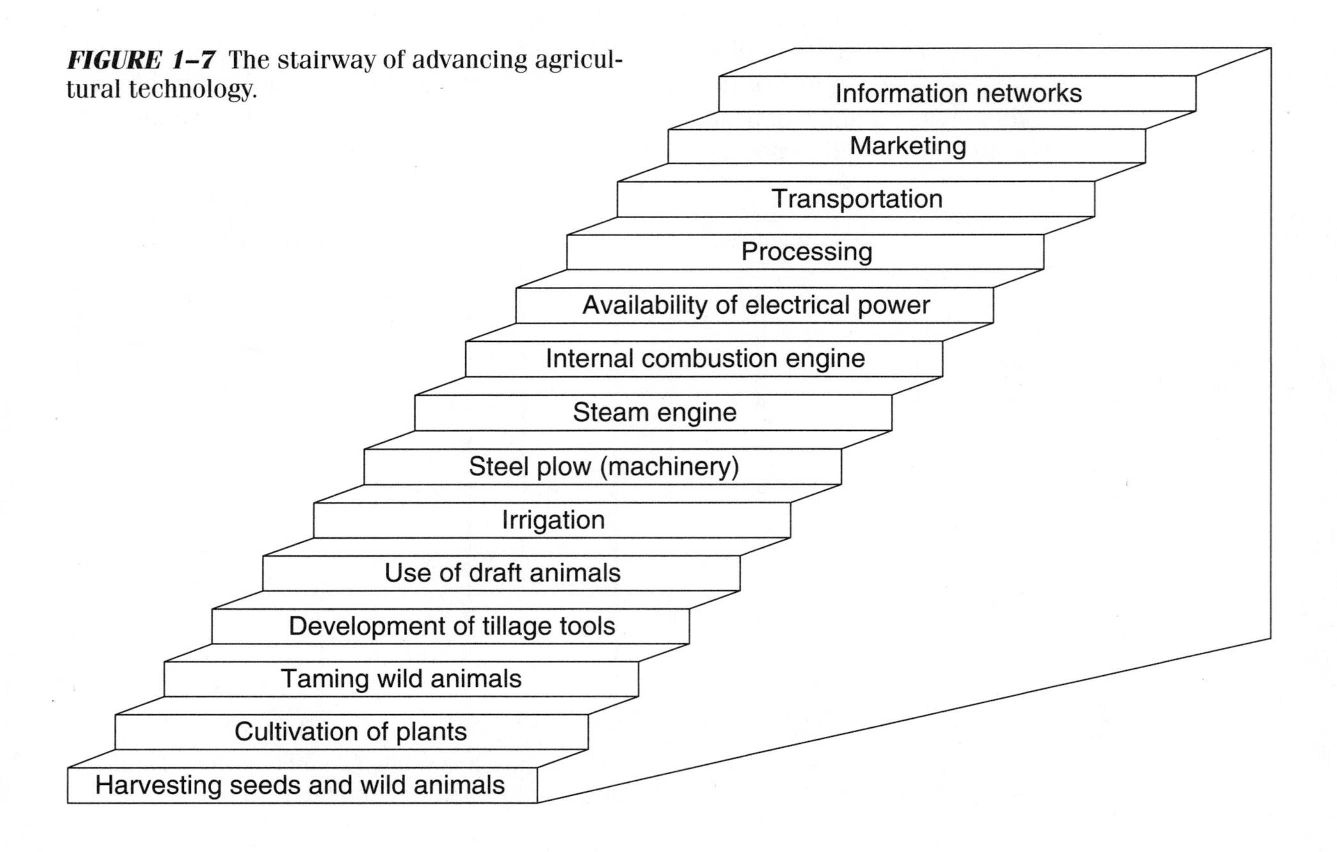

FIGURE 1–7 The stairway of advancing agricultural technology.

FIGURE 1–8 The steam-powered tractor was one of the developments that followed the invention of the steam engine.

The invention of the steam engine by Thomas Savery, and the introduction of the Watt steam engine in 1769 signalled the beginning of tremendous growth in the manufacturing industries. This came to be known as the **industrial revolution**. Steam power was soon harnessed to machines that performed many kinds of tasks, Figure 1–8. As mechanized manufacturing processes were developed to take advantage of this new resource, many new industries were born. Among these was the **textile industry** which developed processes for making cloth. Improved spinning and weaving technology was successfully combined with new concepts in mass production and the result was the textile mill.

In 1793, Eli Whitney invented the **cotton gin** to remove cotton seeds from the cotton fibers. This technology was important to the success of the textile industry, but Whitney's innovations dealing with the factory concept of mass production were equally valuable. He developed the idea of mass production of products—each worker performing the same small task on each item on the production line. This same process is widely used today.

The development of the steam locomotive in 1829 soon resulted in the use of steam power for transport of raw materials and manufactured goods. Previous to this time, the only method of mass transport had been along the coasts, rivers, and canals on boats and barges.

The invention of the **Bessemer process** for making steel had a great impact on technology, Figure 1–9. Steel is harder than iron and is more dependable in some uses than iron. Steel became an important construction material and was widely used in construction of buildings and machinery. Agriculture came to depend on this durable, tough material to build the machinery used on the farm and in the agricultural support industries.

The **land grant university** system was implemented in 1862 when Congress granted land and money to the states to establish agricultural colleges and uni-

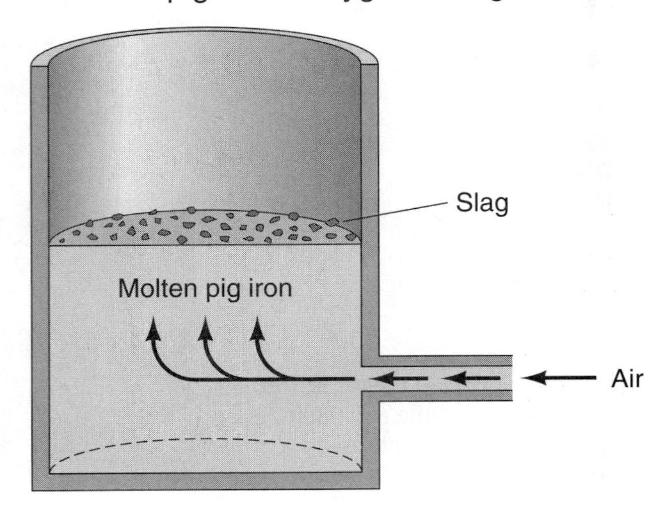

FIGURE 1–9 The Bessemer process injected air into a container of molten pig iron. The presence of oxygen allowed impurities to burn off forming slag. The slag was skimmed off, and the product that remained was called steel.

versities, Figure 1–10. The agricultural experiment stations were authorized by Congress in 1887. Information from these sources proved difficult to implement, however, until agricultural education programs were started. The Agricultural Extension Service was established in 1914, and secondary school vocational education in agriculture was authorized in 1917. Both of these organizations have made significant contributions in bridging the information gap between agricultural research and the farm.

The steel moldboard plow invented by John Deere and the reaper invented by McCormick played significant roles in the early mechanization of agriculture. No single event was of greater significance, however, than the invention of the **internal combustion engine**. This type of engine burns fuel in a closed chamber and generates power from the expansion of heated gases.

FIGURE 1–10 The Morrill Land Grant Act of 1862 led to the establishment of Land Grant Universities. Morrill Hall is a name that can be found on agricultural buildings on many of the Land Grant University campuses today.

CAREER OPTION

Agricultural Education

A person who chooses agricultural education as a career will need a professional agricultural education degree from an agricultural college or university. Most land grant universities and other agricultural universities offer B.S. and graduate degrees in agricultural education.

Teachers of high school agriculture classes are usually hired when they complete their B.S. degrees. They will be teaching classes about agribusiness, agriscience, horticulture, animal science, plant science, mechanics, natural resources, and leadership skills. They are expected to work well with young people, since they will be advisors to the youth organization called the FFA, Figure 1–11.

Agricultural educators with M.S. or Ph.D. degrees often continue to work as high school teachers, but they are also qualified for jobs as **Agricultural Extension Educators** or as professors at agricultural colleges. Agricultural extension educators work as university faculty members in the counties of each state, Figure 1–12. They help solve agricultural problems for people who live in the area, and they manage 4-H club programs for young people.

FIGURE 1–11 Agriculture teachers provide instruction in and about agriculture to students of all ages. *(Photo courtesy of USDA)*

FIGURE 1–12 Agricultural Extension Educators provide technical information on agricultural subjects and help bring new research information to the agricultural industry. *(Photo courtesy of USDA/ARS #027)*

FIGURE 1–13 Early gasoline-powered tractors provided new sources of power to perform farm work.

The gasoline-powered engine developed by Otto preceded the development of the diesel engine by Rudolf Diesel in 1893. Tractors powered by steam and later by fossil fuels (petroleum) gradually assumed dominant roles on farms during the years preceding World War II, Figure 1–13.

Electric power became available to the farm during the 1930s and had an immediate impact on the quality of life and the productivity of farmers. Many of the menial tasks associated with rural living were taken over by machines powered by electrical energy.

Synthetic fertilizers manufactured from nonorganic sources became available during this same time period. Nitrogen fertilizers are a modern by-product of the steel industry and other manufacturing processes. Nitrogen is the main component of the air we breathe, but it is not useful to most plants in its gaseous form. Industrial processes change the form of this important element, making it possible for plants to use it.

Phosphorus and potassium fertilizers are other common products used by plants. These products are mined from rich natural deposits in the crust of the earth. The ore is refined to provide these nutrients in forms that can be easily used by plants.

These fertilizers were used to replace or supplement natural fertilizers such as animal manure. In combination with soil testing procedures, farmers were able to identify and replace nutrients that had been depleted from the soil by crop production.

The ability to test soil fertility levels and correct nutrient deficiencies resulted in sustained crop yields. When these technologies were teamed with hybrid seed, the greatest production increases in the history of agriculture were realized, Figure 1–14.

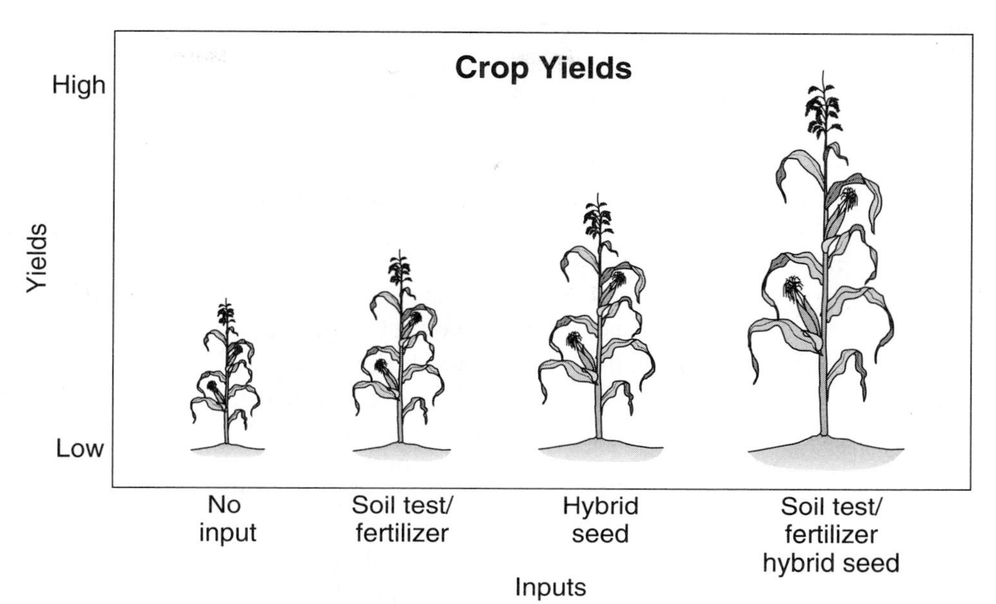

FIGURE 1–14 Three technologies that have resulted in increased crop yields are soil testing, chemical fertilizers, and hybrid seed.

A **hybrid** is obtained by crossbreeding two pure varieties of plants or animals, Figure 1–15. The offspring that result are usually of greater value than either of the parent varieties, Figure 1–16.

FIGURE 1–15 Hybrid seed has boosted the production of crops to record high yields. (*Photo courtesy of USDA*)

FIGURE 1–16 Plant research has resulted in plant varieties and farming practices that produce higher crop yields. (*Photo courtesy of USDA*)

TECHNOLOGY TODAY

Agricultural technology has expanded at a rapid rate since the beginning of recorded history, but no generation has witnessed more rapid change than the present one. Many of today's retired farmers started their careers behind teams of horses or mules. Today, farmers operate air-conditioned tractors, doing more work in a few hours than their parents used to complete in a week, Figure 1–17.

Hybrid vigor or **heterosis** is an increase in productive performance in certain hybrid animals, poultry, and crops. It is the result of many years of work by agricultural scientists, Figure 1–18. The increased production of crop and animal products that has resulted from adoption of such technologies has assured an abundance of food and fiber for the foreseeable future.

Efficiency in production has become a key to survival of farm businesses. Not only are today's farmers in competition with their neighbors, but they must frequently compete with farmers in many other parts of the world. The export of agricultural technology to other nations has made it possible for some of them to become fierce competitors in the world market. Many countries that once imported large amounts of food have become food exporters in recent years.

Biotechnology is the use of engineering techniques to solve problems associated with living organisms. Some of the more commonly known products of biotechnology are antibiotics, enzymes, and amino acids. These products have proved to be of great medical value for both animals and humans. Biotechnology research has led to the development of techniques such as artificial insemination, superovulation, embryo transfer, and hybridization of

FIGURE 1–17 Modern farm tractors are the result of many years of development by agricultural engineers.

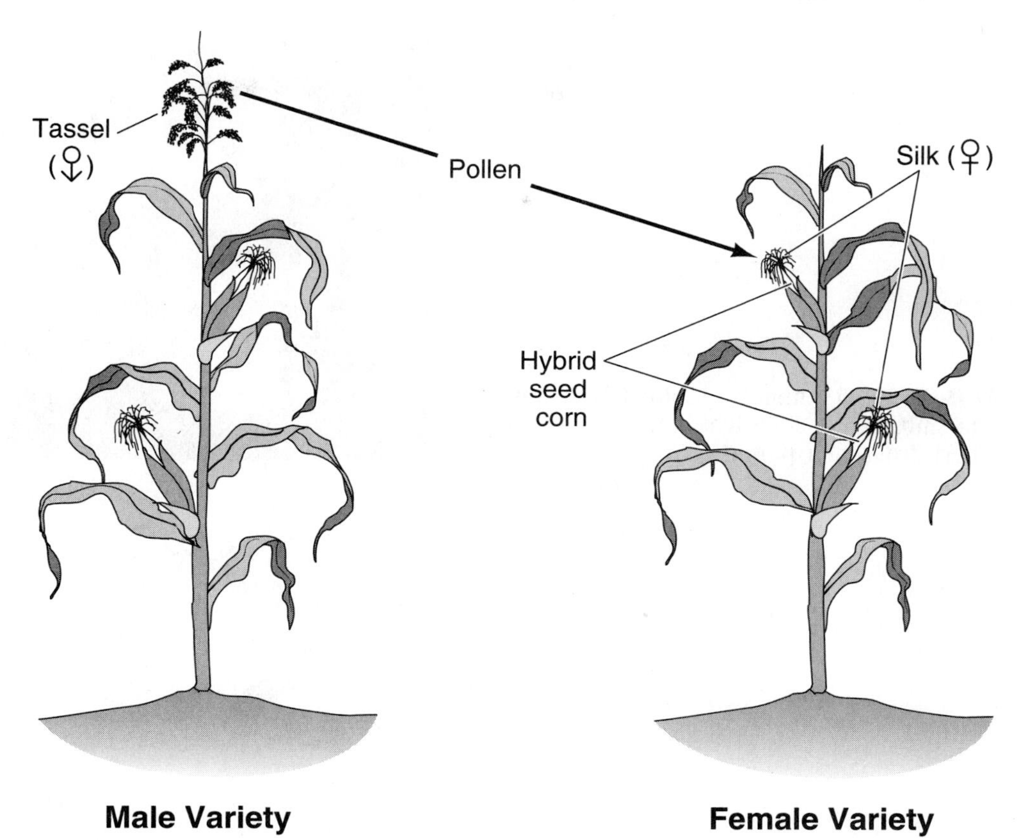

FIGURE 1–18 A hybrid seed is produced when the male flower parts of one variety produce pollen that fertilizes the female flower parts of another variety. Only the seed from the female plant variety is hybrid seed. The tassels are removed from the female variety to prevent self-pollination.

plants and animals. The impact of these technologies on agriculture has been immense, but the possibilities for future biotechnology applications challenge the imagination.

A **clone** has been regarded by many people as a figment of the imagination. In reality, the process has been used for many years to grow new plants from plant tissue. A surgical technique has also been developed that is used to enhance animal reproduction by dividing an embryo in parts to produce multiple individual animals of identical genetic makeup, Figure 1–19.

A **gene** is the cell part that controls heredity. It controls size, sex, and other visible and unseen characteristics. Genes are located on structures called **chromosomes**. Recent discoveries about the chemistry of genes have allowed scientists and technicians to change the **heredity** or genetic makeup of living things. They have discovered ways to cut the genetic structure apart, Figure 1–20. They have also learned to reconstruct the chromosome of an organism

FIGURE 1–19 Animals obtained from splitting embryos are identical in the traits they inherit from their parents.

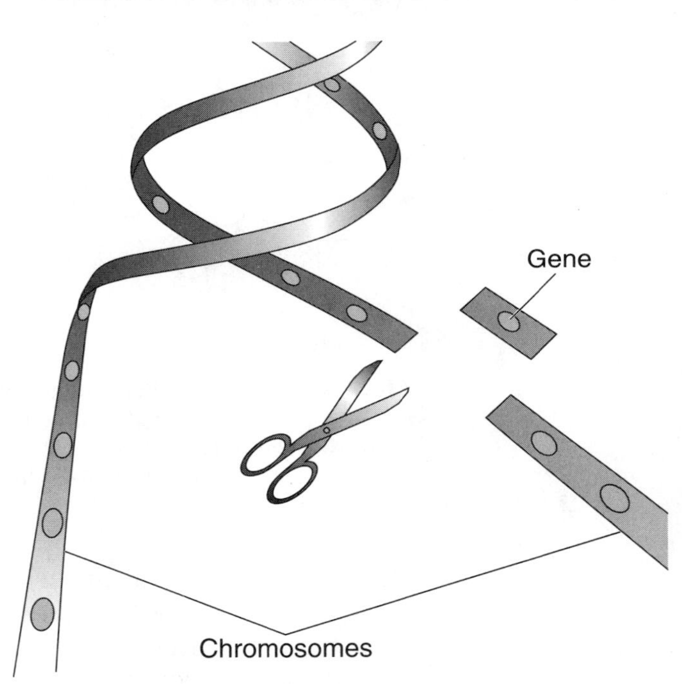

FIGURE 1-20 Scientists have learned to cut a desirable gene from a chromosome and insert it in a chromosome of a completely different organism.

in the desired form. New genes can even be transferred between living organisms to create new life forms. Genetic changes such as this are part of the new technology known as **genetic engineering**.

Some of the new technologies have raised concerns about ethics and morality. The ability to change the genetic makeup of living organisms is thought to be immoral and unethical by some groups of people. Consumers' rights groups frequently oppose the release of genetically engineered organisms into the environment because they fear that the balance of nature may be upset, or that the engineered organism may prove to be harmful. Others may believe that it is morally wrong to alter the genetics of living things. These are some of the social issues that must be settled before genetic engineering is readily accepted in our society.

As science unlocks the secrets of genetics, plants and animals may become resistant to many of the diseases and environmental conditions that restrict their productivity today. Application of the principles of genetic engineering to crop research has resulted in plants that are tolerant to several plant diseases. Some of these new plant varieties produce their own insecticides and herbicides. Some genetically-engineered microbes are even capable of protecting plants from frost damage.

Infrared photography records the amount of heat reflected from plant surfaces in the form of **radiant heat**, Figure 1–21. This technology can be used to detect crops that are showing signs of stress. Soil moisture conditions can be accurately monitored using a variety of instruments. Computers can accurately predict crop requirements for water and nutrients based upon data from a variety of sources. Analysis of plant parts provides information about the

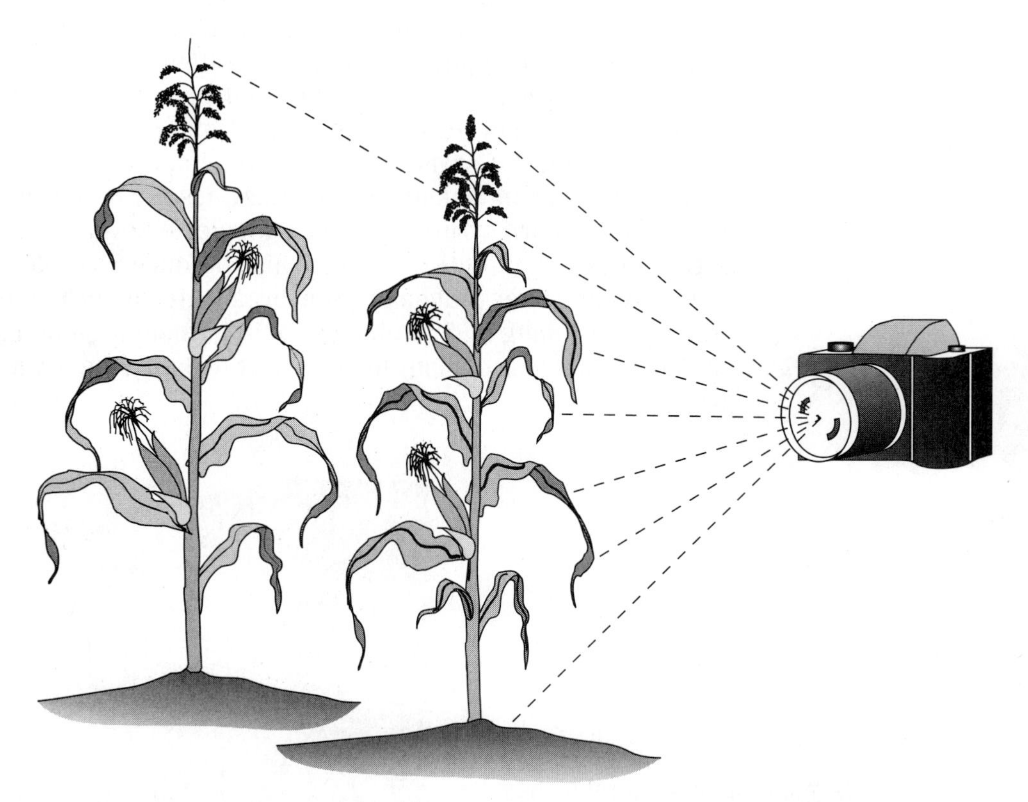

FIGURE 1-21 Infrared photography is capable of recording heat that is radiated from plant leaves. Stressed plants radiate the most heat and appear in different colors in infrared photographs than healthy plants do.

nutrient needs of plants. This is a reliable way of determining when an adequate supply of plant nutrients is available.

Animal agriculture has also benefited from new technologies. Some single-cell organisms such as **bacteria** have been used to produce new vaccines for deadly animal diseases such as tetanus and scours. Use of supplemental hormone materials by animals has proved to stimulate efficient production of milk and meat. Enzymes have been developed that improve digestion of feeds by animals. **Enzymes** are substances that stimulate life processes. Other animal research is leading to superior animal species that are resistant to insects, parasites, and diseases.

New technologies allow livestock producers to scan their herds electronically to detect sick animals. Abnormal body temperatures are detected by special instruments, and the technician uses this information to identify individual animals in need of medication.

Animals can be fitted with transmitters and computer chips that provide the identity of each animal to a computer. As each animal enters a special feeding stall, a special ration is metered into the feeder. The ration content and amount can be adjusted to match the needs of individual animals. This system is also used to separate animals from the herd using automatic gates that are activated by the computer chip that identifies the animal to the computer.

New technologies in packaging and preserving foods have resulted in greater convenience to consumers and increased shelf life of many products, Figure 1–22. Enzymes are used to tenderize meat and to separate milk solids from whey in making cheese. Enzymes are important ingredients in the production of corn syrup and even low calorie beverages. Food fermentation is the basis for an entire industry and produces a wide range of products. Yogurt, wine, beer, sausage, and bread are all products of fermentation processes. Cloning of desirable genes in fermentation organisms is improving production methods of many foods, and stimulating the development of new food products.

FIGURE 1-22 Modern agricultural technology has provided the finest food supply ever available to man. (*Photo courtesy of Utah Agricultural Experiment Station*)

FIGURE 1-23 Perhaps the next century will see agriculture practiced in distant parts of the universe. (*Photo courtesy of Utah Agricultural Experiment Station*)

Pollution of the environment is frequently cited as a negative product of modern agriculture. Among the contaminants of surface and ground water are salts, pesticides, and nitrites. Some of these materials come from applications of farm chemicals. Many of these contaminants can be degraded by soil microorganisms that have been modified to produce specific enzymes. Genetic engineering has great potential for reducing the pollution problem that has plagued the modern world.

Technology in agriculture has played a major role in modernizing the world in which we live, work, and play. It is through new and developing technologies that the food supply has kept pace with the expanding human population. Technological advances have improved both the quantity and quality of food available to a hungry world. Humanity will continue to depend upon technology in agriculture as we advance to the year 2000 and beyond, Figure 1–23.

EMERGING TECHNOLOGIES

Technology, together with the biological sciences, has been projected to be the next source of revolutionary change in the agricultural industry. An equally powerful partner in the biotechnology revolution will be the new generation of computers. Together, these two rapidly advancing technologies could bring more expansive and rapid change to agricultural production, processing, and management than any other development in history.

Animal science may benefit from the ability of technicians to predetermine the sex of animal offspring through artificial insemination using sexed sperm or implanting embryos whose sex is known. Researchers continue to work on new ways to address these technologies.

Scientists and technicians are searching for ways to control diseases and pests by splicing genes. Genetic abnormalities may be corrected using similar procedures.

The ability to produce vaccines, hormones, and other biological products using bacteria as hosts instead of living plants and animals will be enhanced beyond present technology. Cloning procedures for development of new plant species will be refined and expanded into animal reproduction.

Plants of many species may one day be modified to produce their own nitrogen fertilizer through bacteria in the roots, much as legume plants do now. Flavor and nutrient content of many plants now raised for human food may be modified to meet the dietary and health needs of a variety of consumers.

Amid the speculation concerning future scientific and technological advances in agriculture, one certainty exists; technology in agriculture will advance beyond our wildest dreams.

CHAPTER SUMMARY

Agricultural technology is the application of science to agricultural problems. It includes all of the agricultural tools and processes that have been developed since humans began to cultivate the soil and raise animals in captivity. Much has changed since the dawn of agricultural technology. Machines of many kinds have replaced human labor and animal power. Soil testing, synthetic fertilizers, and hybrid varieties of plants and animals have resulted in increased production of food and fiber. Computers and satellites are now used to provide instant data about crops, changes in weather, and markets.

Biotechnology is an emerging scientific field that deals with heredity and life processes. It has been called the "New Biology," and appears likely to result in a period of rapid changes and increased levels of production in agriculture. New technologies are simply new ideas to solve old problems.

CHAPTER REVIEW

Discussion and Essay Questions

1. Explain how agricultural technology is different from agricultural science.
2. List some examples of agricultural technologies and describe how each has improved modern agriculture.
3. Describe some ways that agricultural technologies have contributed to adequate food supplies in America and other countries in the world.
4. Describe how the use of modern agricultural technologies has contributed to new careers outside the field of agriculture.
5. Explain how a career in agricultural education relates to adoption of new agricultural technologies by the industry of agriculture.
6. Explain what biotechnology is, and describe what effects it is likely to have on agriculture.

7. Identify a modern agricultural problem and suggest some possible solutions to it. Explain why you think your solutions will solve the problem.
8. Defend the following statement: "Hybrid seed, soil testing procedures, and synthetic fertilizers are the basis of the agricultural export market."
9. Explain how we can justify the cost of conducting agricultural research.
10. Identify some positive and negative aspects of genetic engineering.

Multiple Choice Questions

1. The most common source of agricultural power in America before 1930 was:

   A. steam engines.
   B. draft animals.
   C. gasoline engines.
   D. electricity.

2. Farm and ranch uses of computers generally *do not* include:

   A. record management.
   B. farm equipment management.
   C. research design.
   D. analysis of farm financial records.

3. The average American farm worker produces enough food and fiber each year to meet the needs of:

   A. 25–35 people.
   B. 40–60 people.
   C. 100–150 people.
   D. over 250 people.

4. The percentage of the population working on farms in America today is about:

   A. 2 percent.
   B. 5 percent.
   C. 15 percent.
   D. 30 percent.

5. Which of the following is *not* considered to be an agricultural product?

   A. soaps
   B. food
   C. clothing and textiles
   D. petroleum products

6. Which of the following manufacturing processes was developed by inventor Eli Whitney?

   A. each worker performs a small task on each item
   B. each worker performs all the manufacturing tasks
   C. each worker performs as a member of a production team
   D. each worker is responsible for improving the design of the finished product

7. Synthetic fertilizers generally consist of:
   - A. chemicals obtained from sea water.
   - B. chemicals recovered from toxic waste dumps.
   - C. nonorganic chemicals that are usually obtained as by-products from manufacturing steel or that are mined and refined.
   - D. chemicals obtained from fermentation of straw and other plant materials.

8. A seed that is produced when the pollen from one plant variety fertilizes the female flower parts of another similar plant variety is called a:
   - A. hybrid seed.
   - B. monoclonal seed.
   - C. biotechnic seed.
   - D. linebred seed.

9. A gene is the basic unit of heredity, and it is located:
   - A. only in reproductive cells.
   - B. on a chromosome.
   - C. in the fluid component of blood.
   - D. in digestive enzymes.

10. Which of the following is a product of a fermentation process?
    - A. monoclonal antibodies
    - B. pesticides
    - C. nitrogen fertilizer
    - D. wine

LEARNING ACTIVITIES

1. Select a person whose ideas had an impact on agricultural technology, research his/her contributions, and report your findings to the class.

2. Identify an agricultural technology that has affected agricultural production, research the subject, and report your findings to the class.

SECTION II

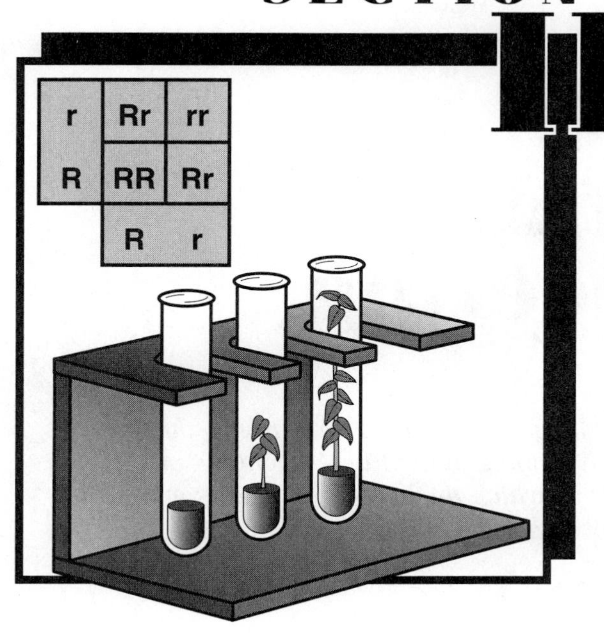

Biotechnology

- Genetic Engineering
- Animal Management Technologies
- Animal Reproduction Technologies
- Biotechnology Applications for Plants
- Plant Propagation Techniques

CHAPTER 2

Genetic Engineering

Modifying living organisms by changing the genetic code of the chromosomes is the basis of genetic engineering. Many leading agricultural scientists are engaged in research to identify methods by which the heredity of plants and animals can be altered to solve agricultural problems.

OBJECTIVES

After completing this chapter, you should be able to:

- define basic genetic terms.
- report situations for which gene splicing may be a reasonable solution.
- identify methods by which the genetic codes of plants and animals may be changed.
- analyze the value of cloning as a tool in plant and animal breeding.
- explain how genetic engineering can be used to enhance the health of plants and animals.
- evaluate the moral and ethical issues associated with developing new life forms.

TERMS FOR UNDERSTANDING

The following vocabulary terms should be studied carefully as you read:

cell	cytoplasm
cell wall	nucleoplasm
cell membrane	vacuole
permeable	chloroplast
nucleus	mitochondrion
protoplasm	golgi apparatus

- mitosis
- interphase
- prophase
- chromatid
- centromere
- metaphase
- centrioles
- spindles
- anaphase
- telophase
- gametes
- meiosis
- homologous chromosomes
- homologue
- spores
- sperm
- egg
- ovum
- haploid
- diploid
- genetic code
- selective breeding
- natural selection
- gene pool
- locus
- gene mapping
- restriction enzyme
- gene splicing
- ligase
- electroporation
- mutant genes
- *Agrobacterium tumefaciens*
- plasmid
- cloning
- callus
- tissue culture
- embryo splitting
- transgenic animal

PLANT AND ANIMAL CELLS

The basic unit of life is the **cell**, Figure 2–1. Cells contain important structures that make life possible. The cells of plants and animals are similar in many ways, but in some ways they are different, Figure 2–2.

The **cell wall** is only found in a plant cell. It is the rigid outer covering of the cell that contains high amounts of fiber. The cell wall makes it possible for

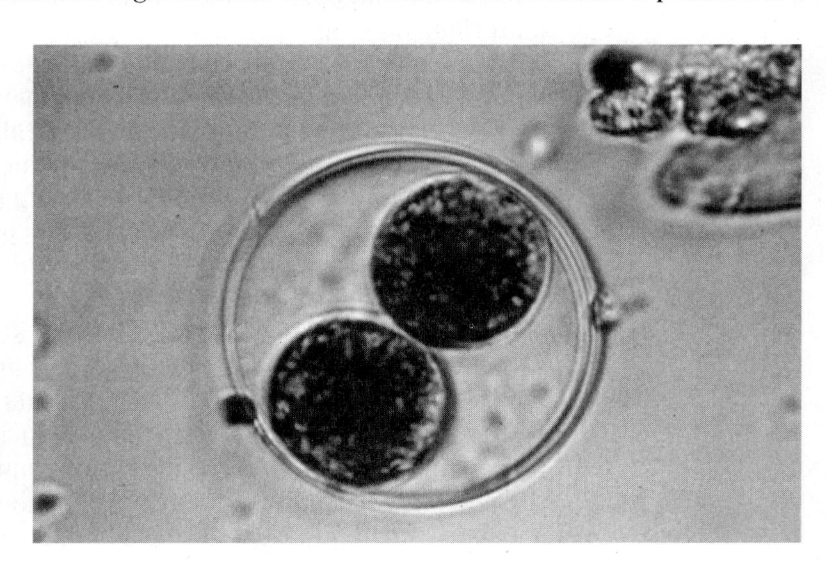

FIGURE 2–1 A living organism begins as a single cell that divides repeatedly, forming different tissues until an entire organism is formed. *(Photo courtesy of Utah Agricultural Experiment Station)*

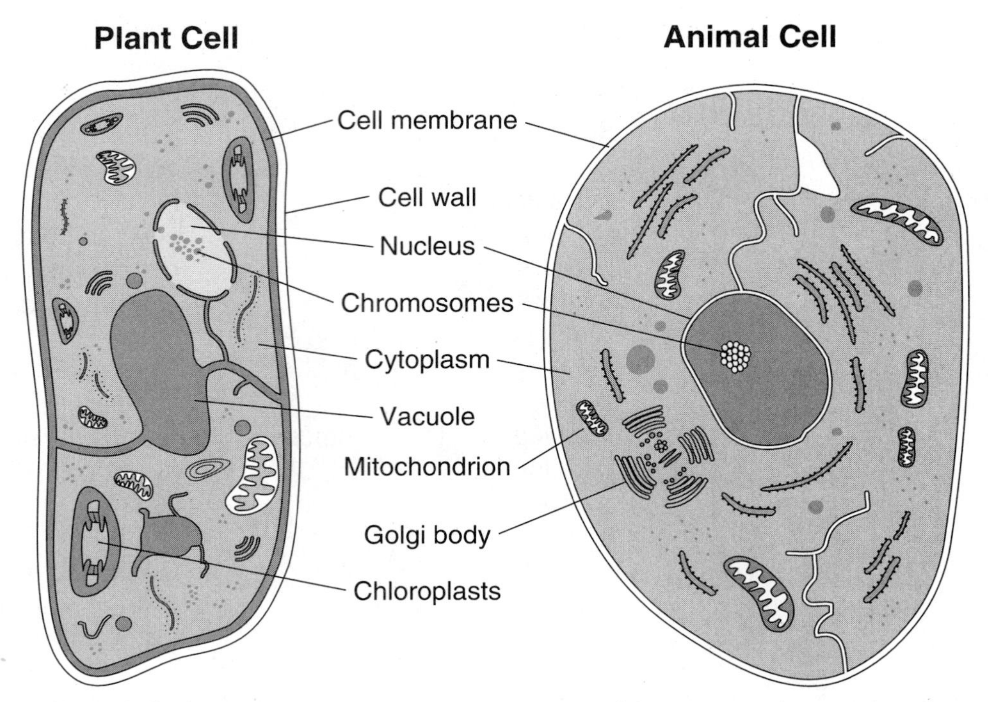

FIGURE 2–2 Plant and animal cells contain some of the same structures, but they also exhibit differences.

plants to maintain their shapes. A **cell membrane** is found inside the cell wall in plant cells, and it is the flexible outer covering of an animal cell. Animal cells do not have cell walls. Cells are usually **permeable**, meaning that body and plant fluids can pass through them to deliver nutrients and to remove waste materials.

Each cell has an oval-shaped **nucleus**. This structure contains the hereditary material through which a living organism passes its traits to its offspring. The chromosome is a structure found within the nucleus. It is a long strand of material on which many smaller structures called genes are located. Chromosomes are found in pairs, and each male and female parent furnishes one of the chromosomes of each pair that is present in their offspring. The genes contain the information that determines the characteristics an organism inherits from its parents.

Protoplasm is made up of all of the structures and substances located within the cell membranes of both animals and plants. It consists of a thick liquid composed of salts, water, proteins, fats and carbohydrates. Two kinds of protoplasm are found in cells. **Cytoplasm** includes all of the cell contents except for the nucleus. The other kind of protoplasm includes the material found in the nucleus of a cell. This material is called **nucleoplasm**.

Plant cells contain some structures that are not found in animal cells. Two of these are the **vacuole** and the **chloroplast**. A vacuole is a round structure that collects excess water and wastes within the cell and discharges them through the cell wall. The chloroplast contains materials that capture energy from the sun as the plant produces sugars and starches. A more detailed discussion of this process is found in Chapter 5.

Mitochondrion are rod-shaped cell structures in which many cell processes are controlled by chemical materials called enzymes. Another structure that is found in animal cells is the **golgi apparatus**. It is a network of fibers, rods, and granules found in the cell cytoplasm.

Cells divide to form new cells throughout the life span of an organism. The type of cell division that occurs to cause growth in an animal or plant is called **mitosis**, Figure 2–3. Several steps occur during mitosis. Cells exist for most of their life spans in a resting or nonreproductive stage. This is called **inter-**

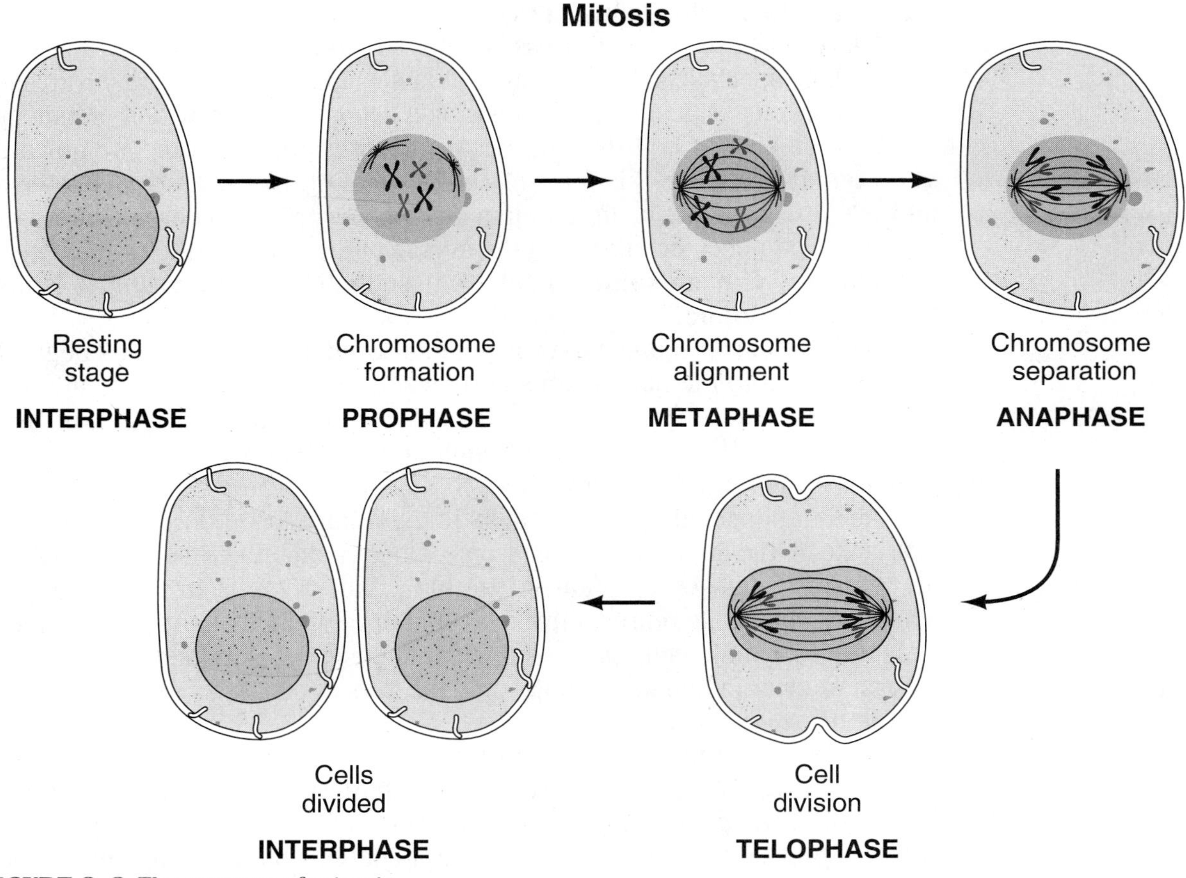

FIGURE 2–3 The process of mitosis

phase. The first stage of active cell reproduction is known as **prophase**. During prophase, the membrane around the nucleus disappears, and the chromosomes appear. Each chromosome has been replicated, and each half of the doubled chromosome is known as a **chromatid**. The point where the chromatids are attached is called the **centromere**.

The next step in cell reproduction is **metaphase**. The chromosomes are pulled to the center of the cell by fibers attached to cell structures called **centrioles** that have migrated to opposite sides of the cell. These fibers or **spindles** are attached to the centromeres that connect the pairs of chromosomes together. The chromatids are pulled apart by the spindles as the cell elongates. This step is known as **anaphase**. A full set of chromosomes becomes evident on opposite sides of the cell during anaphase.

Telophase is the last phase of mitosis. The cell becomes constricted with the cytoplasm shared equally by the two new cells that are forming. A full set of chromosomes eventually becomes separated into each new cell. The membrane around each cell nucleus forms once again, and in plant cells, a cell wall begins to develop between the two new cells.

Cells divide through the process of mitosis to form clusters of cells. Cell clusters become specialized to form different kinds of tissues in the organism, such as lung tissue or heart muscle. Mitosis also accounts for growth of organisms.

Another kind of cell division occurs when reproductive cells known as **gametes** are formed. This form of cell division utilizes a process known as **meiosis**, Figure 2–4. The first step in meiosis occurs when chromosomes are duplicated and become aligned in the middle of the cell with their matching or **homologous chromosomes**. Each of these matching chromosomes is also called a **homologue**.

Once the chromosomes have become aligned, meiotic divisions begin. In the first division, the homologous chromosomes are separated into different cell masses. The division of cytoplasm is equal in the formation of male gametes, but in the female, this division of cytoplasm is unequal. One of the cell masses is larger than the other.

The second meiotic division results in the separation of the two chromatids that make up the duplicated chromosome. Once again, the division of cytoplasm between cell masses is equal in the male, but unequal in the female. In plants, the four male reproductive cells that develop are called **spores**. In animals, the four male cells that result from meiosis develop long, slender tails to form **sperm**. In female animals, only the largest cell mass matures to form an **egg** or **ovum**.

The new cell that is formed through meiosis consists of one chromatid from each original chromosome pair. It is a **haploid** cell because it only contains half of the genetic material of the cell from which it was formed. The parent cell is a **diploid** cell, meaning that it contains both homologues of each chromosome.

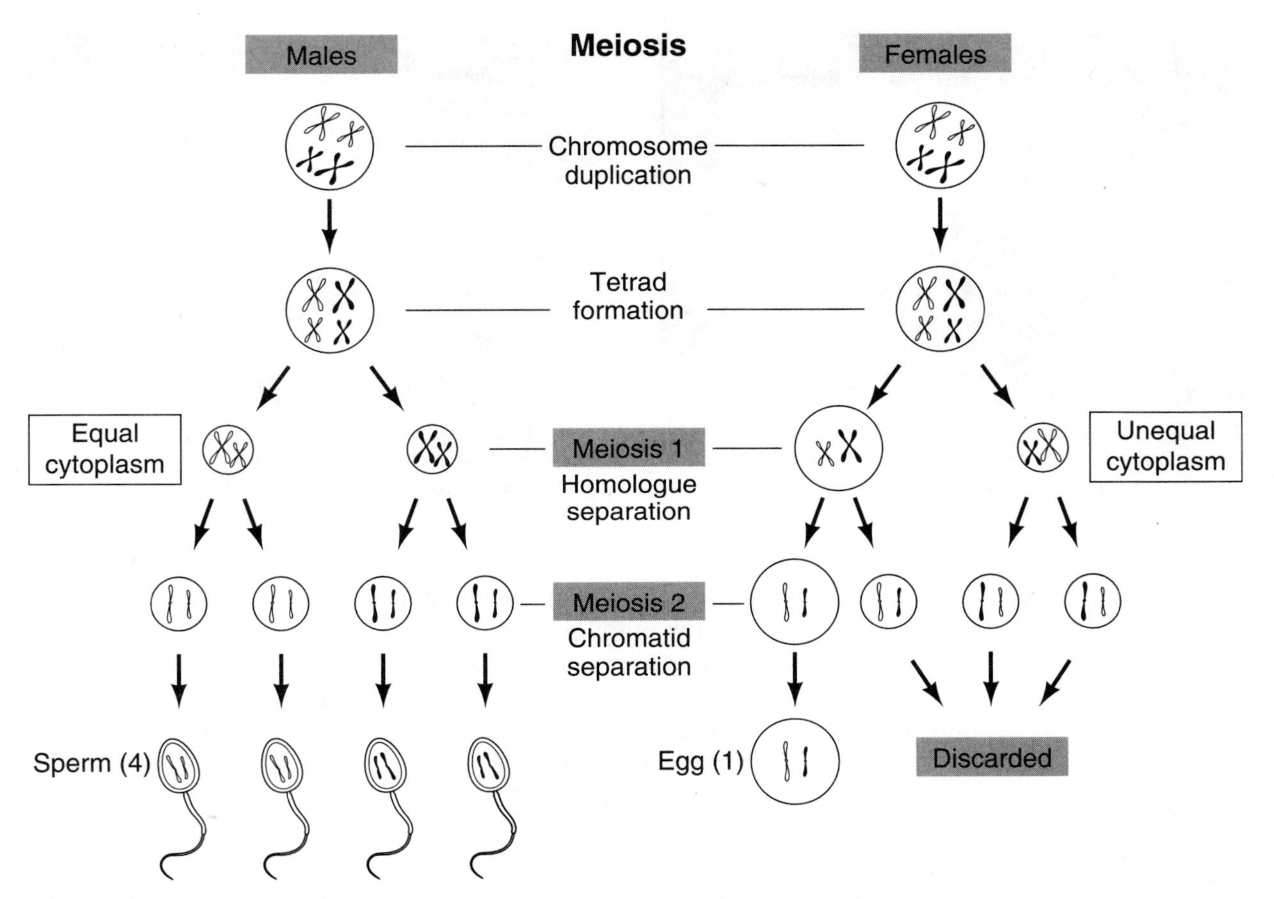

FIGURE 2–4 The process of meiosis

BASIC GENETICS

The cells of every living creature contain chromosomes, Figure 2–5. Genes contain the **genetic code** or the formula that determines each characteristic of an organism. The color of an animal, the shape of seeds, and the mature size of individual organisms are examples of the kinds of traits controlled by genes. Each cell nucleus of an individual organism contains the same genetic code. It contains all of the information needed for its own growth, development, and reproduction.

RECOMBINANT DNA TECHNOLOGY

Humans have manipulated the genetics of domestic plants and animals for a long time, Figure 2–6. They have done this by selecting plant seeds and breeding animals from among those individuals that are most productive under the environmental conditions to which they are exposed. In this manner,

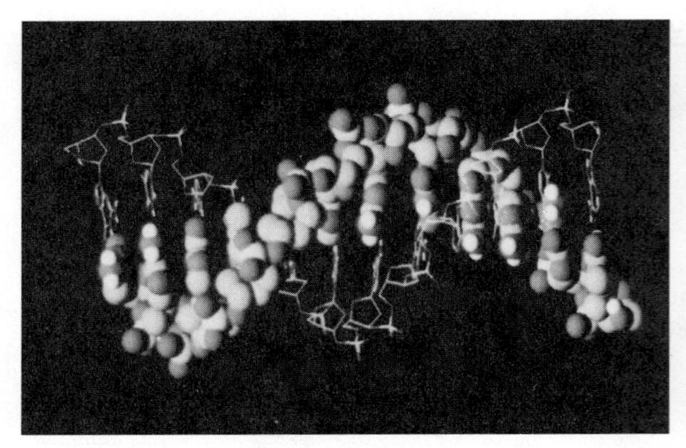

FIGURE 2–5 The chromosomes contain the code of life for every living organism. *(Photo courtesy of Utah Agricultural Experiment Station)*

FIGURE 2–6 Recombinant DNA technology makes it possible to produce an entire plant from a single modified plant cell. *(Photo courtesy of Utah Agricultural Experiment Station)*

it has been possible to develop plant and animal varieties and breeds that are well-adapted to their surroundings and to their intended uses.

For example, beef cattle and dairy cattle have been selected for different purposes. Beef cattle are muscular because they have been selected for their ability to produce meat. Dairy cattle are light muscled and lean. This is because they have been selected for the ability to convert the feed they eat to milk instead of meat, Figures 2–7 and 2–8.

A similar example is the difference between chickens selected for meat production and those selected for production of eggs. Meat breeds are large and muscular whereas egg producing breeds are small birds, Figures 2–9 and 2–10.

Production of hybrid individuals is an important tool to both plant and ani-

FIGURE 2–7 Dairy cows have been selected for the ability to produce large amounts of milk.

FIGURE 2–8 Beef cows have been selected for the ability to produce meat. In comparison with dairy cows, they are thick-bodied and well-muscled.

FIGURE 2–9 Some chicken breeds have been selected for the ability to produce large numbers of eggs with a minimum amount of feed. These chickens are quite small in size.

FIGURE 2–10 Meat chickens are selected for the ability to grow rapidly and to produce a high quantity of meat. They are quite large as mature birds. The chickens in Figures 2–9 and 2–10 are the same age.

mal breeders. Hybrid offspring are the result of mating parent individuals of different genetic makeups. The hybrid offspring are frequently more vigorous and productive than their parents. This condition is called heterosis.

One of the most widely used hybrid animals in history is the mule. The mule combines the strength of the horse with the stamina of the donkey. It was widely used as a draft animal before tractor power became common, Figure 2–11.

Hybrid strains of swine, now commercially available, combine desirable mothering traits from the female line (sows) and maximize the meat production traits from the male line (boars) in the pigs produced from these matings,

FIGURE 2–11 The mule is a hybrid animal that is the result of a mating between a horse and a donkey.

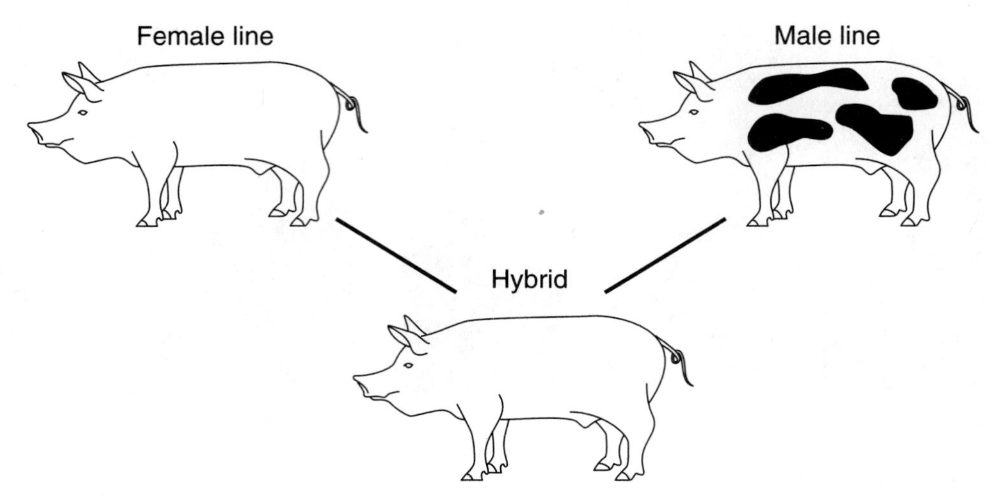

FIGURE 2–12 Hybrid swine production combines desirable traits from both the male and female lines. (Illustration is not to illustrate conformation standard.)

Figure 2–12. As with all hybrids, these offspring are not used for breeding purposes because they do not breed true to type. Some offspring of hybrid matings would resemble each of the original parent types while others would resemble the hybrid line. The hybrid performance advantage and uniformity—achieved by using the original breeding lines—would be lost.

The ability to harvest fruits and vegetables that are easily bruised or damaged by use of mechanical methods has been improved by changing some of the characteristics of the crop. Varieties of tomatoes have been developed that ripen at the same time and have tough skins to withstand the rigors of mechanical harvesting.

Wheat varieties have been developed that are resistant to rust infections and other diseases. Modified stems, strong enough to reduce lodging problems, have been developed in several small grain varieties, Figure 2–13. Each of these problems is capable of inflicting heavy damage on a small grain crop. Similar advances have been made with many crop and livestock species.

The genetic nature of living things has been altered in the past by a process called **selective breeding**, Figure 2–14. Changes in organisms are brought about by mating individuals that exhibit desired traits, Figure 2–15. This procedure resembles the **natural selection** process through which living things have evolved into distinct species. In a natural environment, traits that improve the ability of an organism to survive and reproduce tend to appear in an increasing number of individuals within the population until most individuals exhibit the trait. When selection of breeding stock depends on certain traits being present, this process can be speeded up. Genetic engineering bypasses the selection process by inserting the desired gene into the chromosomes of a living organism, Figure 2–16. The target cells receive the new genes that were obtained from another organism, Figure 2–17.

FIGURE 2–13 Many characteristics of plants, such as length and thickness of plant stems, have been modified by plant researchers. *(Photo courtesy of Utah Agricultural Experiment Station)*

FIGURE 2–14 Plant research has resulted in many new varieties of plants that yield more abundant harvests than their parent varieties. *(Photo courtesy of Utah Agricultural Experiment Station)*

FIGURE 2–15 Many new plant varieties have been produced by controlling pollination. *(Photo courtesy of Utah Agricultural Experiment Station)*

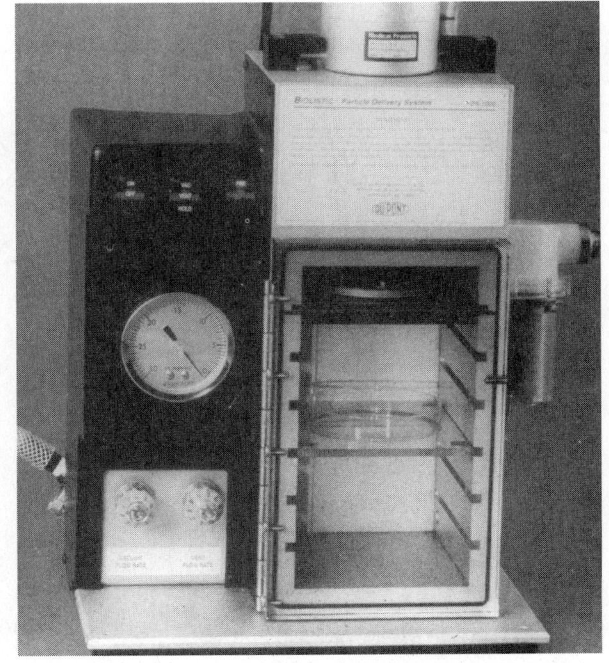

FIGURE 2–16 Using a device known as a "gene gun" (Biolistic particle delivery system), a gene can be propelled into growing tissue by putting material containing the gene on a projectile and shooting it at the target cells. The projectile, which is similar to a BB, is stopped, but the genetic material continues to move and it penetrates the target cells. *(Photo courtesy of DuPont Co. Biotechnology Systems)*

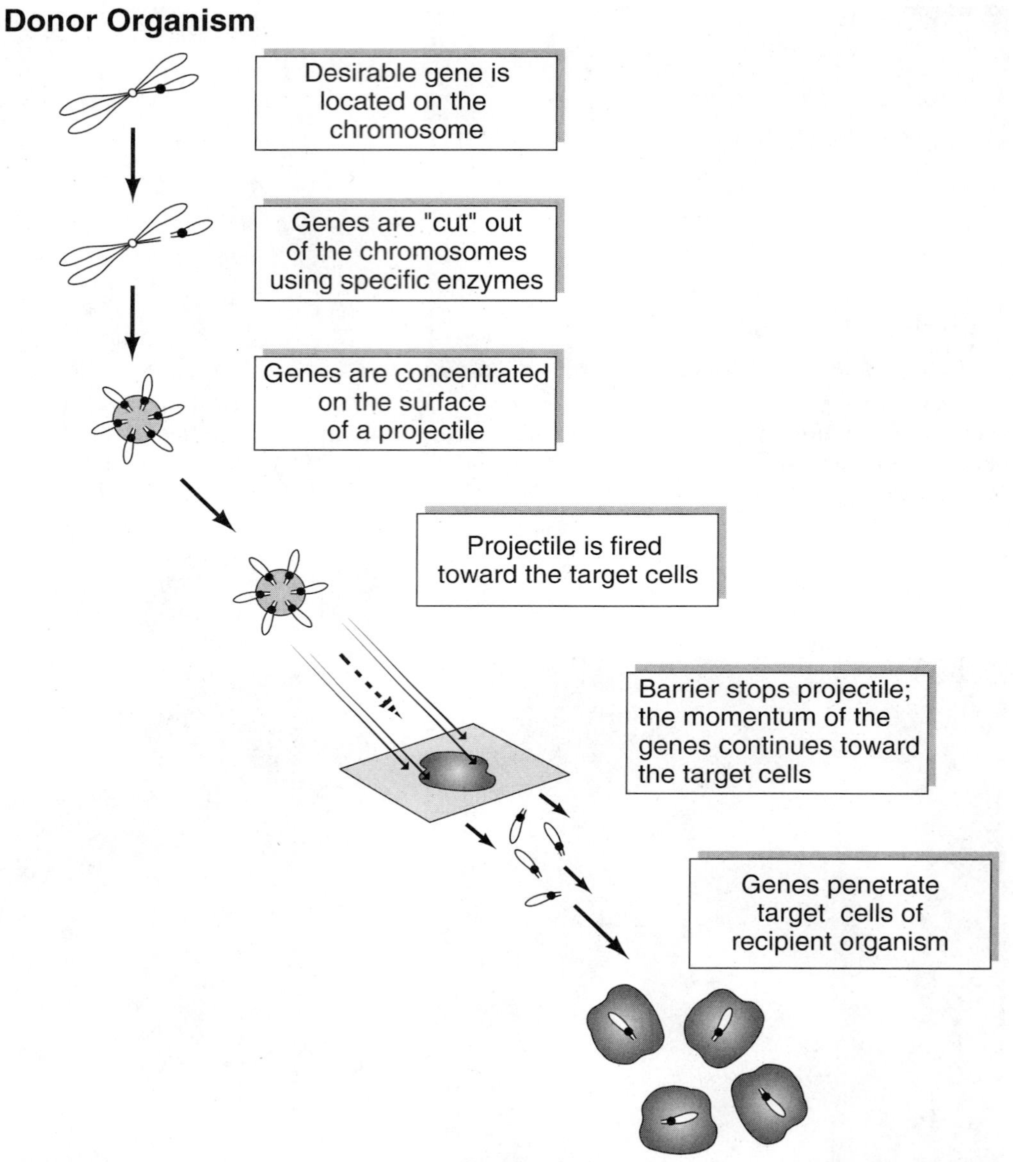

FIGURE 2–17 Desirable genes may be transferred from one organism to another.

Many traits that are evident in an animal or plant result from the genetic information contained in the code of a single pair of genes. An example of this type of heredity is the trait for the horned or polled condition in cattle. Other traits may result from the action of more than one pair of genes.

GENE SPLICING

The **gene pool** for any organism consists of all of the genes that can be found in the total population of genetically similar organisms, Figure 2–18. The more variability there is within the genetic makeup of an animal or plant species, the greater is the potential for developing useful breeding lines to overcome specific problems. When only moderate differences exist, scientists may attempt to introduce more genetic variety into the organism. In some instances, genes from unrelated organisms are inserted into chromosomes to add desired traits to the genetic makeup of the organism.

Chromosomes contain the information that determines characteristics of all living organisms. The gene for a particular trait is located at a specific site or **locus** on the chromosome. This makes it possible for scientists to prepare genetic maps. These maps identify the location of specific genes on a chromosome and identify the trait that the gene controls.

Some of the most important and intense biotechnology research is focused on **gene mapping**. This is the process of finding and recording the location of a particular gene on a chromosome, Figure 2–19. Plant and animal cells contain millions of genes located on many different chromosomes, but only a few gene locations are known. The exact location of a gene on a chromosome must be known before the gene code can be modified.

The scientist must isolate the chromosome upon which the desired gene is located, and identify the locus on the chromosome where the gene is positioned. Once the location of the gene is known, research is conducted to find a **restriction enzyme** capable of cutting a particular gene out of a chromosome.

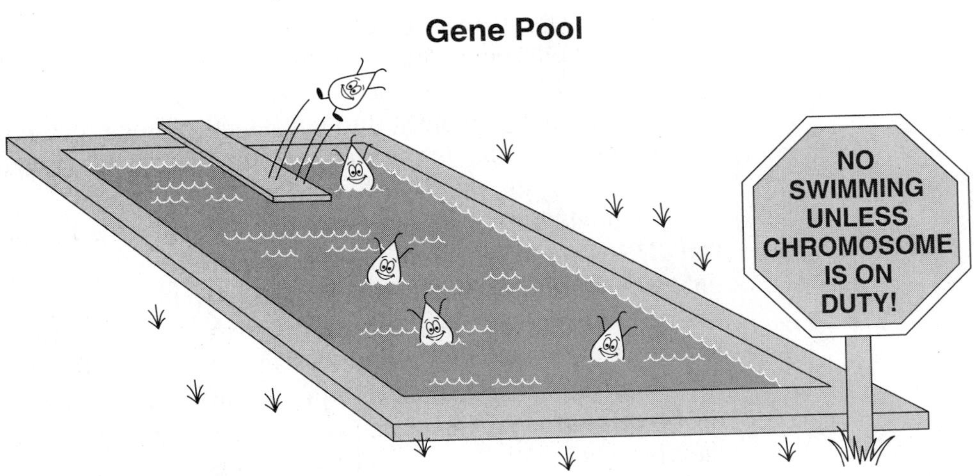

FIGURE 2–18 The gene pool contains all of the genes that exist in the population.

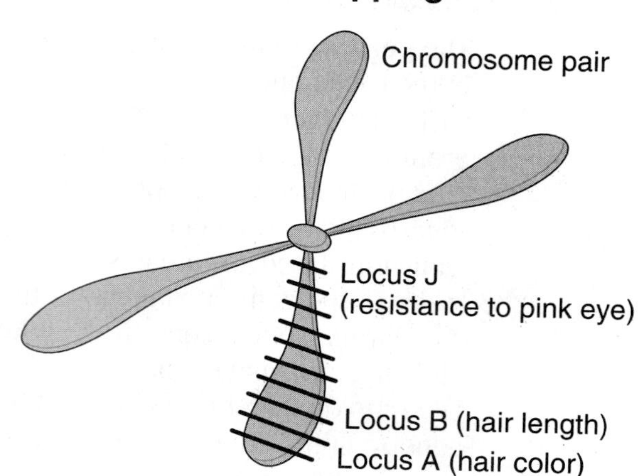

FIGURE 2–19 Gene mapping is a difficult process. Only a few gene locations are known, but scientists have identified important gene locations in both plants and animals. A gene map identifies gene locations on the chromosomes.

A scientist can cut and splice a chromosome using enzymes in much the same way that a person uses a pair of scissors to remove a blemish from an object. Enzyme preparations are very specific and can be targeted to specific genes. Using the proper enzyme, a scientist can remove a gene from its position on a chromosome and replace it with another. This process is called **gene splicing**. When undesirable genes are removed from a chromosome, sticky chromosome ends remain where the cuts occurred. By applying an enzyme called a **ligase**, desirable genes, placed in close proximity to the modified chromosome, become attached at the same location from which the undesirable gene was removed. This enzyme bonds the new gene permanently in place on the chromosome.

When a cell nucleus containing a modified chromosome divides during the reproduction process, the resulting offspring will exhibit characteristics of the new gene. The foreign gene has become part of the genetic makeup of the organism. Genetic changes through selective breeding tend to occur slowly, and the degree of change is restricted by the size of the gene pool for a particular organism.

Scientists have discovered several different ways to remove desirable genes from DNA and to insert them into the DNA of different organisms. Restriction enzymes can now be attached to DNA strands, which allows any gene in the strand to be cut free from the original DNA molecule. Prior to this development, restriction enzymes only worked at approximately 100 specific gene locations.

Genes can be inserted into plant chromosomes using a technique called **electroporation**. This process uses electricity to make tiny holes in cell membranes. This allows genetic materials from other sources to pass into the cells. This process even allows genes from unrelated species of plants to be spliced into the DNA of the treated cells.

CAREER OPTION

Research Geneticist

A person who chooses a career in genetic research will need an aptitude for science and related disciplines plus the ability to pay close attention to detail. A student who aspires to be a research geneticist should pursue an undergraduate degree in a strong science-related discipline. An advanced degree (Ph.D.) from a good university and experience under the guidance of a respected professional in the field are important qualifications.

A research geneticist will do research related to the genetics of plants, animals, and other forms of life, Figure 2–20. He or she may be employed by a university, a company-owned research laboratory, a government supported research center, or an international research organization. Genetic research may be conducted either alone or as a member of a research team. Good communication skills are important. Technical scientific papers are the primary means of learning what has already been done and of reporting research activities to other professionals.

FIGURE 2–20 A research geneticist works with living organisms to discover ways to understand and manipulate the genetic code. *(Photo courtesy of James Strawser, The University of Georgia)*

Before gene splicing techniques were developed, scientists depended on naturally-occurring mutant genes to modify traits of living things. **Mutant genes** are different from the parent genes that produced them. They occur only rarely in nature.

The processes that cause genes to mutate are not well-defined, but it is known that exposure to radiation and certain chemicals can result in gene mutation. Such gene mutations can be either beneficial or damaging. The dark red color of some apple varieties is an example of a mutant gene that causes the fruit to be more attractive. The growth of cancer cells from otherwise healthy tissue is an example of a harmful mutation.

Gene splicing techniques allow scientists to transfer genes within and between species to obtain desired characteristics without waiting for useful gene mutations to occur naturally.

When a gene is selected for transfer to a plant, it is sometimes spliced into the DNA of a special strain of bacteria, Figure 2–21. ***Agrobacterium tumefaciens*** is a naturally-occurring bacterium capable of inserting a portion of its genetic material into a chromosome in a plant cell nucleus. In a natural setting, this bacterium has been observed to cause tumor-like growths on plant stems called Crown Gall. Research has confirmed that this plant disease occurs when Agrobacterium transfers genetic material to the cells of a plant. The bacterium must be altered to eliminate the Crown Gall gene from its DNA before it is used in the gene transfer process.

The gene that has been selected for transfer to the plant cell is then spliced into the **plasmid** ring of the bacterium. This plasmid ring is a chromosome in

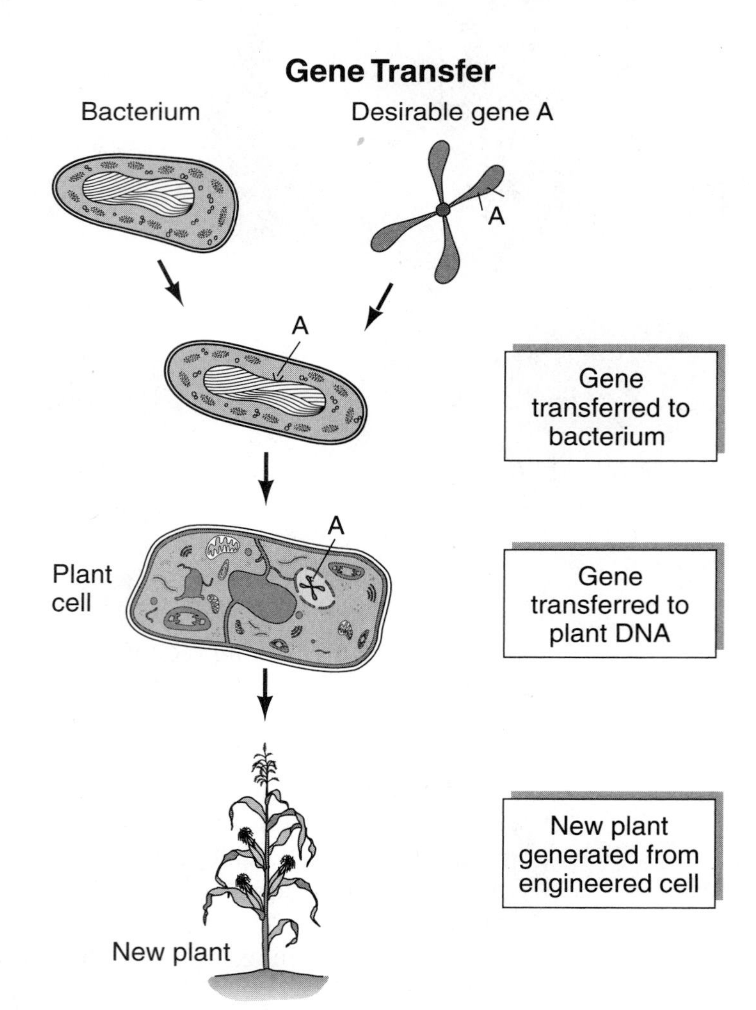

FIGURE 2–21 Genes from unrelated life forms can be transferred into the chromosomes of plant and animal cells.

a circular form that is found in bacterial cells. A culture of the modified bacteria is then prepared, and plant cells or leaf particles are placed in the solution containing the bacteria. The Agrobacterium transfers the new gene into the chromosome of a plant cell. Such cells are then regenerated into complete plants using a tissue culture technique.

Although selective breeding continues to be a valuable tool, it can now be used to propagate individuals that carry modified genetic traits. Gene splicing allows science to control and direct genetic changes in organisms in a much more orderly manner and at a much more rapid rate than was possible using selective breeding alone.

CLONING TECHNOLOGY

Cloning is a process through which genetically identical individuals are produced. By replacing the nucleus of an unfertilized ovum with the nucleus from a cell of the organism being duplicated, individuals of identical genetic makeup can be produced. Each cell composing an organism contains a full component of genes. When different tissues are produced, only a portion of the available genetic information is used. Genes performing other functions are turned off. A major problem with this process is that it does not work with mammals when the nucleus of an adult cell or advanced embryonic cell is used.

Cloning may be accomplished in plants by stimulating cell division of a single plant cell to produce a complete plant. To do this, the cell must be returned to a state in which it is undifferentiated into special tissues. If the cells in the culture medium came from leaves, they must be changed by adding hormones to cause all of the genes to become functional again. The cells resulting from this treatment are called **callus**. Once this is accomplished, the culture medium is changed to include nutrients and hormones that cause the genes to switch on and off in a proper order to stimulate growth of roots, stems, leaves, and other plant organs. This procedure is frequently referred to as **tissue culture**. Ultimately, a new plant is generated and planted in soil. Seed from the plant is tested for the presence of the new gene. The process is valuable as a method of reproducing plants from modified plant cells in which superior genetic qualities, such as resistance to disease or pests, have been introduced.

Plants differ in their ability to regenerate from plant parts and cell culture techniques. Because of these differences, progress has been slow in soybeans and grains. Some plants—such as tomatoes, tobacco, and some ornamental flowers—regenerate easily. With these plants, cloning and tissue culture techniques are more easily applied. New techniques have made it possible to perform genetic engineering and tissue culture practices with grains such as barley, rice, and corn.

Plant breeders have used the cloning process to speed up the development and propagation of new plant varieties. When cloning is used in combination

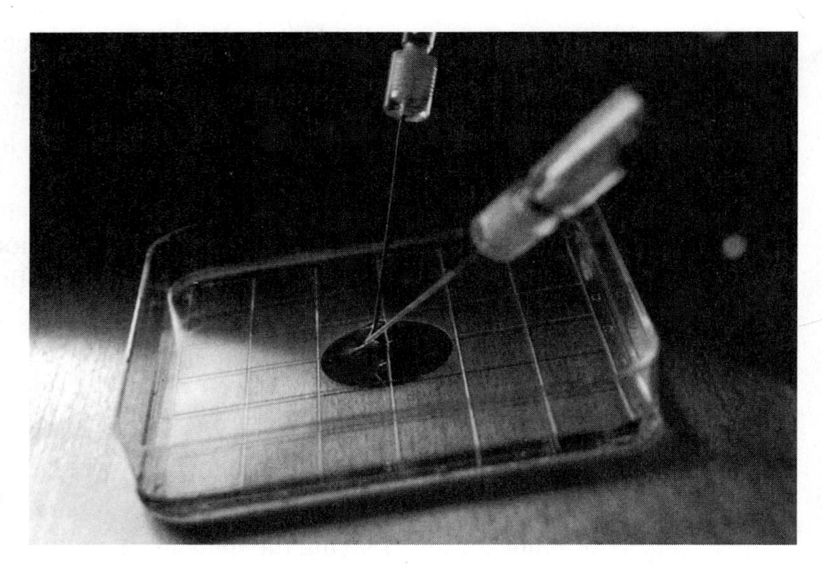

FIGURE 2–22 Animals are cloned by splitting a living embryo into two or more separate cell masses. *(Photo courtesy of Utah Agricultural Experiment Station)*

with gene splicing, it is possible to stimulate the rate of development of useful plant varieties.

Cloning of animal genes has proved to be possible, and techniques have been developed for creating synthetic genes. Cloning of mammals, using techniques effective with plants, has not proven to be successful.

One approach to cloning of mammals has proved to be very effective, however. **Embryo splitting** is a form of cloning in that individuals produced through use of this technology are genetically identical, Figure 2–22. It is accomplished by dividing a growing embryo using a form of microscopic surgery. An embryo in the 16 to 32-cell stage can be successfully split into two or more viable embryos. The procedure is expensive, but identical individual animals can be produced when the embryos are implanted in receptor females. The practice has been most widespread in the dairy and beef cattle industries.

Thousands of calves have been produced from split embryos since the practice of embryo transfer has gained acceptance. If the embryo is transferred without first splitting it, the pregnancy rate is about 65 percent. By splitting embryos prior to transferring them, technicians have been able to obtain success rates approaching one calf produced for each healthy embryo recovered. Half of the split embryos produce pregnancies, but since twice as many embryos are available, the net result is about 35 percent more calves. Genetic progress is enhanced greatly when this procedure is used with superior females.

IMMUNE SYSTEM STRATEGIES

Genetic engineering techniques offer a variety of possible solutions to the problem of disease control in plants and animals. One solution is to identify a gene that is resistant to a particular disease organism and splice it into the appropriate chromosome in place of the less resistant gene. Tissue culture or a variety of other techniques may then be used to produce a disease resistant plant.

The ultimate solution to control of diseases in animals is to alter the genetic makeup of embryos prior to implantation. One method of doing this is to inject resistant genes into a single cell embryo. The transferred gene is then duplicated during the normal process of cell division. An animal that is produced using this method is called a **transgenic animal**. A similar process is known to occur in nature as a result of virus infections.

Disease-resistant genes may be obtained from animals of the same or different species. Another source of resistant genes is available through use of another new technology that involves engineering and production of synthetic genes.

DEVELOPMENT OF NEW SPECIES

Genetic engineering has made it possible to modify the genetic makeup of organisms. This is done by removing undesirable genes and splicing in desirable genes. These techniques have resulted in plants that are resistant to such problems as drought conditions or saline soils. Plants that were damaged by certain herbicides have been altered genetically to show tolerance for the same chemical.

Research into genetic engineering processes has resulted in the abilities of scientists to change the heredity of certain animal species. New species of laboratory animals, altered genetically by this process, have been patented by the United States Patent Office to protect the production and marketing rights of the scientists who developed the new life forms.

As the science of genetic engineering advances, many animals will be genetically altered to increase their usefulness to man. This new knowledge is being challenged on moral and ethical grounds, however, and genetic engineering is having an impact on issues of morality and philosophy in modern society.

CHAPTER SUMMARY

Genetic engineering is the process of modifying living things by changing the genetic code of the organism. In times past this has been accomplished by selecting plant seeds and breeding animals from among those plants or animals exhibiting desired characteristics through the process of selective breeding. It is now possible to directly change the genetic makeup of organisms using recombinant DNA technology. This is a process by which undesirable genes are removed and desirable genes are spliced into the chromosome.

Gene transfer in plants is often done by first inserting the gene in the plasmid ring of a special strain of bacteria. *Agrobacterium tumefaciens* is capable of inserting the gene into the chromosomes of certain plants. Other effective methods for inserting genes in DNA include use of a gene gun or use of the process known as electroporation.

Cloning is a process that results in the production of genetically identical individual plants, animals, and other living things. This can be accomplished in plants by several methods, the most common of which is regenerating new plants from plant parts. It is usually accomplished in animals by embryo splitting techniques.

Genetic resistance to plant and animal diseases can be accomplished by inserting resistant genes into the nuclei of the target cells. In a similar manner, scientists are able to create new species of living organisms.

CHAPTER REVIEW

Discussion and Essay Questions

1. Define the relationship between chromosomes and genes.
2. Describe the gene splicing process and give examples of problems that might be solved by application of this technology.
3. Define what is meant by genetic code and identify two methods used to modify it.
4. In what ways are hybrid plants and animals superior to their parents in modern agricultural production?
5. What are the differences between selective breeding and natural selection? How are they similar?
6. Why is gene mapping important before genetic engineering processes are attempted?
7. Contrast the cloning methods used with plants and the cloning methods used with animals.
8. How have mutant genes affected agriculture and society in positive ways? Negative ways?
9. Explain how genetic engineering techniques can be used to improve the health of plants and animals.
10. Compose a set of regulations to deal with the ethical and moral issues associated with genetic modification of living organisms.

Multiple Choice Questions

1. One structure that is found in the cell of a plant that is not found in the cell of an animal is a:

   A. chromosome.
   B. nucleus.
   C. cell membrane.
   D. cell wall.

2. The most basic unit of life is the:

   A. nucleus.
   B. chromosome.
   C. cell.
   D. cytoplasm.

3. The stage of mitosis in which the chromatids are pulled apart is called:

   A. interphase.
   B. prophase.
   C. metaphase.
   D. anaphase.

4. The structures to which the spindles are attached during cell division are called:

   A. mitochondrion.
   B. centrioles.
   C. chloroplasts.
   D. vacuoles.

5. An example of a diploid cell is a:

   A. brain cell.
   B. spore.
   C. ovum.
   D. sperm.

6. Modification of the genetics of a plant or animal by removing an undesirable gene and inserting a desirable gene is called:

   A. telophase.
   B. selective breeding.
   C. genetic engineering.
   D. natural selection.

7. Which of the following genetic tools and procedures is used for a purpose other than inserting genetic material into plant chromosomes?

   A. tissue culture
   B. *Agrobacterium tumefaciens*
   C. electroporation
   D. gene gun

8. The gene pool for an organism:

   A. consists of the nucleoplasm in a cell.
   B. includes all of the genes in the population.
   C. is a liquid nutrient in which genetic material is stored.
   D. is genetic material that is obtained when mutations occur.

9. A ligase is an enzyme that is used by genetic engineers to:
   - A. prevent a cell from rejecting a new gene.
   - B. destroy undesirable genes.
   - C. attach a desirable gene to a chromosome.
   - D. disinfect plant parts that are used for tissue culture procedures.

10. A transgenic animal is an animal that:
    - A. has been imported from a different continent located across the ocean.
    - B. has an extra chromosome.
    - C. is resistant to genetic diseases.
    - D. has been genetically changed using a genetic engineering technique.

LEARNING ACTIVITIES

1. Separate the class into two groups of students. Assign one group to defend the right of an individual or an institution to develop, patent, and use new life forms in agriculture. Assign the second group to oppose the use of genetic engineering on the basis of moral and ethical principles. Debate the issue in class.

2. Plan and conduct a field trip to a research center that deals with some form of genetic engineering; or invite a person to visit the class who is employed in some phase of research dealing with modification of plants or animals.

CHAPTER 3

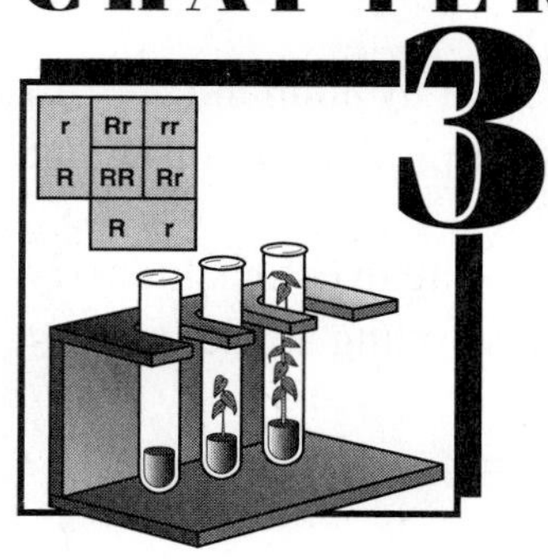

Animal Management Technologies

Animal agriculture exists because of the many technologies developed for animal management. Early humans used technologies such as traps, ropes, and pens to capture and confine wild animals. Other technologies were developed to help care for them in captivity. Methods were devised to cultivate and harvest crops for animal feed. Animal health products were developed, handling equipment was designed, and transportation methods were devised, Figure 3–1. All of these technological advances have played roles in the growth of the livestock industry.

OBJECTIVES

After completing this chapter, you should be able to:

- identify major technologies that have contributed to the development of the livestock industry.
- describe how computers can be used to manage livestock.

FIGURE 3–1 Effective livestock handling systems must be designed to prevent injuries to the animals and to encourage them to move through the handling equipment without resistance. *(Photo courtesy of Utah Agricultural Experiment Station)*

- explain how monoclonal antibodies are used to diagnose and prevent animal diseases.
- list practices that are used to improve digestion by animals.
- discuss the effects of hormone treatments on production of animal products.
- evaluate the importance of electrical power to the livestock industry.
- analyze the advantages and disadvantages of recycling animal wastes.

TERMS FOR UNDERSTANDING

The following vocabulary terms should be studied carefully as you read:

culled
index
scales
yield
vaccines
parasites
monoclonal antibodies
hybridoma
dipstick diagnosis
magnets
hormones
synthetic hormones
bovine somatotropin
electronic sensors
robots
recycled animal wastes
offal
recycled nutrients

SELECTION OF SUPERIOR BREEDING STOCK

The ability to identify individual animals is the first step in developing superior breeding stock, Figures 3–2 and 3–3. For many years, farmers depended on their memories to furnish such information. Livestock producers today often

FIGURE 3–2 Selecting superior breeding stock is an important activity in animal agriculture. *(Photo courtesy of Utah Agricultural Experiment Station)*

FIGURE 3–3 Much attention has been focused on identifying superior animals for breeding purposes. *(Photo courtesy of Utah Agricultural Experiment Station)*

identify each animal with a number tattooed in an ear, or by using permanent ear tags. In some cases animals may be identified using computer chips, electronic codes, and transmitters, Figure 3–4. Computers are used to read electronic codes, and to record production data automatically. A reliable animal identification system is as important as a good set of records for selection of superior breeding animals.

Computers are becoming more important to modern livestock breeders because they make it much easier to maintain good production records. Information—including birth weights, weaning weights, rates of gain, and cost per pound of gain—is easily organized using computers. Individual animals are easily compared using computer printouts which are generated from such information, Figure 3–5.

Computers are valuable tools in determining which breeding animals should be **culled**, or removed from the herd. For example, the milk production of a dairy cow is added to her production record each time she is milked. This gives the farm manager accurate production information that can be used to determine how profitable each animal is. Milk production is only one factor that affects profitability. Computers are capable of analyzing several factors together and combining them into a single number called an **index**. The index number is used to express the value of an animal in comparison with other herdmates, Figure 3–6.

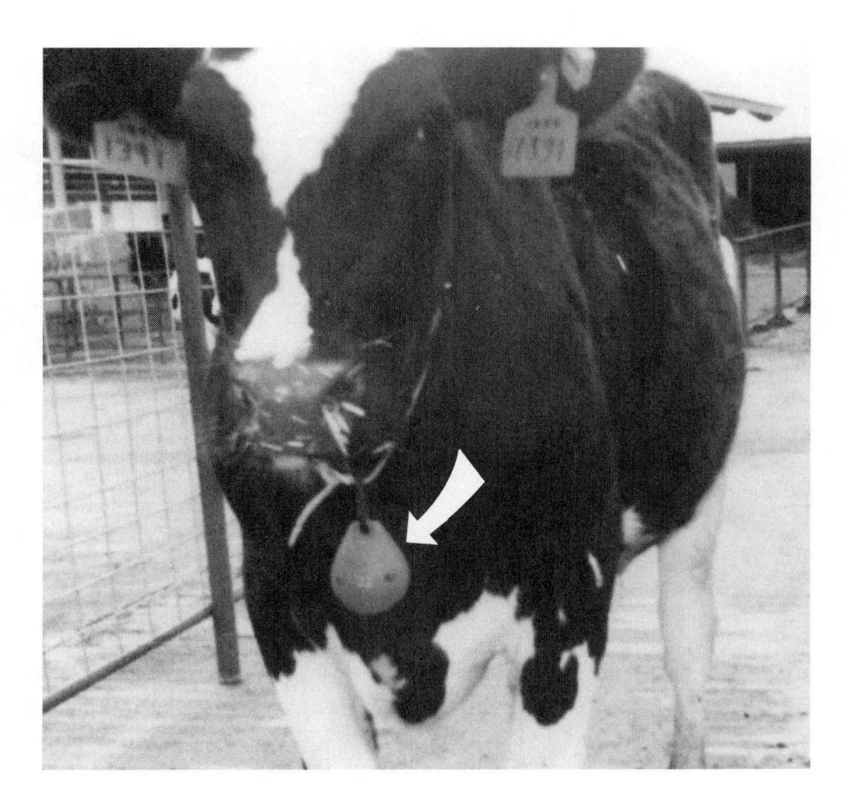

FIGURE 3–4 Modern technology has made it possible to select breeding animals based on facts and records with less reliance on memory, emotion, and hunches. The "necklace" worn by this cow contains a computer chip that identifies her to the computer.

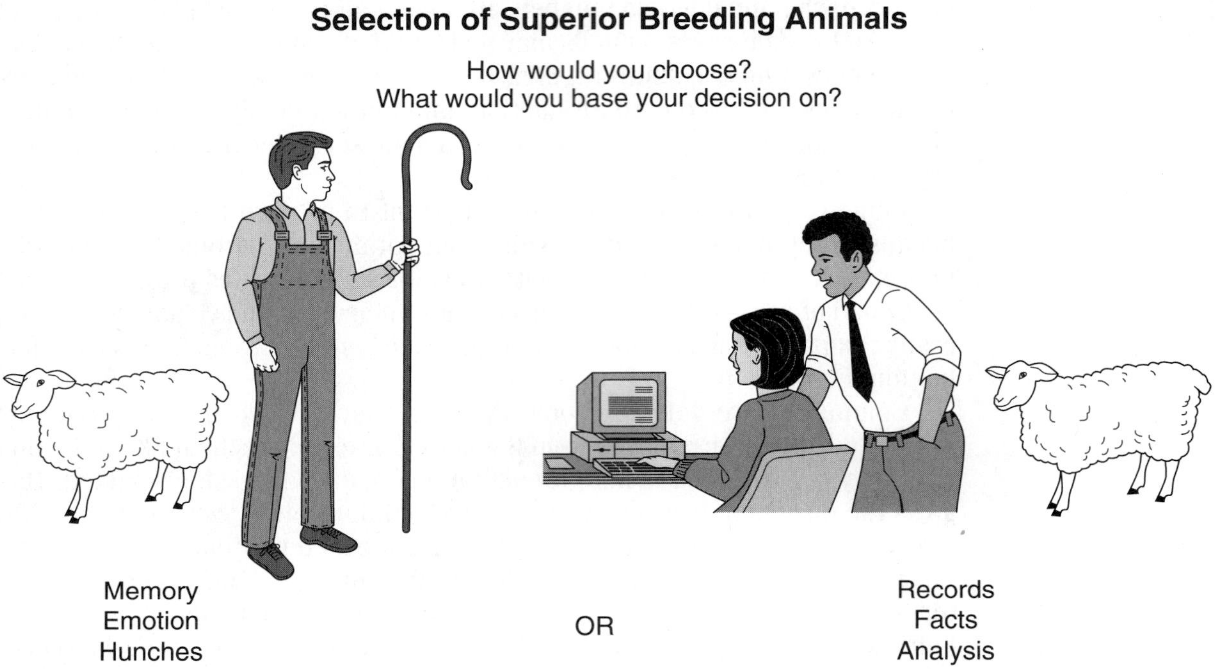

Selection of Superior Breeding Animals

How would you choose?
What would you base your decision on?

Memory
Emotion
Hunches

OR

Records
Facts
Analysis

FIGURE 3–5 Computer printouts containing production records are useful in comparing individual animals within the herd.

Production Record

Herd Ranking	Animal Number	Age	Index (100=average)
1	BR 714	4	107.32
2	BR 307	4	107.13
3	SK 10	5	106.87
↓	↓	↓	↓
39	BR 202	4	100.14
↓	↓	↓	↓
74	SK 19	5	94.19
75	BR 329	4	79.74

FIGURE 3–6 Herd improvement requires culling of inferior animals from the breeding herd. A computer generated index for each breeding female can be used to compare the relative value of an animal with other animals in the herd. Animals with low index numbers are candidates for culling.

One of the most valuable tools available to animal scientists and livestock farmers is a good set of **scales** for measuring production and animal weights. As market animals approach market weights, it is useful to analyze the cost per pound of gain. By comparing this cost to current market prices, a farm manager is able to make more profitable marketing decisions. Market animals can be sold as soon as the cost of additional weight gains approaches the market value of these gains. Computers and livestock scales are excellent tools for calculating these costs.

Electronic scales with digital read-out capabilities have replaced balance beam scales for many uses. Costs of electronic scales are comparable to the cost of balance beam scales. They have the advantage that they can be calibrated electronically, and they are easier to read without making mistakes. Regular use of scales to weigh breeding animals helps identify unthrifty animals because of the tendency of a sick animal to lose weight.

The information gained from weight records has been used to identify animals that have the capacity to grow and produce more efficiently than other animals raised under similar conditions.

Genetically superior animals have had a great impact on the production of most domestic species of livestock throughout the world. The major factors that measure production of animals (rate of gain, weaning weight, days to market weight, pounds of wool or milk, etc.) indicate that the **yield** (ability of farm animals to produce useful products) has improved a great deal in the last twenty years, Figure 3–7.

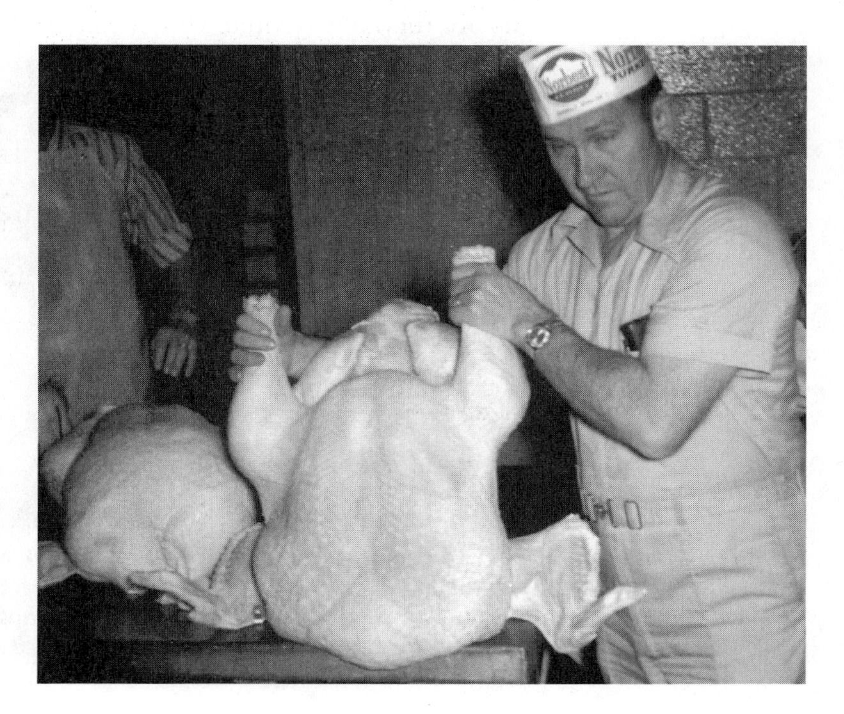

FIGURE 3–7 Production of animal products such as this 58.5 pound turkey depends upon selection of quality parent stock with the genetic ability to grow efficiently. *(Photo courtesy of Utah Agricultural Experiment Station)*

ANIMAL HEALTH

Advances in animal health technologies have made it possible to greatly reduce animal death rates and to improve production, Figure 3–8. Modern animal facilities such as animal shelters, feed storage and delivery systems, livestock corrals, and handling systems have been developed.

Many animal medications are available for use in treating infections and diseases. One of the most important classes of medications is antibiotics. Antibiotics and other medications have been developed for treating many specific livestock ailments. For example, mastitis is an expensive disease that causes the udder of a cow to become swollen and inflamed. Failure to treat this disease results in loss of milk production and poor quality milk. When animals are treated with antibiotics, the milk must be discarded for several days to eliminate the medication from the milk. Total losses from this disease in the United States amounts to around $3 billion each year.

A new protein has been developed using genetic engineering techniques that kills the bacteria responsible for mastitis. It does not contaminate the milk, and the milk of a treated cow may be used during the treatment period. The study of veterinary science for the improvement of animal health is nearly as advanced as the study of medicine is for humans, Figure 3–9. Since modern medicine has become available for animal use, losses due to poor health or death have been greatly reduced.

Animal **vaccines** are available to provide immunity to many diseases that cause death or illnesses in farm animals. **Parasites** that afflict farm animals are routinely treated with chemicals for control purposes. By following

FIGURE 3–8 Innovative research methods such as this lid opening into the stomach of a cow (fistula) have helped scientists study toxic plants and to accurately measure digestion of feeds. *(Photo courtesy of Utah Agricultural Experiment Station)*

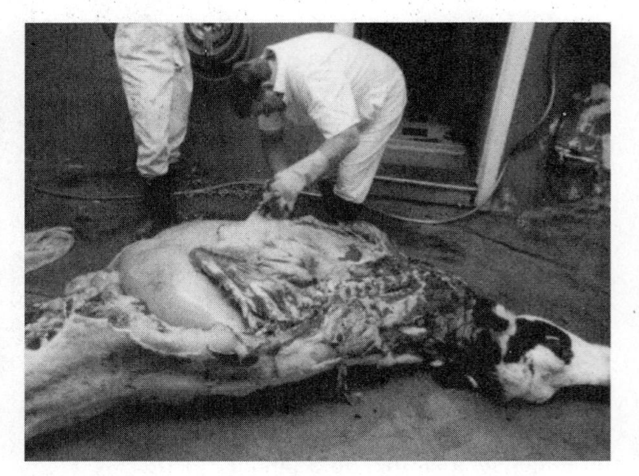

FIGURE 3–9 An animal disease is often diagnosed by performing an autopsy on an animal that has died. *(Photo courtesy of Utah Agricultural Experiment Station)*

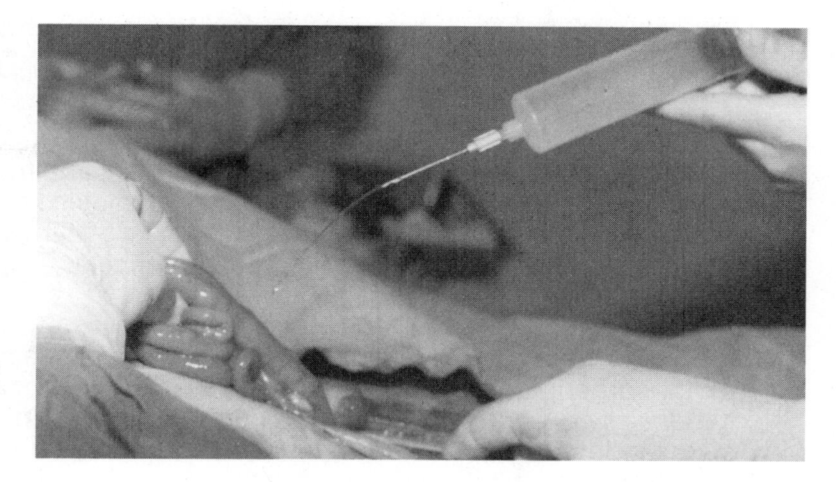

FIGURE 3–10 Veterinary surgery is sometimes used to correct health problems in animals. *(Photo courtesy of Utah Agricultural Experiment Station)*

recommended dosages, the animals are rid of harmful parasites that damage their internal organs, feed on their body fluids, and even chew holes in their hides. Treated animals are much healthier than untreated animals. It is always wise to consult with a local veterinarian to develop a herd health management plan, Figure 3–10.

An emerging field in animal health is the production of **monoclonal antibodies**. These substances are specialized proteins that bind to specific disease organisms and inactivate them. Specific antibodies can now be produced by fusing a cancerous cell with a cell that produces the desired antibody. The resulting cell is called a **hybridoma** and it produces large quantities of the antibody it was designed to produce. The new cell combines the infinite life span of the cancer cell with the ability to secrete antibodies. Feeding small amounts of the proper monoclonal antibody to a newborn calf or pig will provide protection from scours or diarrhea eliminating the need to vaccinate their mothers each year.

Animal diseases can be diagnosed by testing the effects of known monoclonal antibodies on disease organisms found in body fluids of sick or dead animals. By first coating small plastic sticks with specific monoclonal antibodies, the stick can then be tested to determine if the disease organism is present. The exposed stick is placed in a series of fluids. There will be a change in color when the disease agent is found. A similar procedure can be used to test for contamination of milk or other materials. This method is referred to as **dipstick diagnosis**.

Scientists are searching for ways to develop strains of animals that are genetically resistant to diseases. To do this, they must identify the genes that are responsible for resistance, or that improve the ability of the animal to respond to treatment. It is hoped that gene transfer will prove effective in placing genes that resist specific diseases into the chromosomes of animals in such a manner that the gene expresses itself in future generations. It must also be expressed in the proper place in the body and at the proper time.

CAREER OPTION

Veterinarian

A person who chooses a career in veterinary medicine must enjoy working with animals and with people, Figure 3–11. He or she must also be a strong science student with a high level of academic achievement during high school. A degree in veterinary medicine takes nearly as much college experience as a degree in human medicine requires.

A degree in veterinary medicine can lead to careers in research, teaching, or the practice of veterinary medicine. At the beginning of their careers, many veterinarians work for someone else. One reason for this is that they can learn from their more experienced colleagues. Working for an established clinic also gives them a chance to save money for the equipment they will need to set up their own practices.

A veterinarian who goes into private practice needs strong business skills in order to manage the clinic in a businesslike way. Strong communication skills are also needed as he/she meets clients each day.

A veterinarian must be prepared to respond to calls for help at any hour of the day or night. An experienced veterinarian can usually expect to earn an above average income, Figure 3–12.

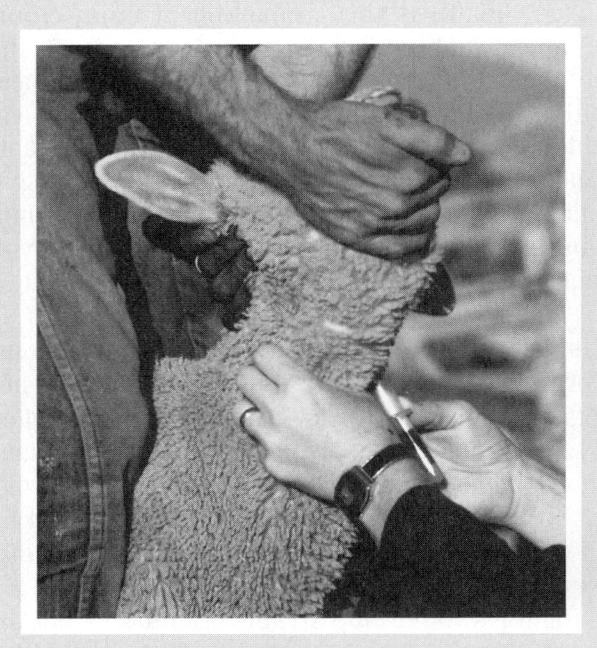

FIGURE 3–11 A veterinarian is an animal doctor who performs medical procedures on animals and/or conducts medical research using animals. *(Photo courtesy of Utah Agricultural Experiment Station)*

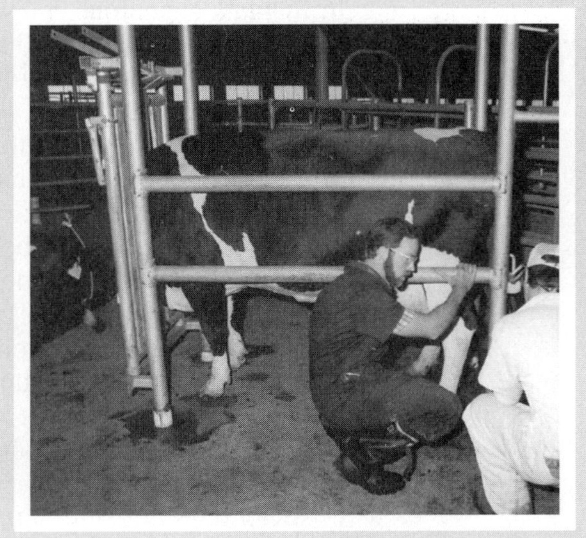

FIGURE 3–12 Large animals are constrained in strong holding stalls while examinations and treatments are performed by veterinarians. *(Photo courtesy of Utah Agricultural Experiment Station)*

FIGURE 3–13 Feed analysis is a service provided by many university laboratories and some private laboratories. The nutrients present in a sample of feed are isolated and measured carefully using high-tech lab equipment. Recommendations are provided for balancing animal rations.

Animal nutrition used to be concerned mostly with the amount of feed that an animal ate. Animal rations today are carefully analyzed to be sure they contain all classes of required nutrients. Procedures have been developed that make it possible to study the quality and amounts of minerals, vitamins, amino acids (components of protein), and other feed ingredients.

The nutritional needs of different classes of animals have been carefully studied, Figure 3–13. Many of the livestock rations used today are formulated by computer technicians. They carefully balance the nutritional needs of animals by blending feed materials from many sources in mixed rations. This approach to nutrition has become quite popular in confinement livestock farming enterprises such as dairy, feedlot beef, swine, aquaculture, and poultry production.

Grasses and other forage plants contain large amounts of fiber that can only be digested by the bacteria that are present in the stomach of an animal. Scientists have learned that they can improve the digestibility of forage plants by zapping these feeds with a special laser beam. The digestibility of most grasses can be improved by more than 10 percent using this treatment.

A new animal health industry, which has emerged in recent times, is the culture and sale of bacteria and other microbes to stimulate the digestive processes of animals. Large numbers of microbes are raised in laboratories, freeze dried, and packaged for sale as livestock feed supplements. When the microbes are fed to an animal, they become active in the digestive tract and stimulate the process of digestion.

Animal health technology includes simple devices such as cow **magnets** to prevent hardware disease in cattle. They function by attracting and immobilizing nails, wire, and other metal objects that contaminate feed to prevent them from passing through the digestive tract. The magnet is swallowed by a cow, and it holds metal objects in an area where they are unlikely to puncture the stomach wall.

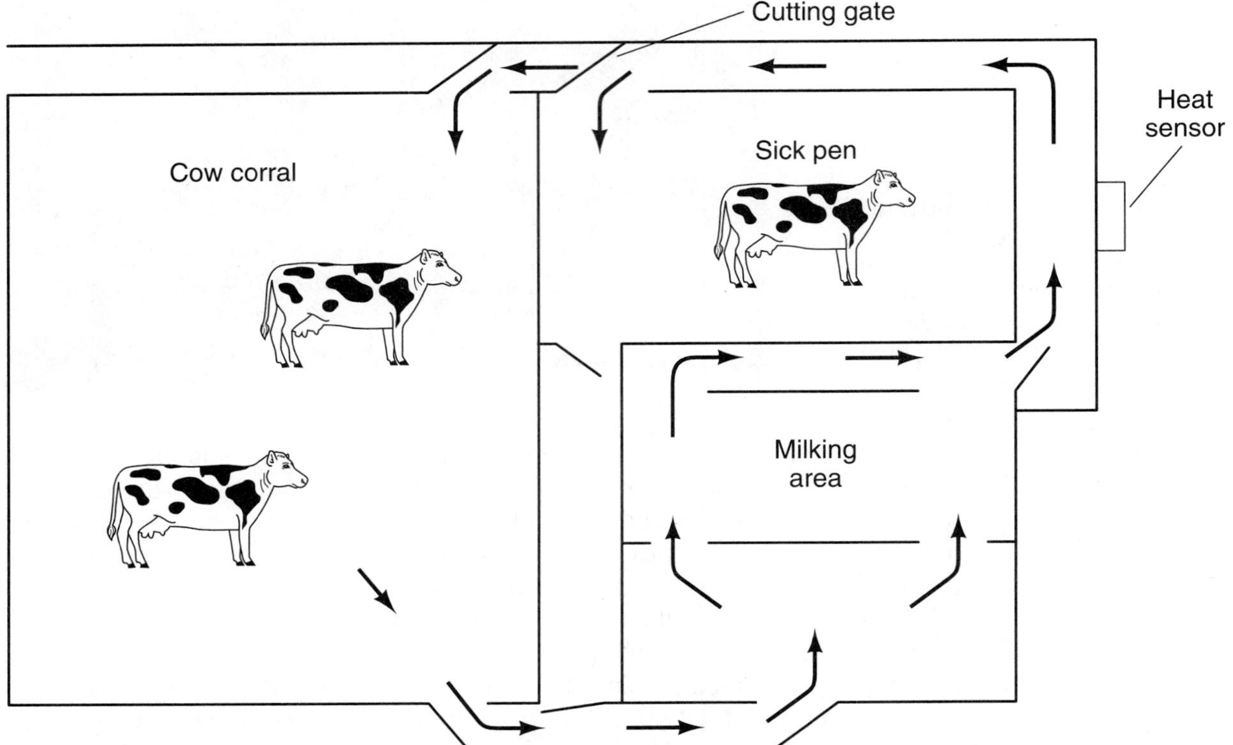

FIGURE 3–14 The body temperature of an animal is recorded as the animal passes near the sensor. Animals with abnormal temperatures are automatically separated from the herd by electronic cutting gates.

Animal health technology also includes complex devices such as remote temperature sensors, which measure radiant body heat of animals. This machine may be linked to a cutting gate that is capable of sorting sick animals with abnormal temperatures into special pens for medical care, Figure 3–14.

Modern equipment has been developed for handling and transporting animals. Some animal stress is always present when animals are moved or handled, but equipment that is designed to accommodate animal behavior patterns reduces animal stress and stress-related illnesses.

HORMONE INDUCED YIELDS

Hormones are substances that are formed in the organs of the body. They are carried by body fluids to other tissues where they cause specific body functions to occur. Scientists have learned how to make or refine some important hormones in the laboratory. Hormones obtained in this manner are often called **synthetic hormones**. Animal research scientists have learned that some kinds of hormone treatments increase the rate and efficiency of growth in meat animals. Sometimes the hormones are provided in very small amounts in the feed of animals. Other kinds of hormones are injected into animals on a regular basis. Hormones that are injected into the tissues of an animal are absorbed rapidly. Some hormones are formed into tiny pellets and injected under the skin in the ear of an animal. When hormones are provided in this manner, the materials are absorbed slowly.

In some instances, genes that cause extra growth hormones to be produced have been transferred into the cells of embryonic animals using genetic engineering techniques. At the United States Experiment Station in Beltsville, Maryland, a strain of hogs has been developed that produces much larger amounts of growth hormones than are usually present. The result of this experimental work is that the genetically engineered hogs grow much more rapidly than would be expected. They also produce less fat that normal pigs.

Production of milk can be increased as much as twenty to forty percent by treating dairy cows with growth hormones, Figure 3–15. This hormone material is also known as BST or **bovine somatotropin**. This hormone treatment has been cleared for use by the United States Food and Drug Administration and the United States Department of Agriculture. BST is a protein that occurs naturally in milk. It is a hormone substance that stimulates milk production. BST can be produced in the laboratory using a fermentation process, and when it is injected into dairy cows, they become more efficient at converting their feed to milk. Milk from treated cows cannot be distinguished from the milk of untreated cows.

Because BST is a product of biotechnology, it has been criticized as a product that is potentially unsafe for human consumption. Research indicates that BST is safe for consumers. Its value to the dairy industry comes from the fact that fewer cows are needed to produce a given amount of milk, or that treated cows will produce significantly more milk than the same number of untreated

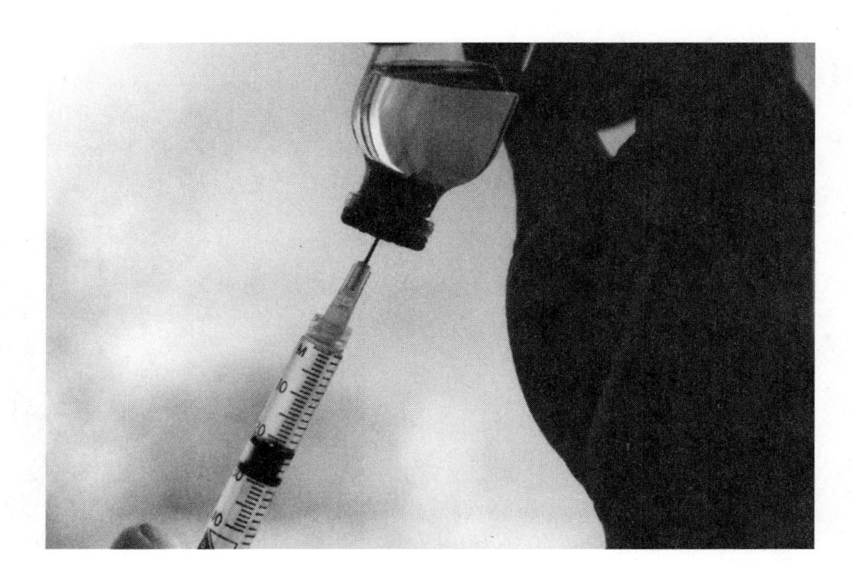

FIGURE 3–15 Hormones such as bovine somatotropin (BST) are sometimes injected in animals to stimulate the production of milk and meat. *(Photo courtesy of Utah Agricultural Experiment Station)*

cows. Perhaps one day we will have dairy cows that carry the genetic code responsible for producing large amounts of growth hormone naturally. Such cows would bring a new level of efficiency to the dairy farms of the world.

The use of hormones, whether by injection into animals or by adding them to the feed that animals consume, has caused controversy in recent years. Some hormones originally approved for use have been taken off the market due to improper use by some producers of animal products. Other materials have been resisted by some consumer groups. A few producers are voluntarily avoiding the use of hormones and other chemicals in the production of livestock and poultry in an effort to fill the market niche these consumer groups have created.

Scientists in Canada have proven that fish injected with growth hormones grow 50 percent faster than other fish. Growth hormones were developed using genetic engineering procedures.

Scientists in the United States have injected a growth hormone gene obtained from trout into carp eggs resulting in transgenic fish that grow 20 percent faster than other carp.

ELECTRONIC ANIMAL MANAGEMENT SYSTEMS

Electric devices have done much to eliminate the manual labor associated with livestock chores. Very few cows are milked by hand these days due to the development of electronic milking systems, Figure 3–16. Many other electrical systems are found on modern livestock farms performing tasks from cooling milk, eggs, vegetables, and fruit to heating livestock facilities. They reduce expensive hand labor and make it possible for a farmer and his family to manage many more animals than was possible just a few years ago.

Feeding systems make it possible to store, process, and deliver many forms of feed to animals and poultry. Storage structures are capable of drying high

FIGURE 3–16 A common example of an electronic animal management system is this modern milking parlor. *(Photo courtesy of Utah Agricultural Experiment Station)*

FIGURE 3–17 Electricity is an energy source that is used to move grains and other similar materials from one bin to another through a series of augers.

moisture grains or preserving moist feeds. Auger systems deliver feed to milling facilities or directly to the feeding areas, Figure 3–17. Even the water supply frequently depends upon electricity for pumping water and to prevent freezing of water lines.

Electronic sensors control livestock ventilation systems in confinement livestock farming. They may also be used to activate sprayers used for applying insecticides or delivering a cool mist to animals on a hot day.

Electronic sensors may be programmed to deliver specific amounts of a ration to individual animals. This technology is especially useful in feeding dairy cows. Animals that are high producers require larger amounts of feeds that are rich in energy to maintain proper health and body fat reserves than less productive cows. Identification sensors that are worn by cows can identify each cow that enters the feeding stall. A computer link to the stall turns on the feed delivery system to provide a custom ration based on the production performance of the cow.

The use of **robots** is beginning to play a role in the production of animal products. Routine tasks, performed the same way time after time, are being performed in some research facilities by specially designed robots. Milking cows is a task capable of being solved by robotic devices. One new experimental facility is developing a modern milking system that can wash, milk, and feed a cow. The computer records milk production and feed consumption for each cow. The cow enters the milking area voluntarily several times each day. The milking machine is attached to the cow by a robotic arm, and it is automatically removed when the cow has been milked.

The task of shearing sheep may soon be performed by a robotic arm that first measures the sheep, calculates the movements of the shearing tool, and then removes the wool. Human labor will still be required to put the sheep in a restraining device, but the tedious labor of removing the wool will be performed by the robotic arm.

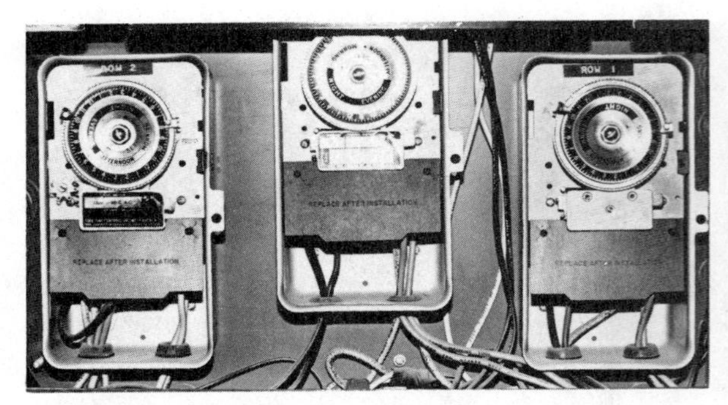

FIGURE 3–18 Light-sensitive devices and/or timing devices are used to turn on supplemental lights.

FIGURE 3–19 A simple, but very important, technology for farms was the development of grain augers for moving grain. This implement has eased the labor intensive chore of shoveling grain on most livestock and crop farms.

Electrical timing devices perform many chores in modern agriculture, Figure 3–18. They are used on many farms to turn on ventilation systems to provide fresh air at frequent intervals. They are used to turn on the lights to artificially lengthen the day for laying hens to stimulate egg production. Even the breeding cycles of sheep may be modified using electric time switches to artificially adjust the length of the daylight period. Electrical power is probably the most widely used technology in modern agriculture. Many other modern agricultural technologies depend upon electricity as an energy source.

Remote television monitors are used in agriculture to observe animal behaviors, provide security, and to monitor machinery in automated agricultural processing facilities. They are especially useful to livestock farmers for observing animals during the birth process (parturition). This allows the farm manager to keep track of potential problems during the day or night without causing a disturbance or having to leave his home or office except to assist with problem births.

The use of electrical energy on the farm has done much to ease the burden of hard manual labor. The distribution of electrical power to rural areas during the 1930s was one of the greatest contributions to efficient farm operation in history. New labor-saving devices continue to be developed, and the technology trend seems likely to continue, Figure 3–19.

RECYCLED ANIMAL WASTES

Recycled animal wastes include such products as manure and animal body parts **(offal)** from processing plants, Figure 3–20. Such materials have often been of little value in the past, but new uses have been found for many waste products from animals.

FIGURE 3–20 Animal manure can be composted to create a useful product for use as a fertilizer or mulching material.

During the process of digestion, animals tend to use fewer nutrients than they have available to them in their feed. **Recycled nutrients** are feeds that use these wasted nutrients in livestock diets. For many years, farmers have recycled wasted nutrients by raising a few hogs in the pens of fattening cattle. The hogs gathered and ate kernels of corn that had gone through the digestive tracts of the cattle without being digested. Small farm flocks of chickens were common sights on most farms until recent years. They were allowed to run freely around the farm to gather their own feed from any source they could find.

Recycled animal wastes have recently become much more important as sources of income. Animal wastes can be recycled as livestock feeds, or recovered as methane gas, or dried manure as sources of heat. Heat from such sources may be used to generate steam for use in the generation of electricity. It can also be used for commercial processing or manufacturing.

Several methods have been used to recycle animal waste on the farm. Large cattle feedlot operators have constructed large concrete slabs on which feed bunks are placed. Manure is "harvested" each day from these areas and mixed with fresh chopped forage. This mixture is then placed in a silo for curing and storage. By the time the feed has fermented, the unpleasant odor is gone, and the mixture of animal manure and fresh forage has been transformed into pleasant smelling, palatable cattle feed.

Waste nutrients from hogs have been recycled by pumping liquid from the manure pit to pipe fittings from which the hogs can drink. The manure pit is stirred frequently to keep solid materials suspended in the liquid. Hogs tend to play in the liquid nutrient suspension and much of the hog's daily requirement for water is provided from this source. Waste nutrients that are suspended in the liquid are recycled in this manner.

A growing source of recycled nutrients is obtained from poultry processors. Feathers provide a source of protein and other nutrients for ruminant animals. Once the feathers have been processed and pelleted, they become palatable

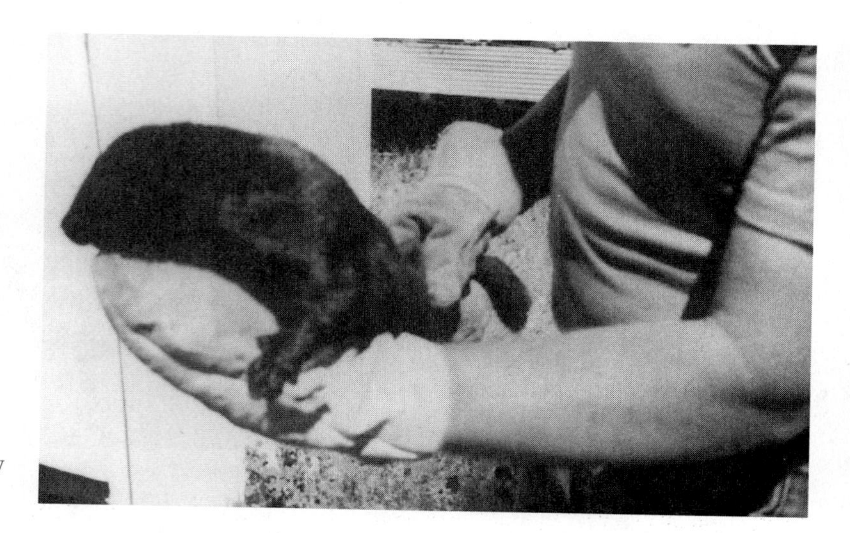

FIGURE 3–21 The fur-bearing industry is one that depends heavily on recycled animal protein.

to cattle, and provide a reasonable source of protein and other nutrients. Poultry manure, which is high in nitrogen, can also be used by the rumen bacteria of cattle as a source of protein.

The agricultural industry has used the offal from processed fish, poultry, and livestock as part of livestock rations for many years. They also depend on waste animal protein from slaughter houses for components of fish rations. Another industry heavily dependent on recycled animal protein is the fur-bearing animal industry, Figure 3–21. Recycled nutrients from animal sources are very important to the livestock industry.

CHAPTER SUMMARY

The advance of agricultural technology has provided many new management tools for management of farm animals. Computers have emerged in nearly every phase of farm management. They are used to balance animal rations, to keep records of animal production, and to select quality replacement animals. They are also valuable in making livestock marketing decisions.

Animal health technologies include all of the advancements that have come about in the sciences of nutrition and veterinary care. They include methods of testing for diseases, development of vaccines, and production of other animal care products. Machinery and equipment for analyzing and processing feed, and the development of structures and devices that may be used to control the environment of animals are also health-related technologies.

Hormones are substances that are produced by animals to control body functions. Synthetic hormones have been used to increase the productivity of farm animals. They are useful in improving the rate and efficiency of gains, and they can be used to stimulate milk production.

The use of electricity on the farm has greatly reduced hand labor. It provides energy to operate sensing devices used for purposes such as environmental control, identifying sick animals, and automatically dispensing individual animal rations.

Nutrients that remain in animal wastes can be recycled for use as animal feeds. Ruminant animals, such as cattle, are able to use recycled animal wastes quite effectively. Offal is a valuable source of protein for fish, poultry, and many classes of farm animals.

CHAPTER REVIEW

Discussion and Essay Questions

1. Identify five technologies that have resulted in improved yields of animal products.
2. Describe two kinds of computer generated information that is useful in selecting superior breeding animals.
3. Discuss two forms of animal health technology that have resulted in higher yields of animal products and lower animal death rates.
4. Explain how monoclonal antibodies are used to diagnose animal diseases. How may they be used to prevent animal diseases?
5. Outline the procedure for diagnosing an animal disease using the dipstick procedure.
6. Suggest two practices that can improve the digestion of feeds by farm animals.
7. What are some positive aspects of hormone use in the production of animal products? Negative aspects?
8. Justify the use of hormones in animal production, taking into account the resistance to hormone use by some consumers.
9. List some major uses of electricity in managing livestock.
10. What are some advantages and disadvantages of recycling waste products from animals?

Multiple Choice Questions

1. Which of the following methods for identifying breeding animals is considered to be a product of modern technology?

   A. computer chip
   B. ear tag
   C. tattoo
   D. brand

2. The most important function of a computer as a tool in livestock production is to:

   A. organize information.
   B. create new information.
   C. cull inferior animals.
   D. pay the bills.

3. An accurate scale with which individual animals can be weighed is useful in obtaining all of the following production information *except*:

   A. weaning weight.
   B. rate of gain.
   C. conformation score.
   D. pounds of feed per pound of gain.

4. A production index number for an animal provides the following information:

   A. the page in the herd book on which the animal is listed.
   B. the value of an animal in comparison with animals that belong to other producers.
   C. the identity of the animal based on an electronic code imbedded in a computer chip.
   D. the relative value of the animal in comparison with its herdmates.

5. Which of the following materials is used to identify disease organisms using a dipstick analysis procedure?

   A. ligase
   B. monoclonal antibody
   C. hybridoma
   D. antibiotic

6. The modern electronic device classed as a remote temperature sensor performs which of the following functions in a livestock facility?

   A. maintains the drinking water at a constant temperature
   B. records the high and low temperatures in buildings where animals are confined
   C. identifies unhealthy animals that have unusually high or low body temperatures
   D. identifies females that are ready for breeding

7. The growth hormone known as bovine somatotropin or BST is used in dairy cows to:

   A. stimulate milk production.
   B. increase the protein content of milk.
   C. increase the body size of small cows.
   D. improve the breeding efficiency of the cows.

8. Most automated management systems operating on modern livestock farms and ranches are activated by devices called:
   A. hydraulic rams.
   B. electronic sensors.
   C. pneumatic switches.
   D. condensors.

9. A very important energy source that became available to most farms and ranches during the 1930s is:
   A. solar power.
   B. wind power.
   C. geothermal heat.
   D. electricity.

10. Offal is a form of recycled agricultural waste material that is composed of:
   A. animal body parts.
   B. livestock manure.
   C. vegetable waste.
   D. compost.

LEARNING ACTIVITIES

1. Conduct a feed trial using rations deficient in different nutrients. Purchase some day-old chicks to be used as subjects; tag them for absolute identification; pen them in separate groups; and feed all but one group of chicks a ration deficient in an essential nutrient. Record the weights of the chicks each week for 4–6 weeks. Graph the results and discuss how each group performed in comparison with the other groups.

2. Conduct a mock trial in the classroom with students serving as judge, attorneys, jury, court officers, witnesses, plaintiff, and defendant. Charge the defendant with failure to hold an animal product off the market for a long enough period of time after injecting antibiotics to control an infection in the animals.

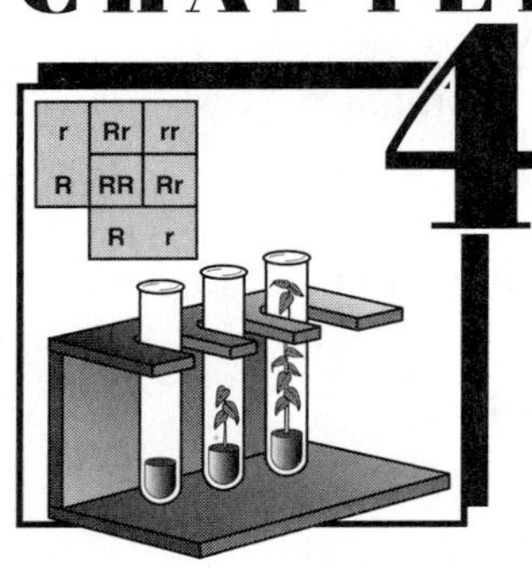

CHAPTER 4 Animal Reproduction Technologies

Farm animals are considered to be one of our most important renewable resources. They are not used up as they are consumed, but continue to produce more of their kind each year. The ability of animals to reproduce themselves makes them different from many resources such as oil or mineral deposits that become depleted with use.

The rate and frequency with which farm animals reproduce varies greatly from species to species, Figures 4–1 and 4–2. By improving the efficiency of reproduction, it becomes possible to maintain breeding herds and still increase the supply of animal products available for consumption. Animal reproduction technologies make it possible for animals to reproduce more efficiently than they could in a natural environment.

FIGURE 4–1 The reproductive rates of farm animals vary greatly. Swine are among the most efficient producers of offspring, producing large litters of young two or three times per year. *(Photo courtesy of the American Landrace Association, Inc.)*

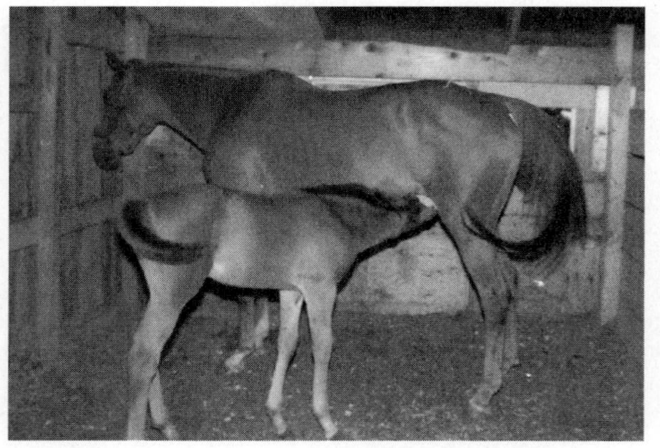

FIGURE 4–2 Mares rarely give birth to twins (1 percent–5 percent); and the period between conception and birth (gestation period) ranges from 334–344 days. *(Photo courtesy of Barbara Lee Jensen)*

OBJECTIVES

After completing this chapter, you should be able to:

- describe reproductive technologies associated with farm animals.
- estimate the rates of genetic improvement for specific traits in farm animals based on the heritability factors associated with the traits.
- explore the advantages and disadvantages of using artificial insemination as a management tool in animal breeding.
- list advantages and disadvantages of embryo transfer as a livestock reproduction tool.
- propose methods to modify the reproductive cycles of farm animals.
- recommend ways to control the time of birth for farm animals.
- describe how the gender of an animal is determined.
- report research efforts to control the gender of farm animals.
- define the steps involved in embryo transfer technology.

TERMS FOR UNDERSTANDING

The following vocabulary terms should be studied carefully as you read:

fertilization
zygote
sexual reproduction
prostate
bulbourethral glands
semen
copulation
urethra
oviduct
vagina
cervix
uterus
placenta
fetus
dystocia
pelvic measurement
proof
progeny
genetics
inheritance
heritability factor
organism
birth rate
artificial vagina
rectal probe
extender
embryo
embryo transfer
donor
recipient
synchronize
FSH
ova
estrus
anestrus
photoperiod
northern hemisphere
southern hemisphere
implant
pessary
conceive
parturition
environmental conditions
gender selection
autosomes
sex chromosomes
flow cytometry

REPRODUCTIVE MANAGEMENT

Reproduction in farm animals is a sexual process. Gametes consisting of the sperm from the male and ova from the female (haploid cells) are produced through the process known as meiosis that was discussed in Chapter 2. The sperm and ovum join together in the process known as **fertilization** to form a diploid cell called a **zygote**. The production of gametes and the process by which they join together is **sexual reproduction**.

All farm animals reproduce in this manner. In most cases, fertilization occurs internally after the male deposits sperm inside the reproductive tract of the female. Fish are notable exceptions in that the females lay eggs in nests where the males deposit sperm on the eggs.

The male and female reproductive tracts are quite different from one another, Figure 4–3. Reproduction in the male begins with the production of sperm cells in the testes. As the sperm passes through the male reproductive tract, fluids are added by other male reproductive organs such as the **prostate** and the **bulbourethral glands**. This combination of sperm and fluids is called **semen**. During **copulation** or breeding, semen passes through the

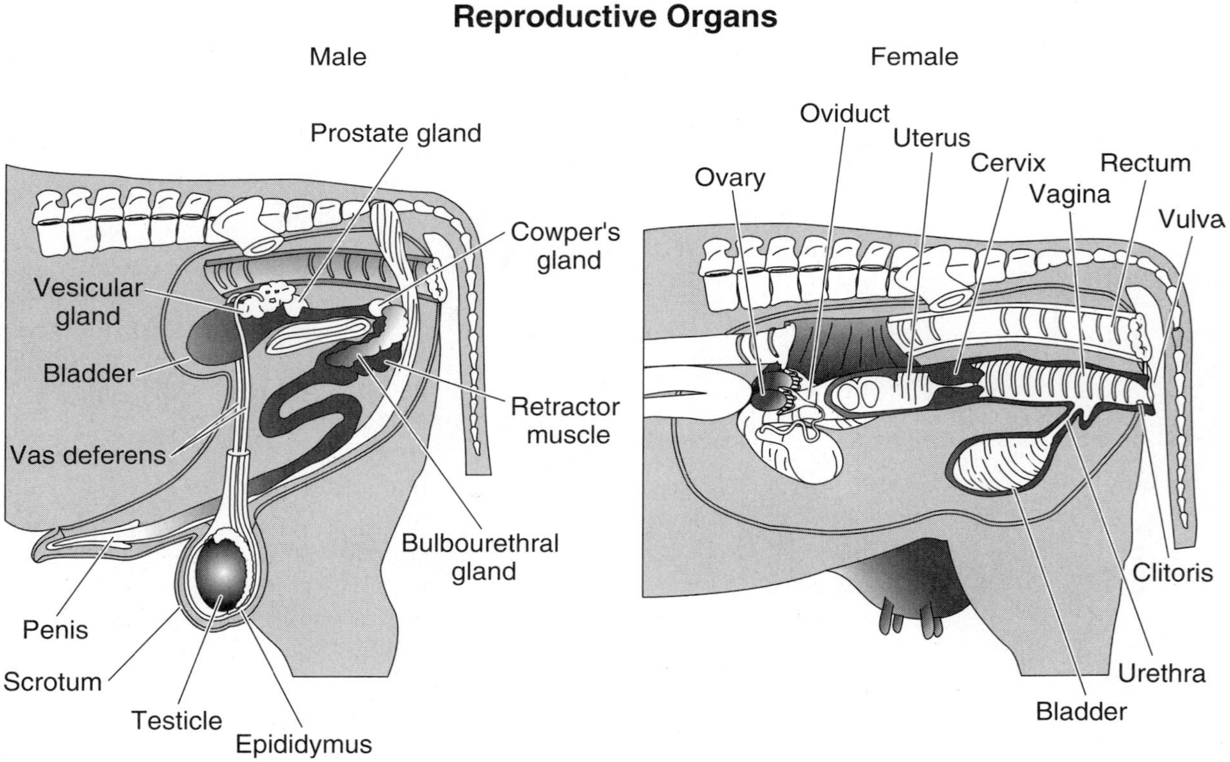

FIGURE 4–3 The male and female organs for reproduction develop from the same tissues, but major differences are evident by the time the animal is born.

tubular **urethra** located in the male penis as it is deposited in the female reproductive tract.

Female reproduction begins with the formation of an egg or ovum in an organ called an ovary. When the ovum is mature, it is expelled from the surface of the ovary, and it enters a tube called the **oviduct**. It is common for fertilization to occur in this location. Sperm are deposited in the female organ called the **vagina** from which they migrate through the **cervix**. This is a muscular organ that seals off the opening and protects the zygote during pregnancy.

The zygote eventually becomes attached to the inner wall of the female organ called the **uterus**. In this location, nutrients from the mother's blood supply pass into the blood supply of the zygote through an organ called the **placenta**. Nutrients and waste materials pass back and forth through the walls of the blood vessels of the uterus and the placenta. There the zygote matures and grows into an immature "baby" animal called a **fetus**.

Reproductive management is a system of selective breeding in which animal matings are planned and not left to chance. Factors that affect the ability of a breeding female to successfully conceive, carry to term, deliver, and nurture her offspring are all important, Figure 4–4. Management of reproduction also includes identifying and culling breeding animals that do not possess these traits or that tend to be weak in them.

A major problem associated with most livestock species is **dystocia** or difficult birth problems due to a variety of reasons. One reproductive management technique that has been proven to be effective in cattle is **pelvic measurement** in both males and females to predict ease of calving. Heifers at the first calving risk reproductive injuries when they have dystocia. Damage to the reproductive tract often results in infection and may delay or prevent future pregnancies. This problem can be reduced by selecting replacement heifers and bulls with large pelvic measurements.

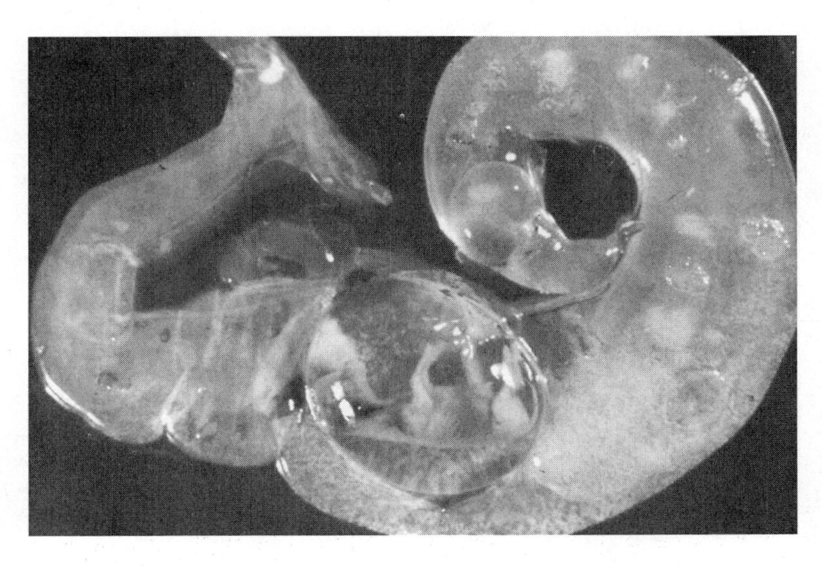

FIGURE 4–4 Reproductive management includes all of the events associated with producing and nurturing a living fetus. *(Photo courtesy of Utah Agricultural Experiment Station)*

Another management tool that addresses the dystocia problem is the **proof** record of the sire. This record establishes what can be expected from the genetic performance of a sire based on past performances of his offspring, or **progeny**. Most commercial breeding associations for cattle provide calf size indexes for their proven sires. These index ratings are based on the birth weights of calves that the bull has sired.

The science of **genetics** includes the study of **inheritance**, or the manner in which traits found in parent animals are passed to the offspring. A **heritability factor** is a measurement of the tendency for a live **organism** to pass physical characteristics on to its offspring. It is expressed as a numerical value. A low heritability rating would include values ranging up to .3 (30 percent heritable). A higher value indicates that genetic improvement for that trait will be achieved more rapidly, Figure 4–5.

The ability to measure heritability factors of different animal characteristics is very important in selection of superior breeding stock. Rapid genetic changes can be achieved for highly heritable traits, but genetic changes are

Trait	Percent Heritability					
	Cattle	Sheep	Swine	Poultry	Rabbits	Horses
Fertility	0–15	0–15	0–15	0–15	—	Low
Number of young weaned	10–15	10–15	10–15	—	3	—
Weight of young at weaning	15–25	15–20	15–20	—	35	—
Postweaning rate of gain	50–55	50–60	25–30	—	60	—
Postweaning gain efficiency	40–50	20–30	30–35	—	—	—
Fat thickness over loin	45–50	—	40–50	—	—	—
Loin-eye area	50–60	—	45–50	—	60	—
Percent lean cuts	40–50	—	30–40	—	60	—
Milk production, lb	25–30	—	—	—	—	—
Milk fat, lb	25–30	—	—	—	—	—
Milk solids, nonfat, lb	30–35	—	—	—	—	—
Total milk solids, lb	30–35	—	—	—	—	—
Body weight	—	—	—	35–45	40	—
Feed efficiency	—	—	—	20–25	—	—
Total egg production	—	—	—	20–30	—	—
Age at sexual maturity	—	—	—	30–40	—	—
Viability	—	—	—	5–10	—	—
Speed	—	—	—	—	—	25–50
Wither height	—	—	—	—	—	25–60
Body length	—	—	—	—	—	25
Heart girth circumference	—	—	—	—	—	34
Cannon bone circumference	—	—	—	—	—	19
Points for movement	—	—	—	—	—	40
Temperament	—	—	—	—	—	25

FIGURE 4–5 The heritability factors for inherited traits of domestic animals vary greatly. *(Courtesy Instructional Materials Service, Texas A&M University)*

achieved at a slow rate when selection of breeding animals is based on traits that are low in heritability. Examples of traits that are rated high in heritability are postweaning rate of gain and loin-eye area. A well-muscled meat animal tends to pass superior muscling on to its young. Fertility is an example of a heritability factor that is considered to be low. Progress in improving the **birth rate**, or the number of young that are born from each pregnancy, is slow in most domestic animals. Many generations of animals may be required to increase the number of offspring born from each pregnancy.

Hormones are important management tools in animal reproduction. They are natural substances that are found in the body fluids of animals and the juices of plants. These substances control or regulate functions of the organisms such as metabolism of food, reproduction, and growth. They are secreted by glands located throughout the body. The most significant hormone producing glands include the ovaries, testes, pancreas, and adrenal cortex.

Hormones are so powerful that only small amounts are needed to stimulate changes in body processes. When large amounts of hormones are found in the body, they can disrupt body processes or even damage organs. Some hormones can be produced using laboratory processes. These materials are called synthetic hormones. Many of these materials are as effective as naturally-produced hormones.

ARTIFICIAL INSEMINATION

The process of artificial insemination involves the use of specialized equipment to place sperm from the male in the reproductive tract of the female at a time when pregnancy can occur, Figure 4–6. When this method is used, semen collected from a male animal is placed in the reproductive tract of a female animal using a slender tube or a gelatin capsule.

Artificial insemination of all major species of farm animals has become a reality. It is practiced frequently with cattle and hogs, and some artificial insemination has occurred in the purebred sheep industry. Even the turkey industry has adopted this important technology. Sale of semen from valuable sires has become a big business. This technology has made it possible for livestock farmers throughout the world to have access to proven sires without owning them or managing them on the farm. This technology has helped improve the quality of farm animals much more rapidly than it could have been done using natural breeding practices.

The process involves several steps beginning with collection of sperm from the male animal. This is usually accomplished by using a soft rubber cylinder called an **artificial vagina**. An electric **rectal probe**, which applies an electrical shock to the male reproductive tract, can also be used for this purpose.

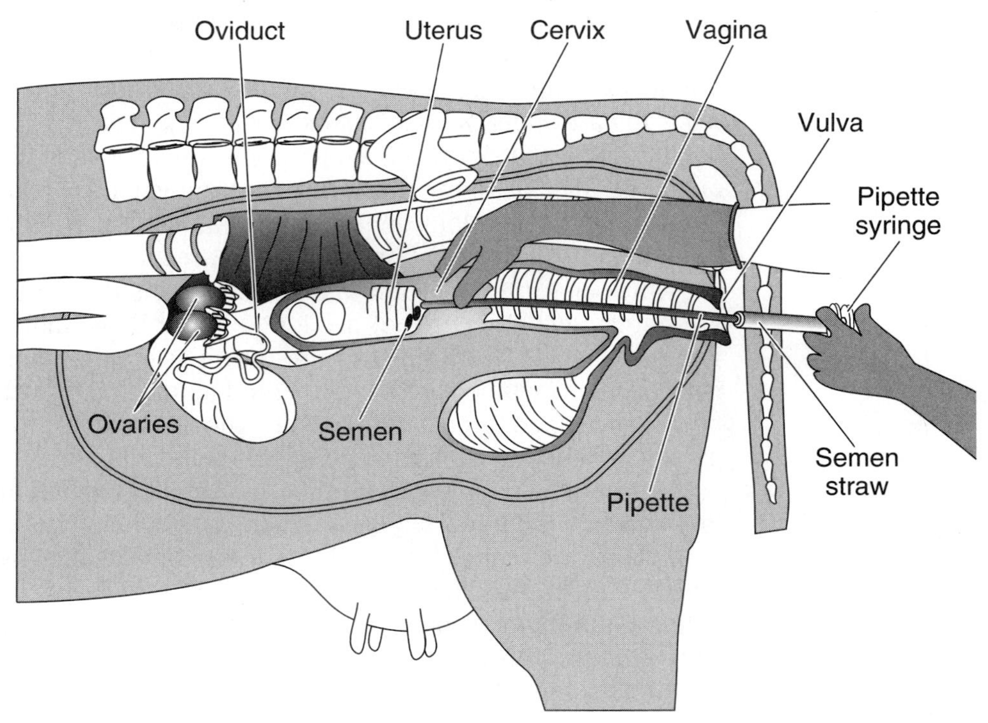

FIGURE 4–6 Artificial breeding of cows involves the use of a hollow stainless steel or plastic pipette. It is used to place semen from the bull in an appropriate location within the reproductive tract of the cow to cause conception and pregnancy.

CAREER OPTION

Animal Breeding Technician

A person who chooses a career as a technician in animal breeding will provide livestock breeding services for clients who operate livestock farms. He/she will need a strong interest in animals and the ability to communicate well with people. The technician operates a business that requires daily travel to the area livestock farms that request breeding services. In most cases the technician will use artificial insemination techniques to breed domestic animals. Other technicians perform ova (embryo) recovery and transfer services, Figure 4–7. Still other technicians work at stud farms where semen is collected, processed, packaged, and stored.

Education for this career usually requires completion of a short-term program requiring several days to several months depending upon which animal breeding options are chosen. The work schedule is demanding and lacks flexibility, because each breeding animal must be serviced at the appropriate time if she is to become pregnant.

Once the semen is collected it is examined to check for abnormal sperm and to determine how many sperm cells are present, Figure 4–8. **Extender** is a nutrient solution that is added to the sperm to dilute it and to increase its volume. The diluted semen is then packaged and frozen for storage in liquid nitrogen or used fresh. Most semen used in cattle is stored in straws and frozen

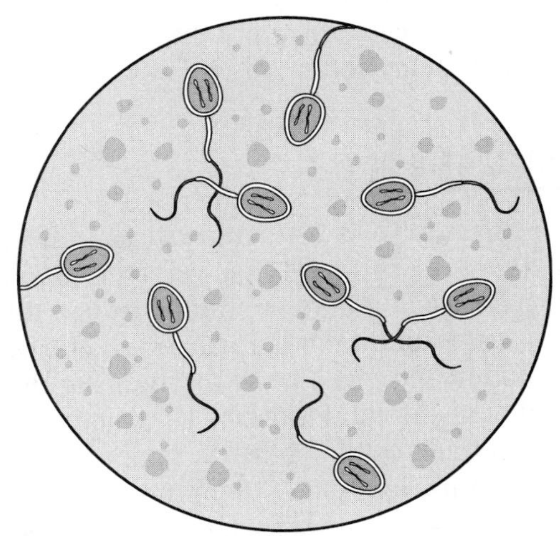

FIGURE 4–8 Semen that is collected for use in artificial insemination must be checked using a microscope to determine its concentration and quality. The semen is then diluted and used fresh, or it is frozen for later use.

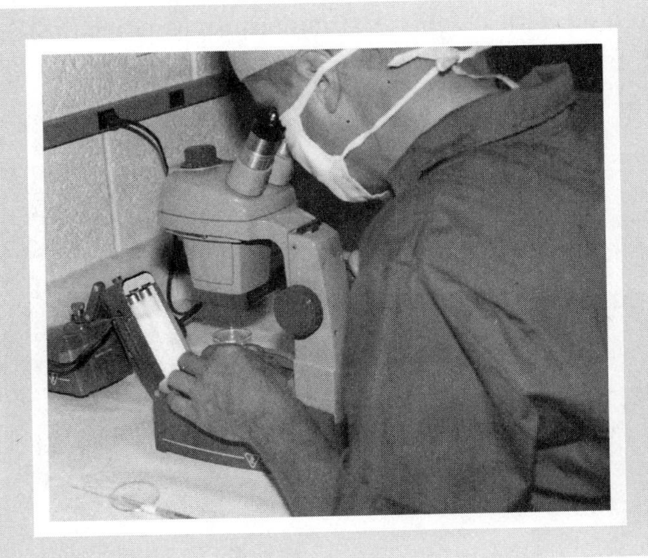

FIGURE 4–7 Animal breeding technicians provide several important services. They include collection, evaluation, and processing semen, as well as providing artificial insemination and/or ova transfer services. *(Photo courtesy of Utah Agricultural Experiment Station)*

until it is needed. A large amount of the semen used in the horse, hog, and sheep industries is still used within a few hours as fresh semen.

The impact of artificial insemination on the domestic livestock industry has been immense. Very few dairy bulls are found on farms anymore. The technology has been adopted throughout the dairy industry. In addition to higher production of dairy products, deaths of farm workers due to attacks by bulls have been greatly reduced. An additional benefit is that secure facilities for herd bulls are no longer needed, and the cost of feeding and caring for dairy herd bulls has been nearly eliminated.

EMBRYO TRANSFER

An **embryo** is a fertilized egg that has divided to form new cells. **Embryo transfer** is sometimes referred to as ova transfer. It is the practice of placing living embryos obtained from a **donor** animal in the reproductive tract of a **recipient** animal, Figure 4–9. The second animal usually has a normal pregnancy and gives birth to offspring that are genetic sons or daughters of the donor animal, Figure 4–10. One of the most significant technologies of modern times is the ability to remove several living embryos from valuable female animals, and place them in the reproductive tracts of less valuable animals. This allows a livestock farmer to produce many more offspring from his best females than he/she was able to produce under natural conditions.

The process begins by treating donor females and recipient females with specific hormones to synchronize their estrus cycles. This is necessary to be sure that the environments in their reproductive tracts are similar, and will not injure the embryos when they are transferred. Just before the donor female is due to ovulate, she is treated with a follicle stimulating hormone (FSH), increasing the number of ova or eggs that are produced. The purpose of this treatment is to produce several embryos, Figures 4–11 and 4–12.

FIGURE 4–9 Embryo transfer makes it possible for female breeding animals to give birth to offspring whose genetic makeup is different from their own. This Mouflon lamb was born to a domestic ewe. *(Photo courtesy of Utah Agricultural Experiment Station)*

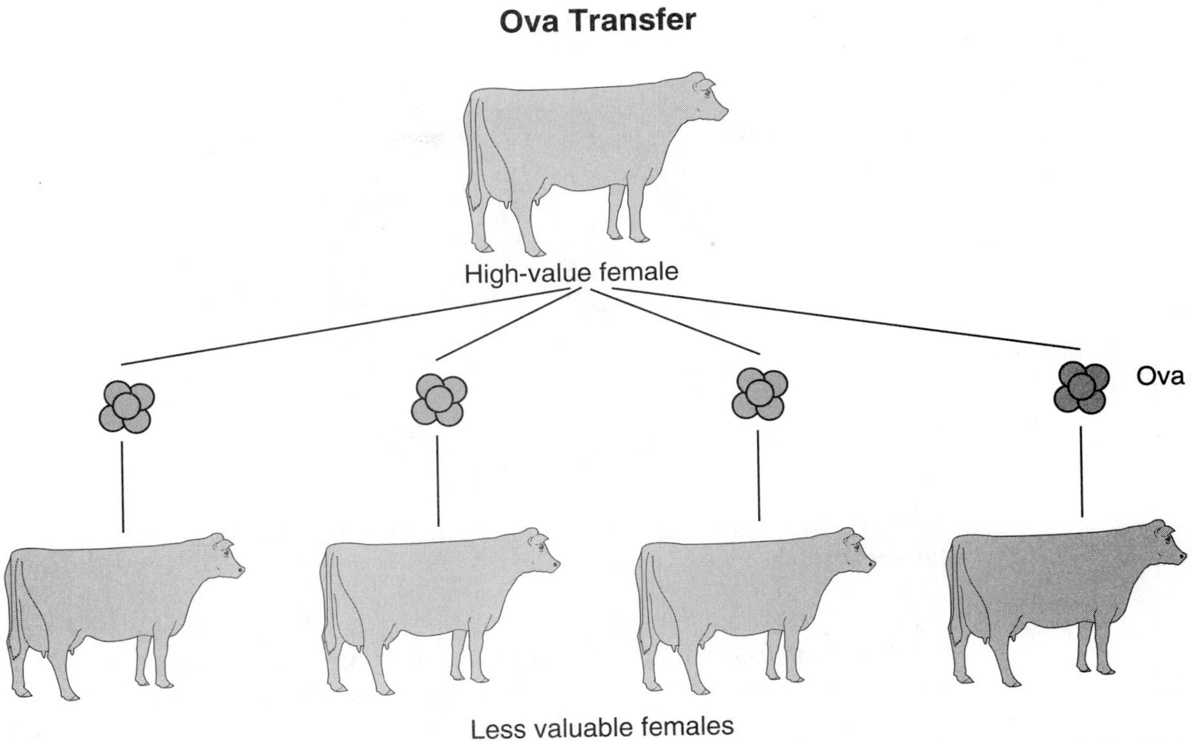

Ova Transfer

High-value female

Ova

Less valuable females

FIGURE 4–10 Valuable female animals are used as ova donors. Ova or living embryos are implanted into less valuable female animals and the offspring are all full genetic sons or daughters of the donor females.

FIGURE 4–11 Several offspring from a single mating can be produced in species of animals that give birth to a single offspring under normal conditions. *(Photo courtesy of William Grange)*

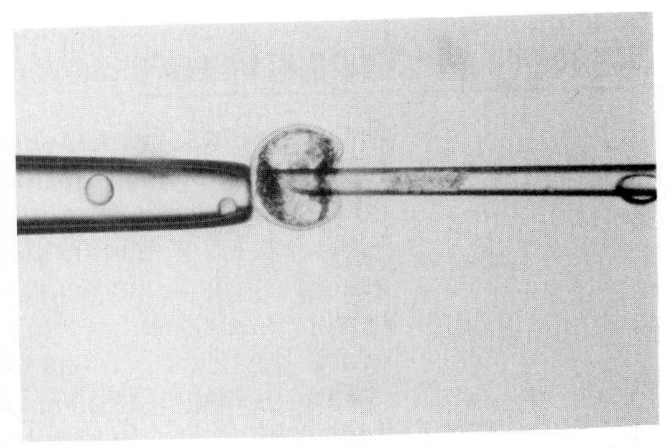

FIGURE 4–12 Several identical individual animals can be produced by separating individual cells from an embryo and injecting each of them into an ovum from which the genetic material has been removed. *(Photo courtesy of The University of Idaho)*

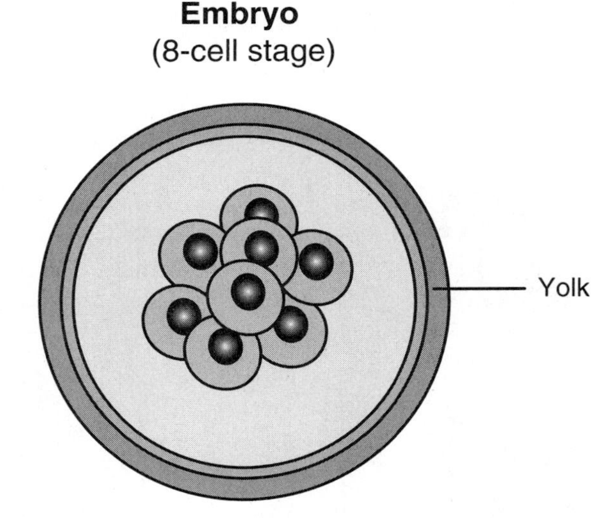

FIGURE 4–13 The ovum divides to form two cells, then four, then eight, sixteen, thirty-two, and so forth.

At the appropriate time the donor female is bred. The growing embryos are recovered from the donor female six to eight days after breeding. By this time, the embryos are in the 8–32 cell stage of development. They are recovered by flushing the reproductive tract of the donor female with fluids. Using a microscope, the embryos are then located and checked visually to be sure that they are healthy, Figure 4–13. An embryo may then be transferred to the uterus of a recipient female, split to form multiple embryos, or frozen for later use.

ESTRUS MANIPULATION

Estrus occurs when a female animal produces an ovum or egg and becomes receptive to the male. Most animals behave differently during estrus than they do at other times. Sexual behavior is evident during this period of time, Figure 4–14. Some species of animals such as horses, cattle, and swine exhibit estrus at any season of the year. Other animals such as sheep tend to exhibit estrus only in certain seasons. **Anestrus** is the period of time when the **ovary**, the primary female sex organ, stops producing eggs.

The breeding season of some species of animals is controlled by the amount of light to which they are exposed each day. The length of this light exposure is called the **photoperiod**. The breeding season for most breeds of sheep begins several weeks after the photoperiod begins to decrease. In the **northern hemisphere**, i.e., that part of the world located north of the equator, the photoperiod begins to decline in late June, but in the **southern hemisphere** this date occurs in December. By restricting the amount of light

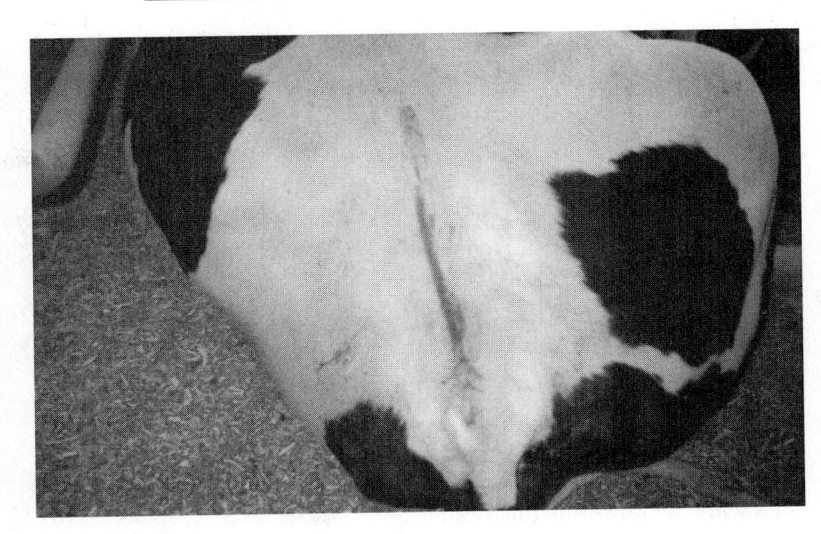

FIGURE 4–14 Several technologies have been developed to identify the correct time to breed cows. Sexual behavior such as cows mounting other cows will cause the colored chalk on the rump of this cow to be spread across the rump of the cow.

to which sheep are exposed, the normal breeding season can be changed or modified.

The reproductive organs of animals are controlled by specific hormones. In most cases, sheep and several species of wild animals begin production of the hormones responsible for estrus in response to a reduced photoperiod (light exposure). Other animals such as cattle and hogs are not affected to the same degree by the length of the photoperiod.

The estrus cycles of all species of farm animals can be modified by changing the hormone balance. This is usually accomplished by injecting carefully measured amounts of hormones into the animal's body, or by placing a hormone **implant** under the skin of the animal. Some hormone treatments are applied by placing a sponge **pessary** containing the hormone material in the female reproductive tract. Hormones are absorbed slowly from implants and pessaries over a period of several days before breeding occurs. The pessaries are removed prior to breeding.

One advantage that may be gained from controlling estrus in farm animals is that large numbers of animals may be bred in a short period of time, Figure 4–15. This is important, for example, when artificial insemination is practiced with cattle in the range areas of the western United States. Cows become scattered over great distances when they are turned out to the open range. By controlling estrus, the majority of the herd may be bred in a period lasting only a few days instead of the three weeks that would be required for all of the cows to exhibit estrus naturally. The cows that do not **conceive**, or become pregnant, will exhibit estrus again about every three weeks. When this occurs, the cows may be bred again by artificial insemination or bred naturally. Bulls are usually turned out with the herd on the summer range for natural mating with cows not settled by artificial breeding.

FIGURE 4–15 Cows that calve over a short time interval produce a uniform calf crop in comparison with traditional breeding practices. A uniform calf crop allows more animals of similar quality to be marketed at the same time, giving the owner a marketing advantage.

Sheep that are managed in confinement systems benefit from estrus control by breeding for lamb production outside the normal breeding season. Sheep are capable of lambing every eight months, producing three lamb crops in two years instead of a single lamb crop each year. With this management system, the buildings and equipment are in use during a greater portion of the year, and lambs can be marketed at different times during the year to level out seasonal price differences.

PARTURITION MANAGEMENT

Parturition is the procss of giving birth. The ability to control the time of parturition is an important management tool because it makes it possible for a farmer to control some of the conditions under which young animals are born. Many of the deaths that occur among farm animals take place during the first few hours following birth. Frequent causes of death among newborn animals include starvation, exposure to harsh weather, and poor **environmental conditions**. These problems can be prevented by giving added attention to the needs of baby animals during this critical period.

Parturition induction is a method of controlling the time when pregnant female animals give birth to their young. It is accomplished by injecting pregnant females in the final days of gestation with special hormones. It is a process that causes birth to occur within a few hours after the treatment is given. When this treatment is used it is possible to plan for births of young animals at times when plenty of help is available. Reliable breeding records are needed to determine which animals are capable of responding to the treatment. During the hours following treatment, the treated animals can be brought into protected areas for close observation and care.

GENDER SELECTION

Gender selection is the ability to control the sex of offspring at the time of mating. Animal scientists have attempted for many years to learn to control the sex of newborn animals. The ability to produce males in the beef industry would allow the cattleman to raise steers from most of his cows. Since steers usually grow more rapidly than heifers, more meat could be produced per cow. Dairy farmers would be able to obtain replacement heifers from their best cows every year.

Animal cells contain two kinds of chromosomes. Those called **autosomes** carry the genes that control all of the characteristics of the animal's body. The **sex chromosomes** control the sex of an animal. The sex chromosomes consist of two types: X chromosomes or Y chromosomes. Animal cells always contain one pair of sex chromosomes. In female animals the two halves of the sex chromosome are always X chromosomes (XX). In male animals the two halves of the sex chromosome consist of one X chromosome and one Y chromosome (XY).

During the cell division process of meiosis, each ovum or egg provided by the female contains an X chromosome. Each sperm cell from the male consists of either an X chromosome or a Y chromosome. If a sperm cell containing an X chromosome fertilizes the ovum, a female offspring (XX) will be produced. If a sperm cell containing a Y chromosome fertilizes the ovum, the result is a male offspring (XY).

Scientists have made many attempts to separate sperm containing X chromosomes from that containing Y chromosomes. The physical differences are so small however that past attempts to separate them have not proved to be effective. Fertile sperm from farm animals cannot be separated using the methods available at the present time.

A method that has shown promise for separating sperm bearing X and Y chromosomes has been the **flow cytometry** process, Figure 4–16. This method uses florescent dye to stain the chromosome contained in the sperm cell. Because the X chromosome is slightly larger than the Y chromosome, it absorbs more dye. A laser is then used to measure the amount of light given off by each sperm cell. Through a complicated process, each sperm cell is charged with a positive or a negative charge based on whether it contains an X or a Y chromosome. It can then be sorted according to its charge. The problem with this process is that it removes the tails from the sperm and the separated sperm are not fertile. Scientists are trying to find a way to use this process without damaging the sperm.

One form of gender selection that exists in the livestock industry today is accomplished by collecting embryos from donor females, sorting them on the basis of their sex, and implanting those of the desired sex, Figure 4–17. A high-quality embryo can be divided into as many as four identical embryos. This is done before they are implanted. Although this type of gender selection is very

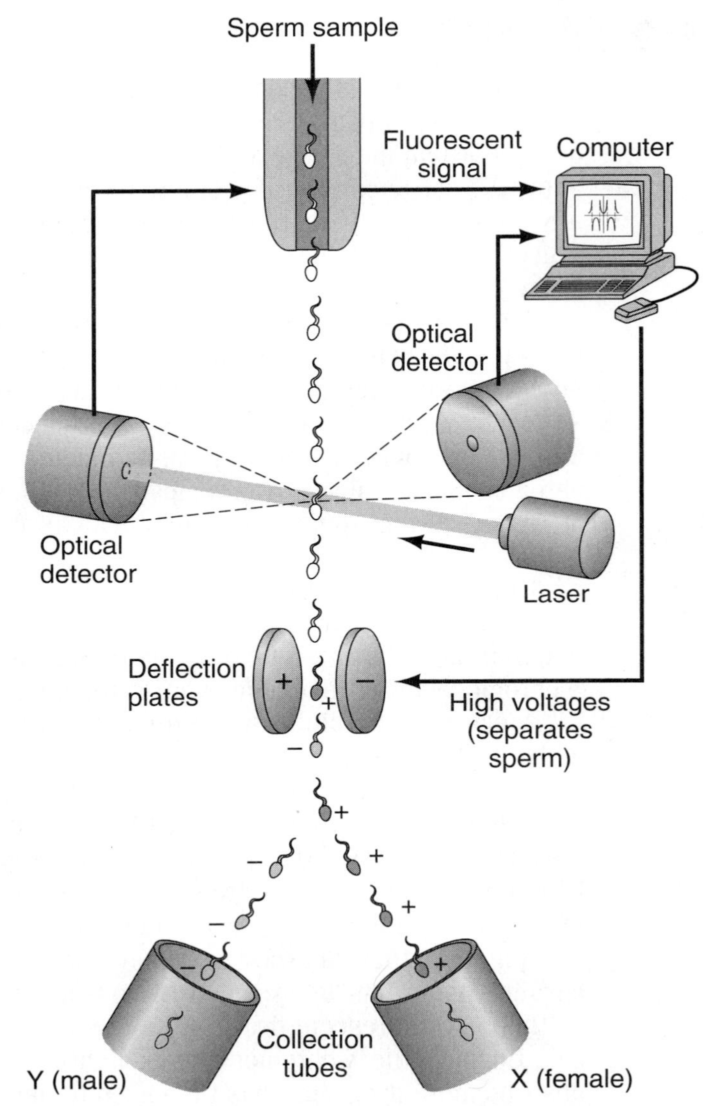

FIGURE 4–16 Flow cytometry is a method of sorting cells which differ only slightly in their structure. Scientists hope to use this process to sort sperm cells into those bearing male (Y) chromosomes, and those containing female (X) chromosomes. *(Illustration courtesy of the USDA)*

limited in the number of offspring that can be obtained, it is the only method available to livestock producers at this time.

The fish industry has overcome many of the problems of gender selection by modifying the process of meiosis. Scientists have discovered that they can modify the sperm obtained from male fish using ultraviolet light to make it incapable of combining with the chromosomes found in the eggs from the female. Eggs fertilized with the treated sperm are capable of developing however, and

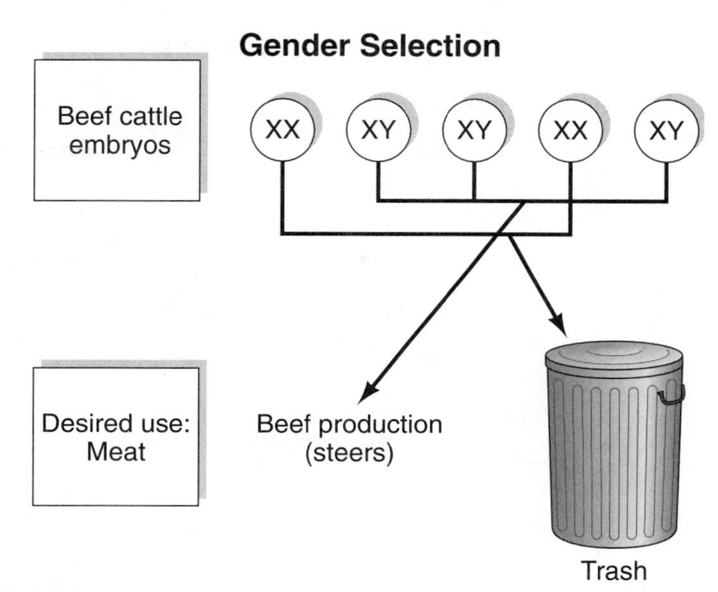

FIGURE 4–17 The only practical form of gender selection available today is to sort embryos according to their sex into male and female groups. If the desired product is market beef, male embryos would be implanted and female embryos would be discarded, because steer calves grow faster than heifer calves.

the result is a hatch of female fish. These females are then treated with male hormones to modify their sex characteristics to that of males. The chromosome makeup of the sperm produced by these modified males consists entirely of X chromosomes. Since the eggs of all fish contain only X chromosomes, all of the fish eggs that are fertilized will produce female fish, Figure 4–18. In the trout industry, female fish are desirable because they produce higher quality meat at market weights than male fish do.

CHAPTER SUMMARY

Reproductive management is a system of selective breeding in which animal matings are planned and not left to chance. Technologies that are useful tools in managing reproduction in farm animals include pelvic measurements, sire performance records, heritability factors, artificial insemination, estrus manipulation, and embryo transfer.

Parturition management is a method of controlling the time when pregnant female animals give birth to their young. This is done by using hormone injections to initiate the birth process. It is used to reduce exposure to hostile environments that often lead to deaths in newborn animals.

Efforts to control the sex of animals by separating sperm bearing X and Y chromosomes have not proved to be effective. Gender selection may be accomplished by using embryo transfer after the embryos are sorted by sex.

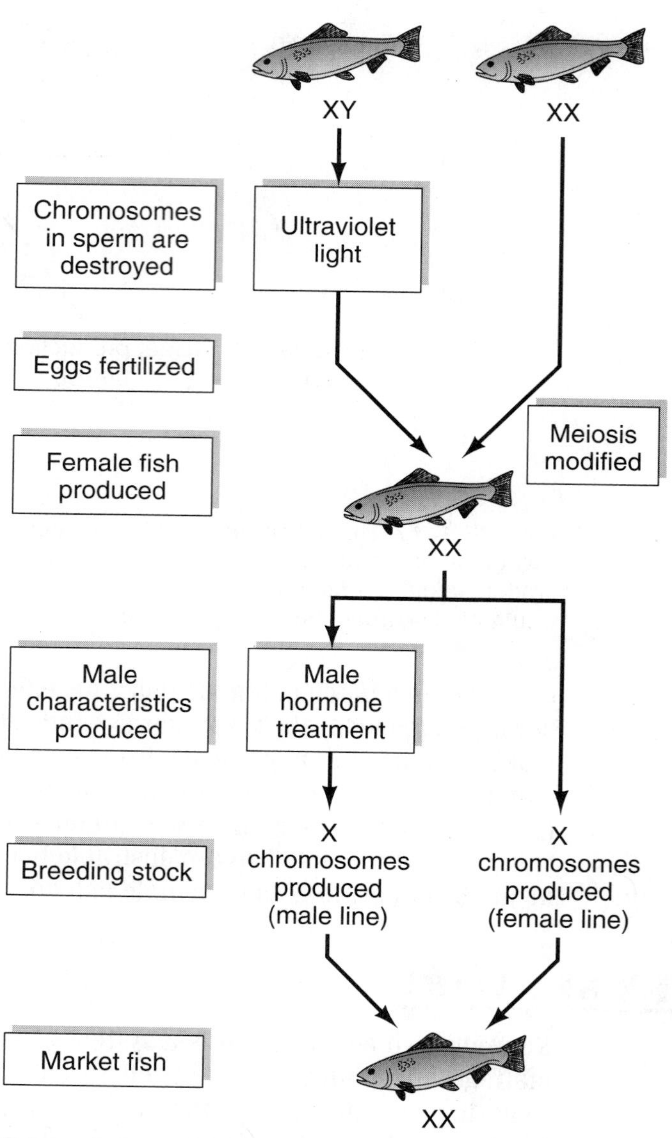

FIGURE 4–18 Female trout are more desirable meat fish at current market weights than male trout. Researchers have modified the meiosis process to commercially produce only female fish.

The fish industry has been able to produce female fish by modifying the process of meiosis.

Numerous new technologies are used in the reproduction processes of farm animals. Those tools have improved production by speeding up the rate of genetic improvement in productive traits.

CHAPTER REVIEW

Discussion and Essay Questions

1. List and describe six technologies used in the reproductive management of farm animals.
2. Compare the advantages and disadvantages of using artificial insemination as a breeding management tool.
3. Describe two ways that reproductive cycles of farm animals can be modified.
4. Suggest conditions that would justify controlling the time of birth for farm animals.
5. Explain how the birth process may be initiated in farm animals.
6. Describe the most promising research efforts to control the gender of farm animals.
7. Outline the steps involved in embryo transfer technology.
8. Justify the use of embryo transfer technology in reproductive management of farm animals.

Multiple Choice Questions

1. The gestation period of an animal refers to:
   A. the number of days between conception and birth.
   B. 344 days.
   C. the period of time when a female animal is receptive to the male.
   D. the time period between pregnancies.

2. The farm animal with the highest reproductive rate is the:
   A. sheep.
   B. horse.
   C. pig.
   D. cow.

3. The study of inherited traits is known as the science of:
   A. reproduction.
   B. genetics.
   C. embryology.
   D. botany.

4. Heritability factors for domestic animals are highest for:
   A. fertility.
   B. percent lean cuts.
   C. milk production.
   D. loin-eye area.

5. A female animal can give birth to offspring to which she is not genetically related by using a procedure called:
   A. embryo or ova transfer.
   B. genotypic modulation.
   C. artificial insemination.
   D. anestrus.

6. A hormone that stimulates female farm animals to produce increased numbers of ova is called:
   A. an enzyme.
   B. FSH.
   C. extender.
   D. semen.

7. Embryos are usually recovered and transferred from donor to recipient females when the embryo cell mass consists of about:
   A. 2–4 cells.
   B. 8–32 cells.
   C. 64–128 cells.
   D. 1/4-inch in diameter.

8. Estrus is a reproductive event that occurs:
   A. only in male animals.
   B. when eggs are no longer produced.
   C. when an ovum is produced.
   D. when the photoperiod is increasing.

9. Parturition induction is:
   A. controlling the time of birth using hormones.
   B. an initiation ceremony for expectant mothers.
   C. treating pregnant females with hormones to produce multiple births.
   D. dividing embryos to produce genetic clones.

10. Autosomes are chromosomes that:
    A. determine the sex of an animal.
    B. carry genes that control the inheritance of all the physical characteristics of the animal's body.
    C. include both X and Y chromosomes.
    D. always produce male offspring.

LEARNING ACTIVITIES

1. Identify a livestock farmer who uses embryo recovery and transfer technology, and arrange a visit to the farm to observe the process.

2. Obtain female reproductive tracts for swine, cattle, and sheep. Observe the specimens in class to identify the various parts and discuss the functions of each.

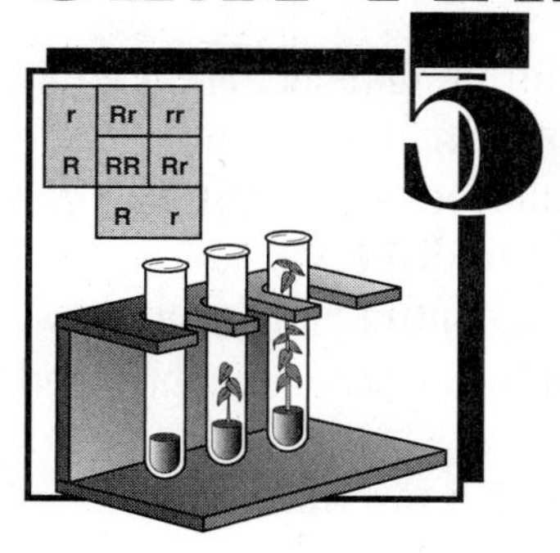

CHAPTER 5
Biotechnology Applications for Plants

All higher forms of life are dependent upon plants for survival. Plant leaves are living factories. They trap and store energy from the sun in forms that other living things can use for food. Plants also provide shelter for many kinds of living organisms. When the ability of a plant to survive and produce is improved, the benefits filter down to all of the organisms that depend on the plant for survival. Plant research has the potential to affect nearly all forms of living things, including humans, Figure 5–1.

OBJECTIVES

After completing this chapter, you should be able to:

- describe ways that biotechnology may be used to improve production of plants.
- understand the effects of improving the process of photosynthesis on the world supply of agricultural products.

FIGURE 5–1 Research facilities have been established by Land Grant Universities and the United States Department of Agriculture. They research many problems associated with plants.

- identify alternative methods of reducing frost damage to crops and explain how each method works.
- distinguish between several methods of controlling insects that damage crops, and describe the mode of action of each method.
- distinguish between the use of cultivars and recombinant DNA procedures to develop plants that are resistant to diseases.
- appraise the use of plant growth regulators to control plant processes.
- discuss the implications of using plant growth regulators on plants used for food production.
- describe research efforts to develop plants that are tolerant to salt.
- assess the advantages to be gained by developing new plants that are capable of nitrogen fixation.

TERMS FOR UNDERSTANDING

The following vocabulary terms should be studied carefully as you read:

photosynthesis
chlorophyll
light reactions
ATP
hydrogen ions
NADPH
Calvin cycle
respiration
cryoprotectant
ice-minus bacteria
entomology
entomologist
pollinators
pollen
biological control
IPM
pheromone
economic threshold
cultivar
herbicide
plant growth regulators
cytokinin
auxin
gibberellin
abscisic acid
ethylene
callus
saline soils
nitrogen fixation
symbiotic relationship
symbiosis
nodules
Azolla

PHOTOSYNTHESIS RESEARCH

Photosynthesis is a chemical reaction that captures energy from the sun, and combines it with carbon dioxide from the atmosphere to form plant tissues. Oxygen is a waste product from this process, but it is a life sustaining element for many other forms of living things, Figure 5–2. Photosynthesis produces two of the most basic needs of animals: food and oxygen.

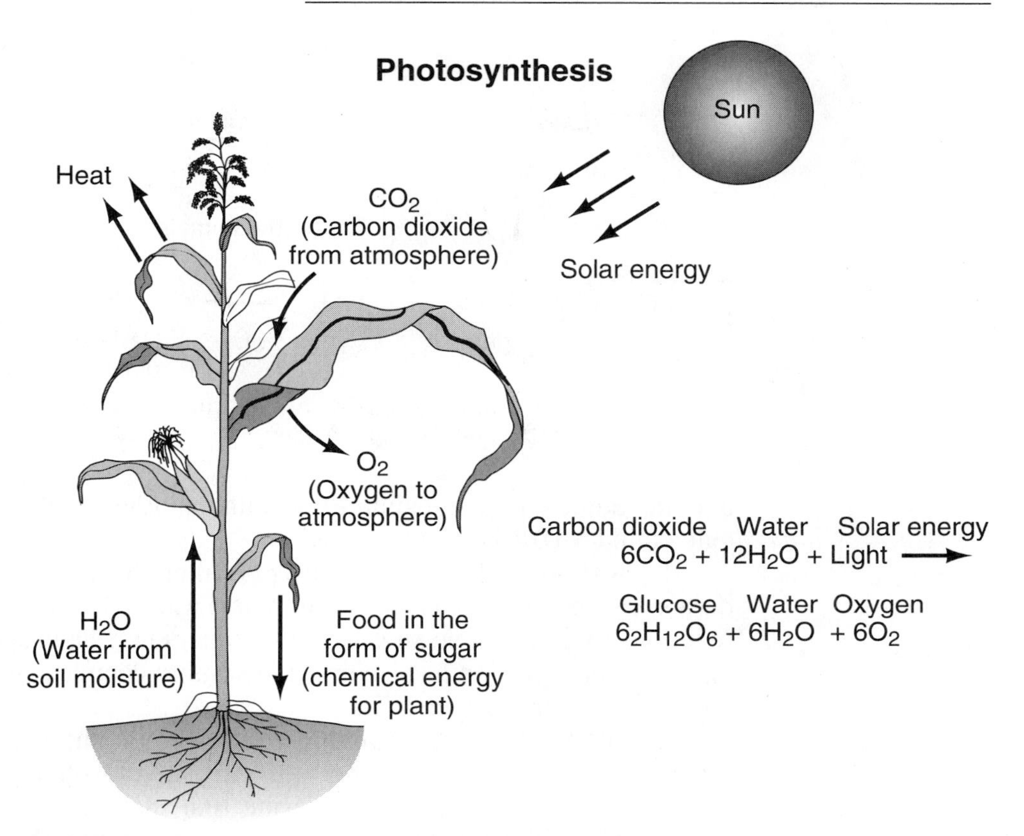

FIGURE 5–2 Photosynthesis is the process by which carbon dioxide is combined with water to store energy obtained from sunlight. Chlorophyll supports this reaction in which sugar and oxygen are produced.

Chlorophyll is a green substance that is capable of capturing energy from sunlight. Plants use this energy to make sugar from carbon dioxide and water. Chlorophyll is located in plant cells in the chloroplasts. It is most abundant in the leaves of plants, and it is required in two important chemical reactions that convert light energy to chemical energy. These two reactions are called **light reactions**.

The first reaction occurs when light is absorbed by chlorophyll molecules. This is followed by several steps that lead to the formation of a high energy molecule called **ATP** (adenosine triphosphate). ATP stores electrons in the form of chemical energy in the electrical bonds that attract and hold its phosphate molecules together.

The second light reaction occurs at the same time that light energy is converted to chemical energy when ATP is formed. Some of the water molecules located in the leaf are split to form oxygen, electrons, and positively charged hydrogen particles called **hydrogen ions**. The hydrogen ions use energy obtained from electrons to bond with a molecule called NADP (nicotinamide

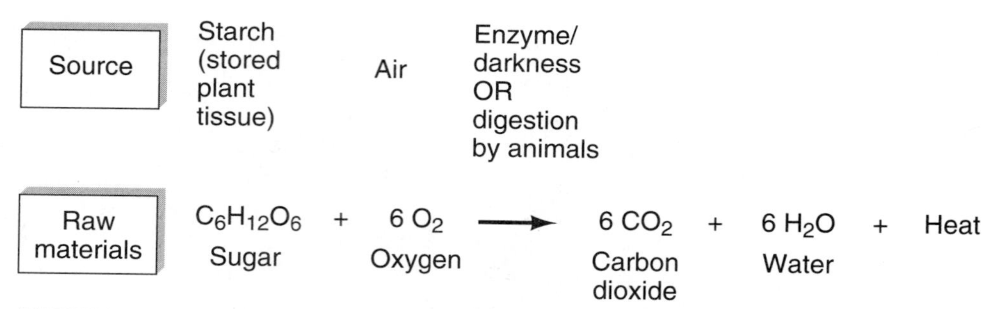

FIGURE 5–3 Respiration is the process by which plant tissues are broken down to produce heat, water and carbon dioxide.

adenine dinucleotide phosphate), forming another kind of high-energy molecule called **NADPH**.

The next phase of photosynthesis is called the **Calvin cycle** during which carbon dioxide from the environment reacts with ATP and NADPH from the light reactions to form simple sugars. These sugars may be used as energy sources for the plant, or they may be converted to more complex sugars, starches, fats and proteins for storage.

One of the most basic applications of biotechnology to plants is the research effort to improve the efficiency of the photosynthesis process. Scientists have discovered an enzyme that plays an important role in the process of photosynthesis. When the enzyme is present in large amounts, greater plant growth occurs. When the supply of the enzyme is limited, plant metabolism slows down.

During hours of darkness stored plant materials tend to decrease. When photosynthesis is interrupted, the enzyme reacts with oxygen and reverses the process of photosynthesis. This process is called **respiration**, Figure 5–3. It occurs in plants when photosynthesis is interrupted, and in animals when stored sugars are digested to release energy. Scientists hope to find ways to prevent the loss of stored plant materials and to stimulate the efficiency of the photosynthesis process. Plants will become much more productive as photosynthesis becomes more efficient in converting energy and carbon dioxide to plant tissue. Perhaps current research into the roles of enzymes in the process of photosynthesis will boost crop yields once again, as agriculture faces the challenge of feeding our hungry world.

FROST PROTECTION

Many fruit and vegetable plants are highly sensitive to freezing temperatures. Some plants sustain severe damage any time freezing occurs during the growing season. Other plants are damaged only when they are flowering or when fruit is immature. The length of time the temperature is at or below freezing, and the number of degrees below freezing makes a major difference in the amount of damage that occurs.

Many attempts have been made to reduce damage due to frost. Some protection can be achieved by irrigating during the coldest hours. Water that is sprinkled directly on the plants tends to maintain the temperature of the plants near 32 degrees even when ice forms on the plant surfaces. The use of microjets to apply a fine water mist to crops that are easily damaged by cold temperatures is a common practice in fruit growing areas, Figure 5–4. Other common practices in fruit growing areas include operating large fans mounted on towers above the crop. The fans are turned on to mix the cold air near the ground with the warmer air from the upper atmosphere, Figure 5–5.

One method used to protect young plants from freezing temperatures is to apply a protective material called a **cryoprotectant**. These products are designed to protect plants from freezing temperatures. One such product is a special protective foam called frost gard. When the material is applied to plants, tiny air bubbles insulate the plants to prevent loss of heat. A narrow band of the foamy material is applied to each row of plants, and it provides frost protection for several hours. If cold weather persists, the foam may need to be applied several times because it melts away when the temperature rises during the day.

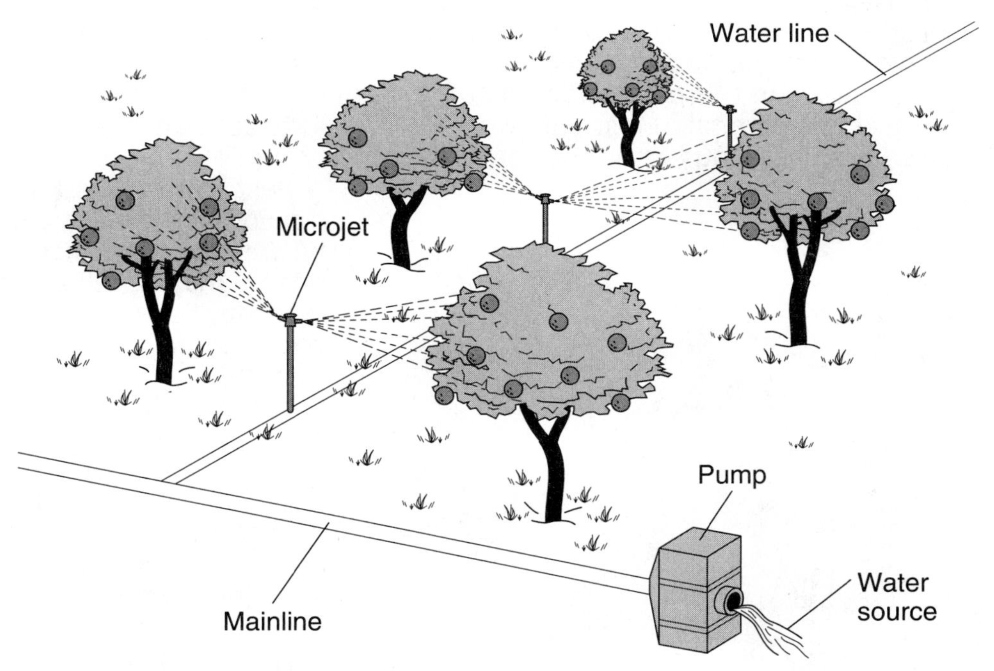

FIGURE 5–4 An irrigation system that delivers a fine water mist or spray to the surfaces of trees and plants is an effective method of protecting valuable crops against freezing temperatures.

FIGURE 5–5 Large propeller blades mounted on engines are capable of warming the air around fruit trees by mixing the comparatively warm upper air with the cold air near the ground.

CAREER OPTION

Pest Control Specialist

A person who chooses a career as a pest control specialist must have an understanding of the science of entomology and a broad knowledge of the habits and behaviors of insects, rodents, and other organisms that cause damage to crops, animals, and property. A pest control specialist must also have a strong background in the safe use of chemicals and poisons used to control pests.

A person in this profession must also understand the laws regulating the use of hazardous materials. Most of the materials used to control insect and rodent populations are hazardous when they are not properly applied, Figure 5–6.

Pest control specialists must learn to recognize the presence of pests by the damage they cause. Insects or small rodents may be difficult to find even after they have caused damage to crops, stored products, livestock, or property.

Careers in pest management include operating pest control businesses or working in government as pest control specialists. Training is available through private pest control service companies, universities and colleges, and government and private training programs.

FIGURE 5–6 A pest control specialist uses knowledge of pest behaviors and habits to control pest numbers.

The biotechnology revolution has given us a new tool to prevent frost damage to plants. It is called the **ice-minus bacteria**. Genetically engineered bacteria are applied to plant surfaces allowing the plant to tolerate temperatures several degrees below freezing. This new technology will be very useful in protecting young seedlings from frost damage.

INSECT RESISTANCE

Entomology is the branch of science that studies insects. A person who works in the science of entomology is known as an **entomologist**. Insects make up the largest group of animal life forms. They have adapted to most climates and environments known to man.

Insects make up a class of living organisms that play many different roles in agriculture. Some insect species cause serious damage to plants. When insects are not controlled, they are capable of completely destroying the crops of large regions of the world. History is filled with examples of famines that have resulted in the starvation and deaths of many people due to uncontrolled invasions by locusts and grasshoppers.

Insect damage to crops costs millions of dollars each year in lost crop yields and damaged products. In addition to these losses, the agriculture industry spends huge amounts of money for insecticides and other materials to control insect populations each year.

Despite massive insect damage to crops, insects must not be eliminated. Many insects are very useful to humans and to plants. Insects provide silk, honey, medicines, and other products for humans, and the fleshy parts of large insects are a source of food for some people. Insect species known as **pollinators** are very important to crop production, Figures 5–7 and 5–8. This is

FIGURE 5–7 The tiny leaf cutter bee is very important to the alfalfa seed industry. It is a pollinator. It accomplishes this task as it gathers nectar from the tiny flowers of the alfalfa plant. *(Photo courtesy of Ron M. Bitner)*

FIGURE 5–8 Honeybees are helpful insects that pollinate crops and produce income from sales of honey. Aggressive use of chemicals to control insect pests can completely destroy a colony of honeybees.

because pollinating insects carry particles containing pollen from one plant to another as they gather nectar. **Pollen** contains the male reproductive gametes of plants. The result is greater production of seeds and fruits. Without insects, many seed and fruit crops would be impossible to produce.

Some species of insects are predators. They are natural enemies to some of our worst insect pests. Use of predatory insects, parasites, or disease organisms to manage harmful insects is called **biological control**. It is neither wise nor possible to achieve 100 percent control of damaging insects using chemical insecticides, because complete reliance on insecticides to control insect pests tends to destroy their natural predators, too.

Some insect pests can be controlled by regulating the transport of fruits, seeds, and other plant materials across state or national boundaries. California uses this strategy to help protect its fruit industry against reintroduction of the Mediterranean fruit fly.

Many harmful insects can be controlled by rotating crops. This kind of insect control is known as cultural control. Cultural control practices work because the environment in the field changes each time a different kind of crop is planted. Many insects are unable to adjust to changes in their environments, and they often fail to survive.

Sometimes it is possible to control harmful insects by removing them from crops by hand or by collecting the insects using machines or other mechanical devices. Either of these methods is usually inefficient and expensive. It is difficult to locate all of the insects that infest a field. It is also impractical to attempt to gather insects that are able to fly, or insects that are very small. Examples of mechanical control of insects include insect traps and vacuum systems that suck up insects from the crop in a similar manner to the way a vacuum cleaner is used to remove dirt particles from a carpet.

Chemical control of insects has been the insect control method of choice on most farms for more than fifty years. This is because chemical control is relatively effective and inexpensive. Some potential problems do exist, however. Failure to follow the instructions on the chemical container can lead to damage to the environment. It can also result in illegal chemical residues on a crop. Many insects are capable of building up an immunity against a chemical that is applied at regular intervals. When this occurs, the chemical in question loses much of its value.

A multipronged attack on harmful insects is the most practical approach to insect control. **Integrated pest management** or **IPM** is a concept for controlling harmful insects while providing some protection for useful insects. It involves use of some chemical pesticides, but relies on natural insect enemies and other insect control strategies to control harmful insects. When insecticides are the only source of control, they kill both harmful and useful insects. Integrated pest management is an insect control program that does not attempt to kill all of the harmful insects. Insect control of this kind also kills the natural enemies. For natural insect enemies to survive, they must have a small population of harmful insects upon which to prey.

Integrated pest management is not a new idea. It was widely used prior to the introduction of modern insecticides. It has emerged in recent years as the best alternative to complete reliance on insecticides. Integrated pest management is an ecosystem approach to controlling insect problems. It takes into account the effects that a particular form of insect control might have on the other living things that are found in the ecosystem.

Two major factors have accelerated the adoption of IPM. They are resistance by some consumers to the use of chemical insecticides on food animals and crops, and the ability of insects to develop genetic tolerances for insecticides. An effective alternative to insect control other than complete reliance on insecticides is needed.

Successful implementation of IPM requires much more attention to detailed observations of the insect populations in the fields. An effective system for sampling potentially harmful insect populations must be developed. Such a system might include using insect traps to which mature insects are attracted through the sense of smell. A **pheromone** is a chemical substance that some kinds of insects use to attract mates. It is an effective "bait" for an insect trap.

Accurate population records of key insect pests are needed in order to determine whether it is cost effective to take control measures. The point at which the losses due to pest damage equal the cost of controlling the pest is known as the **economic threshold**. The use of graphs to record insect populations is a good way to visualize what is happening to the population over a period of time, Figure 5–9.

Reduction of the population of an insect pest may be as simple as introducing predatory insects that eat the key insect pests. This may be all that is

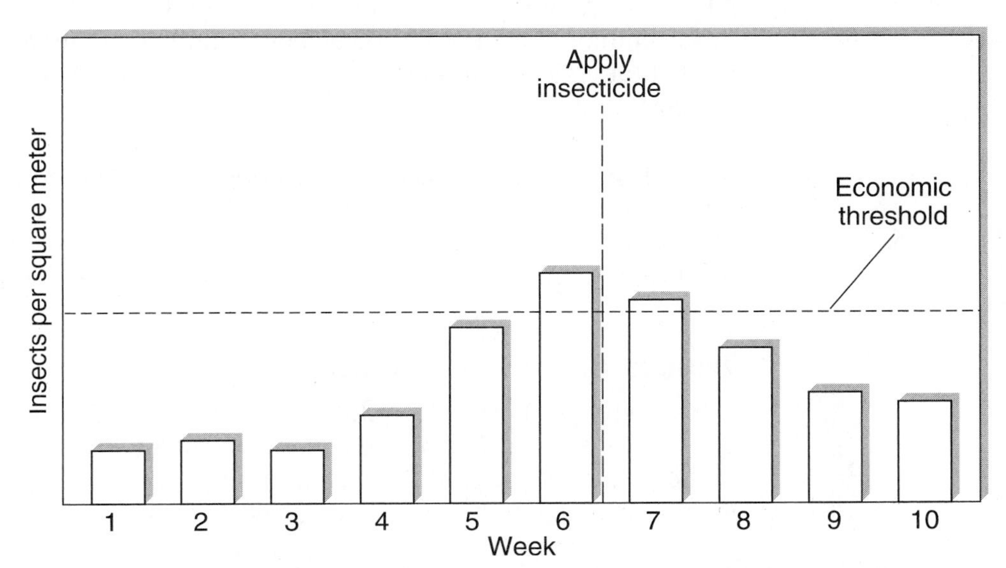

FIGURE 5–9 The population of an insect should be monitored carefully to determine if or when it is cost effective to reduce the insect population using an insecticide.

needed to reduce the harmful insect population below the economic threshold level. In such a case, no further control measures should be introduced. A similar result may be achieved by planting a crop variety that has been genetically engineered to resist insects. Integrated pest management sometimes includes the careful use of an insecticide as an emergency measure to reduce an insect pest population that has become a serious threat to a crop.

Another useful technology for controlling harmful insects is becoming available. It is genetic insect resistance. Some plants are naturally resistant to insects. They may give off an odor that insects avoid. Other plants contain natural insecticides in their plant juices.

Through use of genetic engineering techniques, it is now possible to transfer the genes responsible for producing natural insecticides to plants that have no insect resistant traits. The advantage of this technology is that only those insects attempting to eat the plant are killed. Chemical insecticides are not needed for such a crop. Pollinating insects and natural insect enemies are not subjected to insecticides, so they survive in the field to aid in controlling the damaging insect species.

Insect-Resistant Potatoes

Colorado potato beetles are capable of stripping all of the leaves from large fields of potatoes in a relatively short period of time. These insects are capable of destroying the potato industry if they are not controlled. Plant geneticists at Monsanto Company's research laboratories have isolated a plant gene that is resistant to this insect. These researchers have successfully inserted a beetle resistant gene into the chromosomes of potato plant cells, and they have also regenerated these modified cells into complete potato plants using tissue culture technologies. These new potato plants produce a natural insecticide that is deadly to Colorado potato beetles, but which is also friendly to the environment.

As more major crop producing plants are developed that are genetically resistant to insects, the dependence of the agricultural industry upon chemical insecticides is expected to decline. Production of crops from genetically modified plants should be possible at lower costs and with less damage to the environment than we are now experiencing using nonresistant crop varieties.

Transgenic Cotton Plants

A new cotton plant whose genetics have been changed using genetic engineering techniques is being field tested in Mississippi. A bacterial gene was introduced in the plant causing it to produce a protein that is toxic to armyworms, bollworms, and tobacco budworms. These pests are estimated to cost domestic cotton producers nearly $130 million every year. The source of the gene is a strain of bacteria known as Bacillus thuringienensis that is common in soil.

DISEASE RESISTANCE

Plant diseases are responsible for the loss of millions of dollars each year due to reduced crop yields. In some cases, susceptible crops may not be grown for many years in disease infected areas. Although scientific researchers have succeeded in developing many crop varieties that are resistant to major plant diseases, serious crop losses still occur.

Disease-resistant plant varieties have been developed in several ways. One approach to the problem has been to inspect diseased areas for individual plants, or **cultivars**, that have resisted infection to the disease organism. When the resistance is shown to be hereditary, plant breeders begin the process of developing a new crop variety. The resistant trait must be bred into plants that are adapted to production in the localized areas where they will be grown.

The resistant species is usually tested for several generations to be sure that it is free from undesirable characteristics. In some instances, the development of the new variety is speeded up by raising test plots in the southern hemisphere (south of the equator) during the winter season in North America. This allows the plant breeder to raise two generations of plants in the same amount of time that is usually required to raise a single generation.

The development of recombinant DNA procedures discussed in Chapter 2 has made it possible to transfer disease-resistant genes directly to other plants, Figure 5–10. Transfer of genetic material is possible both within plant species and between plant species. But identifying the genes responsible for resistance to disease can be very difficult.

Resistant genes are usually isolated by identifying those genes that are expressed in resistant plants, but not expressed in the diseased plants. Once a resistant gene has been identified, it is transferred to a suitable cultivar and tested. When it has been determined that disease resistance has been passed to new generations of plants, an adequate amount of seed for the new variety must be produced.

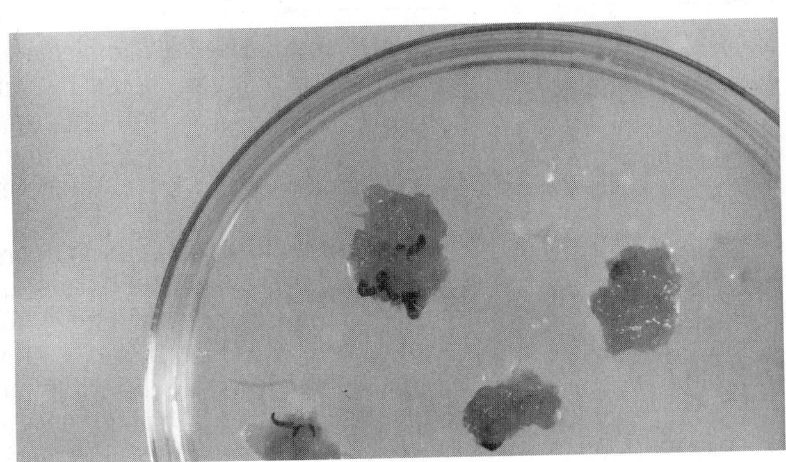

FIGURE 5–10 Damaged plant cells can be penetrated by bacteria that carry genes resistant to plant diseases and pests. When these cells are treated with hormones, they can be stimulated to grow leaves and roots, and to regenerate into complete plants. *(Photo courtesy of R. Zemetra, The University of Idaho)*

HERBICIDE TOLERANCE

A weed is defined as any plant that is growing out of place. Weeds compete with cultivated plants for plant nutrients and water. Some weeds are such strong competitors that the desirable plants near them eventually die or become less productive. Most crops are not capable of producing maximum yields when weeds are present. Some weeds even interfere with the operation of machines that are used to harvest crops.

A chemical compound that is used to kill or control weeds is called a **herbicide**. This class of chemicals is used in many locations by people who have problems with weeds. Large amounts of herbicides are used by farmers to control weeds that compete with crops. Managers of city parks and golf courses use significant amounts of herbicides to kill weeds. Highway and railroad maintenance workers use large amounts of herbicides to control weeds that grow along highways, roads, and railroad tracks. Homeowners also use large amounts of herbicides to control weeds in their yards and gardens.

Herbicides are very useful chemicals, but sometimes they create problems. One problem with some herbicides is that they tend to remain active in the soil for long periods of time. This tendency may restrict the kind of crop that can be grown until the herbicide breaks down and becomes inactive.

Some herbicides are very specific as to the kinds of plants they control. A herbicide that is tolerated by corn plants may be very toxic to strawberry plants, while another herbicide may be tolerated by strawberry plants and deadly to corn plants.

Using genetic engineering and tissue culture technologies, it is now possible to introduce specific genes to plants giving them the ability to tolerate herbicides that formerly killed them before they were genetically changed. This will allow the use of effective herbicides on a much wider range of crops.

PLANT GROWTH REGULATORS

Plant growth regulators are chemical compounds that are found naturally in plants. They act as plant hormones by causing certain genes to be expressed by plants. They play a role in regulating the growth of roots, stems, and leaves, and in controlling the development of fruits and seeds, Figure 5–11. Five growth regulators have been identified, and their modes of action have been determined.

The plant hormone **cytokinin** acts to promote cell division and causes plant cells to develop into specialized tissues such as roots and shoots. **Auxin** is a plant hormone that controls the elongation or lengthening of cells that form plant shoots. The presence of auxin stimulates certain plant genes to be expressed causing growing plant shoots to lengthen.

Gibberellin is a plant hormone that causes the stored food in seeds to be converted to plant nutrients. The nutrients are then available for use during seed

FIGURE 5–11 Once plant cells have developed into tiny plants, they are placed in a special growth medium. *(Photo courtesy of R. Zemetra, The University of Idaho)*

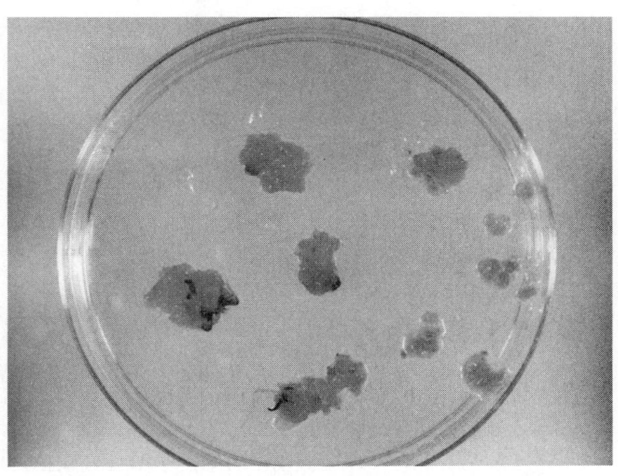

FIGURE 5–12 Callus tissue is plant material that is not differentiated into roots, stems, and leaves. It must be maintained in a sterile environment. *(Photo courtesy of R. Zemetra, The University of Idaho)*

germination and seedling growth. The hormone known as **abscisic acid** acts in a manner opposing the action of gibberellin. It prevents seeds from germinating until they are mature. Another plant hormone, called **ethylene**, is the growth regulator that causes the fruits of plants to soften and become ripe.

Growth-regulating hormones often act together to produce entirely different responses from plants than when they act alone. The interaction between auxin and cytokinin results in the formation of roots and shoots from plant cells called **callus** tissue. This is a mass of similar plant cells that have not specialized into different kinds of plant tissues, Figure 5–12. Callus often forms where an injury to a plant has occurred. Scientists can generate or grow entire plants from individual cells contained in callus tissue using a process known as cell or tissue culture, Figure 5–13.

FIGURE 5–13 Special growth chambers maintain exacting environmental conditions for callus tissue.

It is known that factors in the environment of a plant often interact with the plant growth regulators. The germination of seeds is an example of an interaction between a plant growth regulator and light. Other plant functions are also known to be related to interactions between growth regulators and environmental factors.

SALT TOLERANCE

Adequate supplies of fresh water for irrigation of arid soils are becoming difficult to obtain. Vast tracts of land exist in many areas of the world where annual rainfall is too low to raise cultivated crops. When irrigation water is available, many of these lands are capable of producing excellent crop yields. The oceans of the earth could be used to irrigate all of these lands if our food, forage, and fiber plants were tolerant of salt. A few plants exist in nature that are capable of growth in **saline soils** where high concentrations of salt exist. Some of these plants, such as salt bush and several grass varieties, are capable of growth in or near sea water.

Genetic engineers have been successful in transferring salt-tolerant genes into some species of useful forage crops. Using ocean water as the only source of water available to these new plants, good yields have been produced. As additional plants are developed that are tolerant to salt, crop production using ocean water for irrigation may become a reality on some of our desert lands.

Quack grass is considered by most people to be a weed that is very difficult to control. Now a hybrid plant has been developed by crossing quack grass with bluebunch wheatgrass. The new grass is salt tolerant. Other salt tolerant plants that have been produced include tomatoes and barley. All of these plants have been developed using traditional crossbreeding techniques.

Scientists have recently cloned a salt-tolerant gene that is found in some wild plants. Many new salt-tolerant plants may soon be developed using gene splicing techniques to insert this gene in the chromosomes of other important domestic plants.

NITROGEN FIXATION

The process known as **nitrogen fixation** is very important to agriculture. Large amounts of nitrogen from the atmosphere are converted to nitrogen compounds by nitrogen-fixing bacteria and algae. Some of these bacteria are able to fix nitrogen in their free state, but other forms must first enter into a symbiotic relationship with a plant.

A **symbiotic relationship**, often called **symbiosis**, occurs when two forms of life exist together, and each benefits from the presence of the other, Figure 5–14. This type of relationship exists when a plant becomes a host to nitrogen-fixing bacteria. In legume plants, the bacteria live in structures on the plant

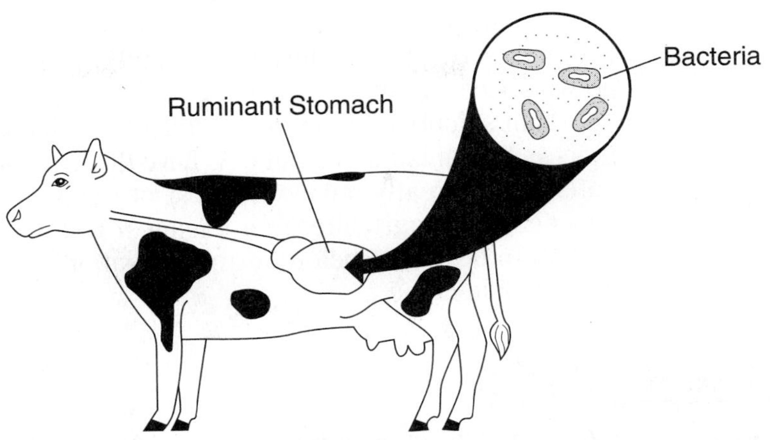

FIGURE 5–14 Symbiosis occurs when two life forms exist together and each provides something beneficial for the other. Rumen bacteria digest fiber in the stomach of a cow before being digested by the cow to provide protein and other nutrients.

roots called **nodules**, Figure 5–15. There they draw nourishment from the plant. The bacteria in turn make nitrogen compounds from nitrogen gas, which nourish the plant. They are also working with a bacterium called *Photorhizobium thompsonum* that is capable of nitrogen fixation using sunlight for energy instead of drawing energy from the nutrients produced by host plants. They hope to create a site for the bacteria on the stems of the host plant where sunlight is available instead of on the roots of the host. This would help the host plant produce more efficiently by using all of its energy for production of crops.

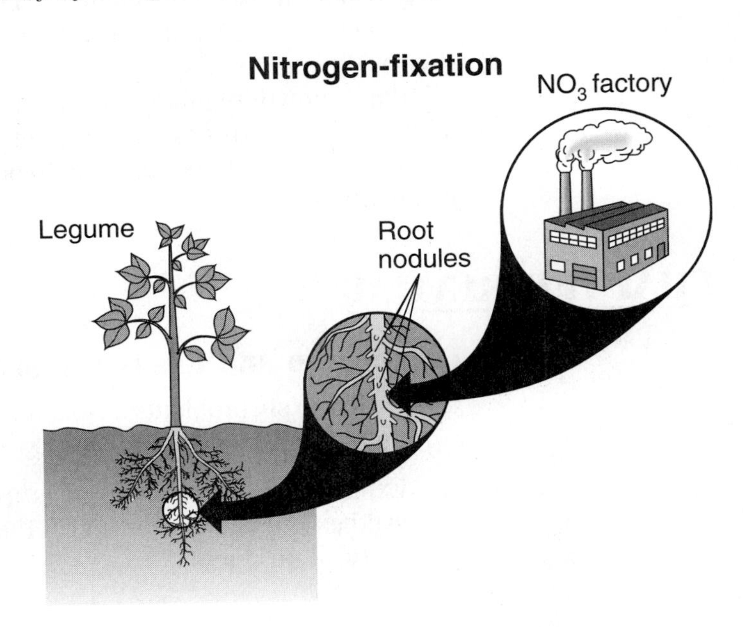

FIGURE 5–15 Nitrogen-fixing bacteria function like tiny living factories located in the root nodules to convert nitrogen gas from the atmosphere into nitrate compounds that plants can use.

Legume plants are not the only plants capable of symbiosis with nitrogen-fixing bacteria. A small fern called **Azolla** is used as a source of nitrogen for growing rice plants.

Plant scientists hope to transfer genes that control the nitrogen fixation process to plants that do not now have this capability. This technology has the potential to greatly reduce the cost of raising crops, and would reduce the dependency of agriculture on chemical fertilizers. Many difficult problems remain to be solved before nitrogen fixation will become a reality with non-legume field crops.

CHAPTER SUMMARY

Biotechnology applications for plants impact other living organisms because they depend upon plants as sources of food and shelter. Research in plant science has resulted in many new biotechnology applications for plants. Some of these new technologies include studies of the basic process of photosynthesis. Work is proceeding to enhance photosynthesis by applying an enzyme that increases the efficiency of the process.

Protection for plants against frost damage has been developed. Ice-minus bacteria prevent damage to plants to which the bacteria have been applied. Foam materials are available that insulate plant leaf surfaces against frost.

Insect damage may be prevented by developing plants that are genetically resistant to insects. Such plants often give off odors that repel insects; or the plant juices may be toxic to certain insects.

Disease-resistant plants have been developed to protect plants against some diseases. The most common methods for providing resistance to disease are selection of resistant cultivars and the use of recombinant DNA procedures to provide genetic resistance to disease.

Plant growth regulators have been identified that are capable of modifying plant growth patterns. Other plant research has resulted in salt tolerant plants. Attempts are being made to develop nonlegume plants capable of nitrogen fixation.

CHAPTER REVIEW

Discussion and Essay Questions

1. Describe six problems associated with plant production for which biotechnology is providing solutions.
2. Explain how biotechnology is used to enhance the efficiency of photosynthesis in plants. How might this research affect the world supply of agricultural products?

3. Name several methods of protecting plants against frost damage, and explain how each method works.
4. Make a chart listing methods of controlling damaging insects, and describe the mode of action for each method.
5. Explain what a cultivar is, and describe how it is used to develop disease resistant plants.
6. Describe how genetic engineering is used to develop disease resistant plants.
7. Name the major plant growth regulators, and discuss the effects of each on plant growth.
8. Discuss the marketing challenges associated with food crops that have been treated with plant growth regulators.
9. Describe research efforts to develop plants that are tolerant of high salt concentrations in soil and irrigation water.
10. What advantages might be gained by developing new nonlegume plants capable of nitrogen fixation?

Multiple Choice Questions

1. Chlorophyll is a substance found in plant cells that is required in chemical reactions that convert light energy to:

   A. sugar.
   B. photosynthesis.
   C. chemical energy.
   D. water.

2. A process by which plant tissues are broken down to produce carbon dioxide, water, and heat is called:

   A. photosynthesis.
   B. respiration.
   C. parturition.
   D. symbiosis.

3. Which of the following management practices is *least* likely to protect a fruit blossom from being damaged by a moderate frost?

   A. applying a solution containing ice-minus bacteria to the trees
   B. applying a fine mist of water to the trees
   C. mixing the upper air with the air at ground level
   D. building fires around the perimeter of the orchard

4. The study of insects is the branch of zoology called:

   A. biology.
   B. pomology.
   C. entomology.
   D. biotechnology.

5. Integrated pest management (IPM) is a method of pest control designed to:
   A. control harmful insects and protect useful insects.
   B. kill all harmful insects.
   C. protect pollinators.
   D. completely eliminate the use of chemical insecticides to control harmful insects.

6. Using the same insecticide to control harmful insects over a period of several years is likely to result in:
   A. a generation of insects that is immune to the insecticide.
   B. an increase in the population of predatory insects that prey on the key insect pest.
   C. an increase in the effectiveness of the insecticide due to increased soil concentrations of the chemical.
   D. the development of a natural immunity in the plants to the insect pests.

7. Each of the following methods of insect management is considered to be a form of biological control with the exception of the following practice:
   A. introduction of insect parasites.
   B. practicing crop rotations.
   C. introducing insect disease organisms.
   D. releasing insect predators in the environment.

8. One disadvantage associated with the use of biotechnology to produce a food plant that is resistant to a plant disease is that:
   A. the food it produces will be contaminated with harmful substances.
   B. the environment will be polluted by the transgenic plant.
   C. the plant will kill useful insects.
   D. consumer resistance to genetically engineered products may have a negative influence on the market appeal of the plant and its products.

9. The plant growth-regulating hormone that prevents seeds from germinating until they are mature is:
   A. gibberellin.
   B. abscisic acid.
   C. cytokinin.
   D. auxin.

10. Nitrogen fixation is a process that occurs in the root nodules of legume plants when:
    A. lightning strikes convert nitrogen gas to nitrates.
    B. a compound called symbiosis is produced.
    C. nitrogen gas is converted to nitrates by specialized bacteria.
    D. a shortage of water exists in the soil profile.

LEARNING ACTIVITIES

1. Develop and conduct a market survey to determine the attitudes of consumers in your area toward the use of plant growth regulators to produce food products. Use specific examples, such as the use of Alar to improve color in apples.

2. Take a field trip to a crop research laboratory or farm, or invite a plant scientist to visit the class to discuss biotechnology research in plants. Ask to see actual examples of plants or plant products that are being developed.

CHAPTER 6

Plant Propagation Techniques

Plants are a versatile class of living organisms in their ability to reproduce their own kind. Many plants reproduce sexually by producing seeds from which the next generation of plants is grown. Some plants tend to reproduce primarily through asexual methods such as the growth of stolons or rhizomes. Some plants reproduce both sexually and asexually.

OBJECTIVES

After completing this chapter, you should be able to:

- distinguish differences between sexual and asexual reproduction in plants.
- describe the process of sexual reproduction in plants.
- identify the parts of a seed and describe the function of each part.
- evaluate the importance of hybrid seeds to modern agriculture.
- define the process of asexual reproduction in plants.
- list four forms of asexual plant reproduction and give examples of each.
- assess the role of tissue culture in the development of improved varieties of plants.

TERMS FOR UNDERSTANDING

The following vocabulary terms should be studied carefully as you read:

propagation
stamen
ovule
anther
receptacle
filament
microspore mother cells
tetrad
microspores
generative nucleus
megaspore mother cells
megaspores

micropyle
polar nuclei
gametophyte
embryo sac
pistil
pollination
cross-pollination
self-pollination
stigma
ovary
style
seed
fruit
variety
seed coat
germination
endosperm
cotyledon
monocot
dicot
regeneration
separation
bulb
bulblets
corm
division
rhizome
tubers
eye
tuberous roots
cuttings
softwood
hardwood
cambium
grafting
scion
rootstock
budding
agar
somatic embryogenesis

SEXUAL PROPAGATION OF PLANTS

Propagation occurs when an organism reproduces itself. Sexual propagation or reproduction is accomplished in plants when the male and female gametes are joined to produce seeds. Seed production occurs when a pollen grain from a male flower part called the **stamen** fertilizes the **ovule** or mature female germ cell, Figure 6–1. After the ovule has been fertilized, it matures into a seed that is capable of growing into a new plant.

The stamen consists of the anther and the filament or stalk. The stamen is the male portion of a flower. The **anther** is the organ in which pollen grains develop and mature. It is supported and connected to the **receptacle**, or base of the flower, by the **filament**. Several anthers are usually present in a flower, and each anther contains four pollen sacs in which pollen grains develop.

Pollen formation begins with the production of **microspore mother cells** inside the pollen sacs. These are diploid cells that contain chromosome pairs. As they begin pollen formation, each of these cells divides through the process of meiosis that was discussed in Chapter 2.

A **tetrad** consisting of a cluster of four haploid cells is formed, Figure 6–2. Later, these four cells pull apart forming four cells called **microspores**. The nucleus of each microspore divides one more time, forming a pollen grain consisting of two cells. One cell contains the **generative nucleus** from which two

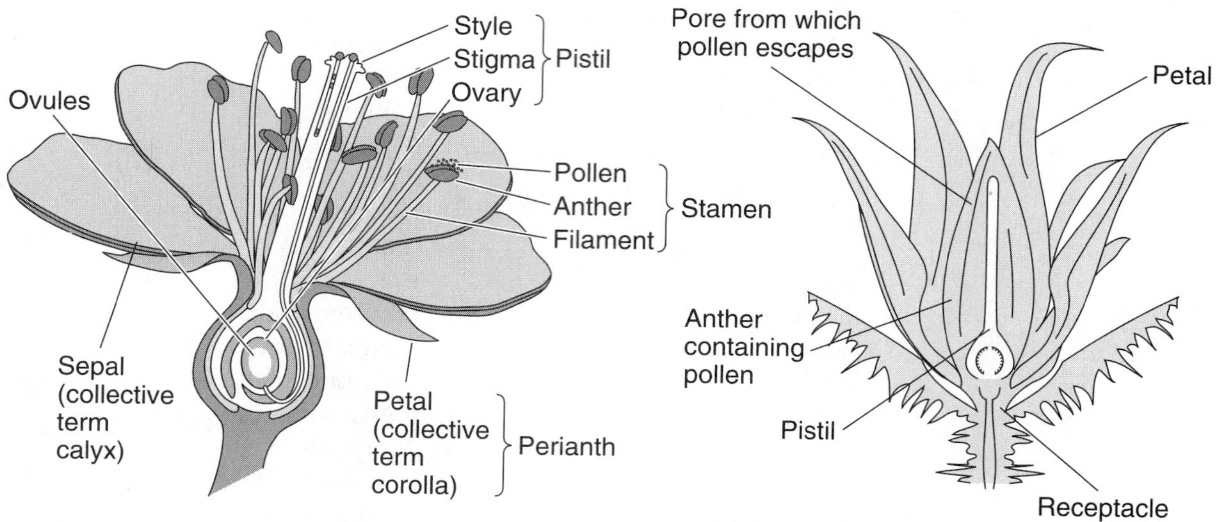

FIGURE 6–1 Major parts of a flower

sperm cells will develop during fertilization of the ovules. The other cell contains the tube nucleus, which is destroyed once it enters the ovule.

Ovule formation begins when a **megaspore mother cell** is produced inside the ovule. It is a diploid cell. This cell divides during meiosis forming four haploid cells called **megaspores**, of which three die. The remaining megaspore grows in size, and its nucleus divides to form two nuclei. Two more divisions occur, and a total of eight nuclei are produced. Each of these nuclei is haploid, meaning that single chromosomes are present instead of paired chromosomes.

Now the eight nuclei move to different locations in the ovule. Three of the nuclei gather near a small opening in the ovule called the **micropyle**. One of these nuclei enlarges and becomes the egg cell. Three other nuclei migrate to the opposite end of the ovule. They will eventually develop cell membranes along with two of the nuclei near the egg cell. The other two nuclei are called **polar nuclei**, and they migrate to the center of the chamber where they form a single cell. They will eventually produce food for the tiny plant embryo.

A total of seven cells are formed, encompassing the eight original nuclei. These seven cells together comprise the female **gametophyte**, or gamete. This structure is also called the **embryo sac**. At this stage of reproduction, the ovule is mature. Once the pollen grains and the ovules have matured, pollination and fertilization can occur.

The female flower parts include the stigma, style, and ovary. These three parts constitute the **pistil** of the flower. **Pollination** occurs when pollen is transferred from the anther to the stigma of a flower. **Cross pollination** is the transfer of pollen from the anther to the stigma of different plants of the same species. When pollen is transferred from the anther to the stigma of the same flower, **self-pollination** occurs.

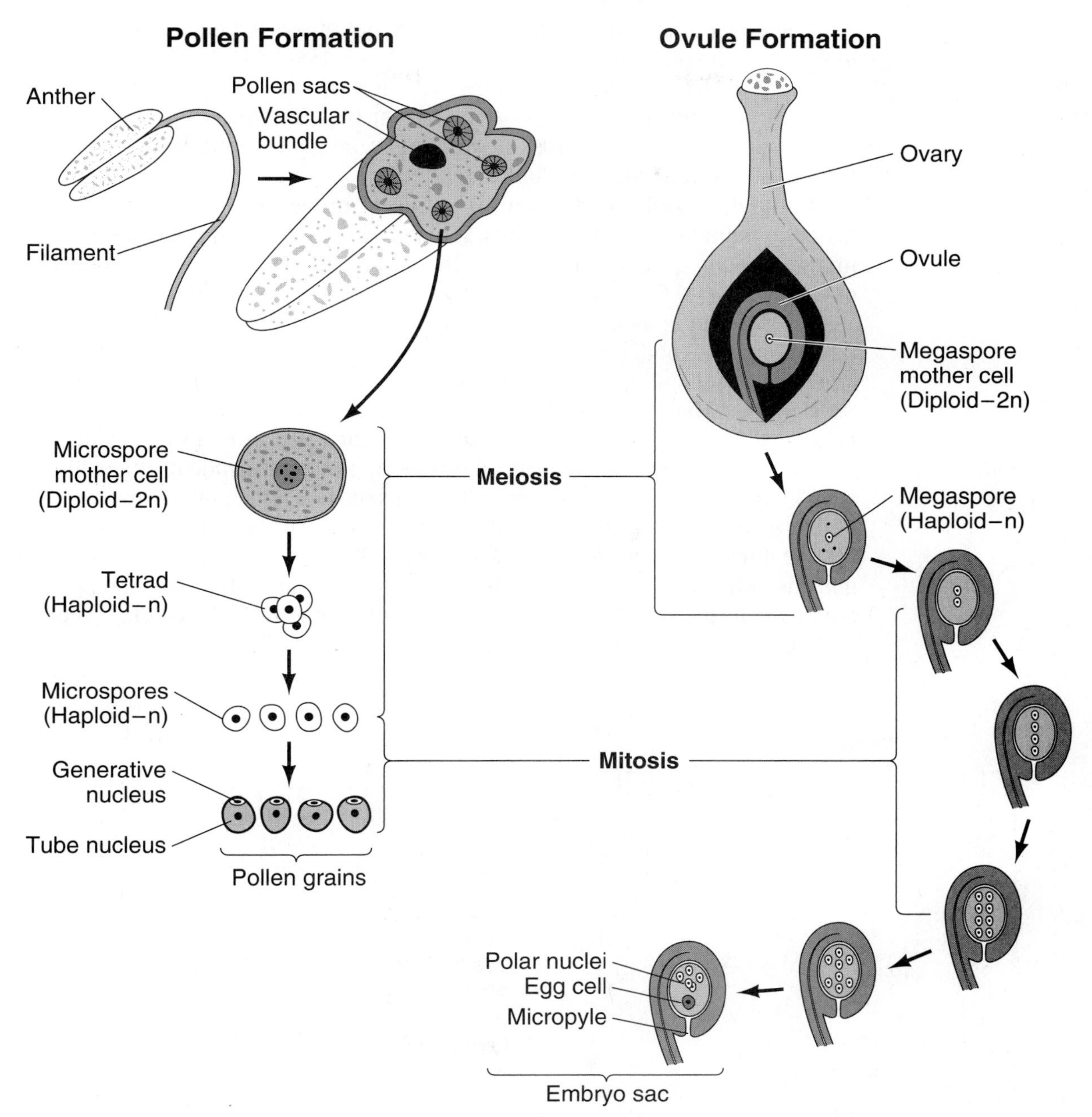

FIGURE 6–2 Gamete production in plants

Cross-pollination is desirable because it promotes genetic diversity. When self-pollination occurs, the plants in the immediate region become very similar to one another, but their vigor sometimes declines. They may become less competitive with other plants in the ability to survive and produce.

Some plants are easily cross-pollinated, but others depend on insects, animals, wind and other forces to carry pollen from one plant to another. Bees and other insects are common pollinators that carry pollen on their bodies from one flower to another as they feed on the nectar that the flowers produce. Some birds, especially hummingbirds, are also important pollinators. Many pollinators are attracted to flowers by their bright colors.

Pollination and fertilization are different processes. Fertilization is the fusion of male and female gametes. It occurs in plants when the male gametes found in pollen grains combine with the female gametes located in the ovules of plants, Figure 6–3.

The **stigma** is an organ located on the female structure of a flower where it functions as a pollen receptor. The **ovary** is the flower part that produces the egg cell, and the **style** is the structure that connects the stigma to the ovary. Soon after a pollen grain lands on the stigma of a flower, a pollen tube begins to develop. After it penetrates the stigma, it grows down through the style until it reaches the micropyle or opening in the surface of the ovule.

The tube nucleus and the generative nucleus enter the pollen tube from the pollen grain and migrate to the ovule. As these structures migrate, the generative nucleus divides forming two sperm cells. The sperm cells and the tube nucleus enter the ovule through the micropyle. One sperm fertilizes the egg cell creating a fertilized cell called a zygote. The other sperm fertilizes the two polar nuclei, and a cell is formed around them.

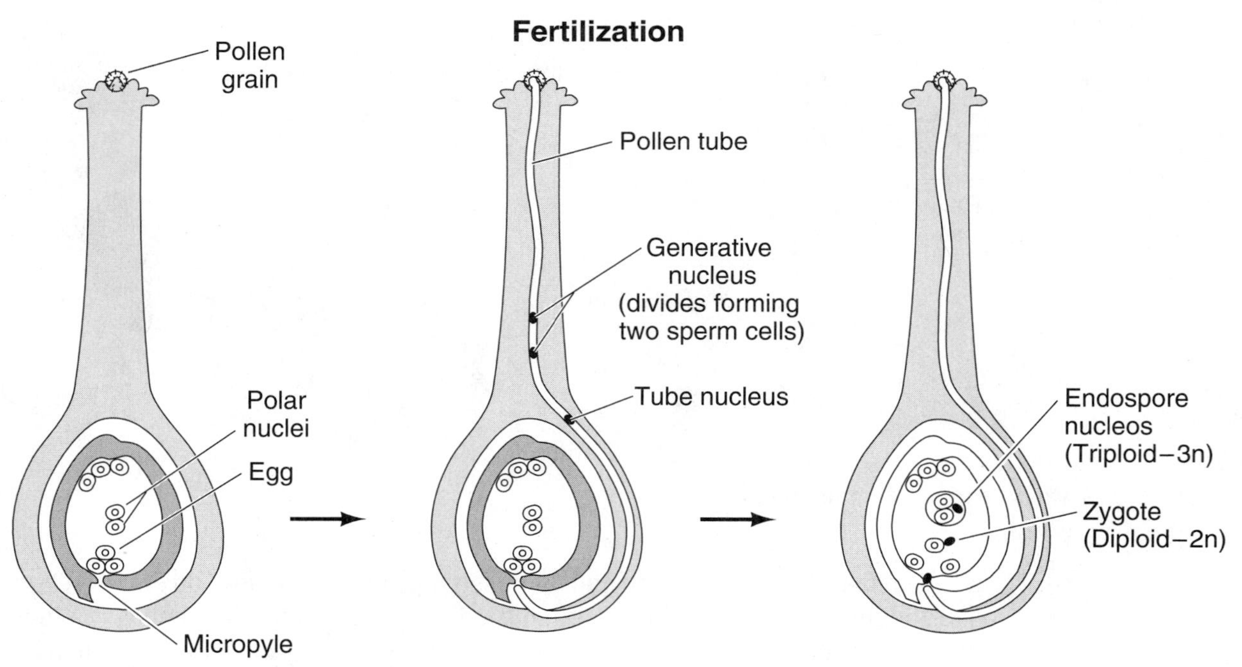

FIGURE 6–3 Fertilization in plants

The development of seeds begins after pollination and fertilization are complete. A **seed** is an ovule that has reached maturity. It consists of a tiny plant embryo along with its food supply enclosed in a protective outer coat. The seed is surrounded by the **fruit**, which consists mostly of the ripened ovary. The fruits are sometimes eaten by birds and animals and dispersed to new sites in their droppings.

Many of the varieties of seed crops produced in agriculture such as wheat, oats, beans, and peas are examples of sexually propagated plants. Plants that have inherited the same genetic traits are said to be of the same **variety**. Most of these plants breed true to type and are therefore very similar in the characteristics they exhibit. The individual plants that make up a variety have descended from the same parent stock.

Plant Production From Seeds

Seeds vary greatly in size, but they are similar in many ways. A seed is made up of three basic parts: (1) the seed coat; (2) the embryo; and (3) the endosperm. The **seed coat** is a protective covering for the embryo and the endosperm. The seed begins to grow when moisture penetrates the seed coat and causes the contents of the seed to swell until the seed coat is broken. This process is called **germination**.

The **endosperm** is the food supply for the embryo. The endosperm consists of starch, protein, and oils. It nourishes the new plant as it emerges from a seed until the root begins to draw nutrients from the soil and photosynthesis begins.

The embryo is a tiny immature plant complete with roots, stems, and leaves. Once the seed coat is broken, the root and stem of the embryo emerge from the seed, leaves begin to form on the stem, and the plant begins to produce its own food supply.

The first leaf of a new plant is the **cotyledon**, or seed leaf. Sometimes the cotyledon stores the nutrients that are needed by the embryo. In other plants, the cotyledon absorbs nutrients that are stored and makes them available to the embryo. When a single cotyledon is present, the seed is called a **monocot**, Figure 6–4. A seed with two cotyledons present is called a **dicot**.

Each seed contains a supply of nutrients that is stored in the tissue of the endosperm. The endosperm was developed from a cell that formed from the fusion of a sperm with two polar nuclei during fertilization.

Germination is critical in producing plants from seeds. Several factors affect germination, but seed quality, moisture, and temperature are of prime importance. Aeration of the soil and soil characteristics also play key roles in germination of seeds.

One of the greatest single advantages of sexual reproduction in plants is that large numbers of seeds can be produced in a growing season. This allows plants to multiply rapidly when growing conditions are favorable.

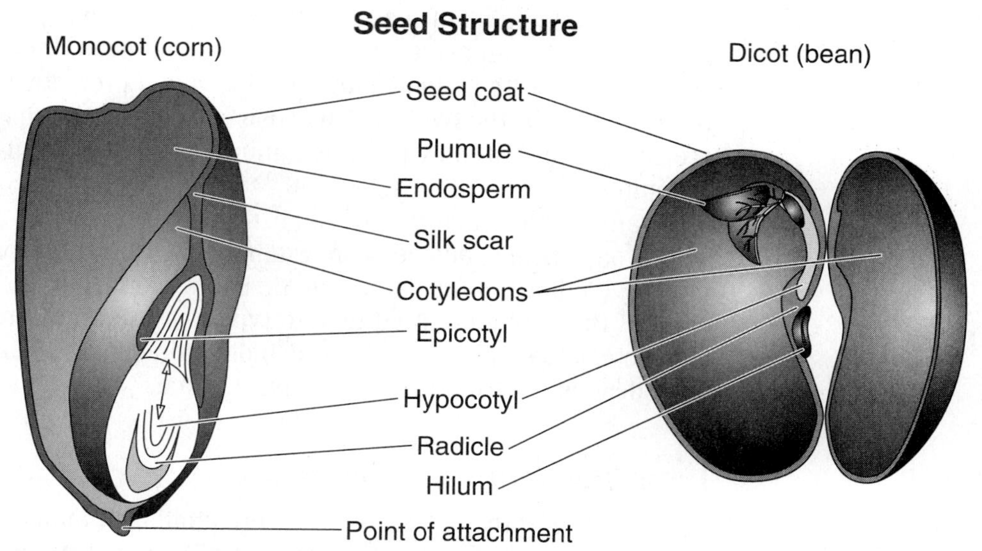

FIGURE 6–4 Distinct similarities and differences exist between seeds classed as monocots and dicots. The embryo is made up of the radicle, hypocotyl, and plumule.

Hybrid Seed Production

Frequently, the seed of a plant is the primary agricultural product for which the plant is cultivated. Seeds are often eaten by humans and by livestock or poultry as major portions of their diets. This makes seed production much more important than it would be if only enough seed production was needed to produce the next generation of plants.

Production of hybrid seeds is a method of plant breeding that crosses two genetic strains of a species of plants to produce seed for commercial production of a crop. Grains and grasses are good examples of crops that use hybrid seeds to stimulate yields, Figure 6–5.

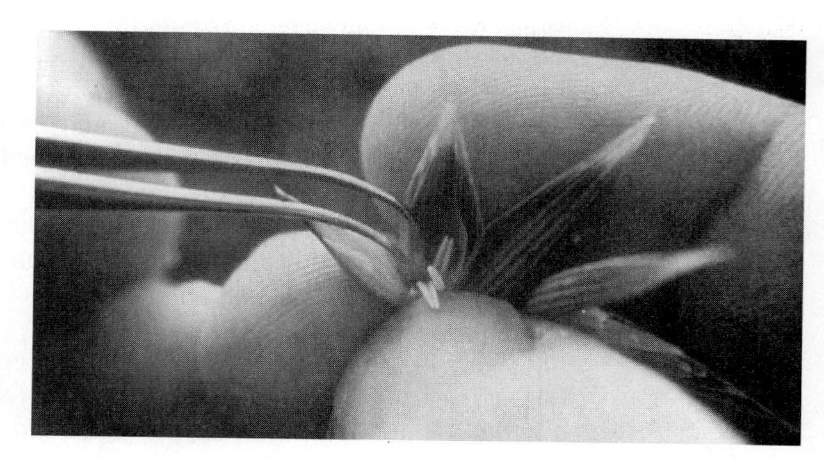

FIGURE 6–5 Hybrid plant seeds are produced by removing male flower parts to prevent self fertilization. *(Photo courtesy of Utah Agricultural Experiment Station)*

FIGURE 6–6 Hybrid seed corn is produced by planting male rows that are evenly spaced among the female rows. The tassels are removed from female rows to assure that only pollen from the male rows will be available to fertilize the crop.

FIGURE 6–7 Pollination in plants is sometimes controlled by placing bags over flowering plant parts. *(Photo courtesy of Utah Agricultural Experiment Station)*

Hybrid seed is produced by preventing the flowers of one plant strain from being pollinated by pollen of the same strain. Instead, it is pollinated by the variety or strain with which it is being crossed, Figures 6–6 and 6–7.

The characteristics of the two parent plants are often very different, and the plants grown from the hybrid seed usually combine the qualities of both parent varieties. Hybrid plants usually appear to be quite uniform, and they often produce greater or higher quality yields than either of the parent varieties.

Hybrid seed must be produced each year because seed produced from the hybrid plants does not breed true. It should not be planted. It produces individual plants that lack uniformity in looks and performance. Some of the plants will resemble each of the parent varieties, but most of the plants will be different from either parent and different from the hybrid plants. To maintain hybrid yields, first generation hybrid seed must be used for each planting.

Hybrid seed is the result of a lot of research. Universities and experiment stations have done much to develop hybrid varieties of plants, but large seed companies also research new crop varieties. They do this in order to compete for profits from the sale of commercial varieties of hybrid seeds.

ASEXUAL PROPAGATION OF PLANTS

One characteristic of plants that is different from many other living organisms is their ability to replace missing parts by growing new parts. This process is called **regeneration**. Some plants are able to regenerate or grow an entire plant

CAREER OPTION

Seed Analyst

A person who chooses a career as a seed analyst works with seeds to make sure that seed quality is high. Seed analysts require various degrees of technical or collegiate-level training prior to employment. They must be patient people who are able to perform many repetitions of the same task.

Most of the major agricultural food and fiber crops are grown from seeds. Because large amounts of quality seeds are needed each growing season, a giant seed industry has developed. Many seed analysts work for large seed companies. They test the seed the company plans to offer for sale to determine how well the seed can be expected to perform in the field, and to ensure high seed quality, Figure 6–8. Some seed analysts work with government agencies that are responsible for making sure that commercial seed meets the standards of quality defined by seed laws.

Seed analysts test seeds to see how well they germinate. They observe the seeds carefully to ensure that noxious weed seeds and foreign materials are not mixed with the seeds of the desired plant variety. Seed analysts play a major role in preventing the spread of noxious weeds by rejecting contaminated seed crops, Figure 6–9.

FIGURE 6–8 The end result of successful plant breeding is the development of healthy and productive plants for crop production. *(Photo courtesy of Utah Agricultural Experiment Station)*

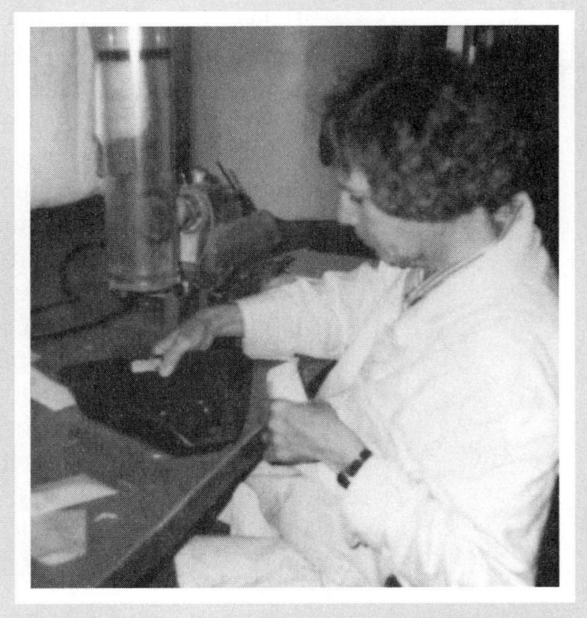

FIGURE 6–9 A seed analyst is responsible for checking seed samples to make sure that the seed is free of weed seeds and other impurities.

from only leaves, roots, or stems. Asexual propagation is the only form of plant reproduction that can be used to retain pure varieties of some plants. Many varieties of fruit and ornamental plants do not breed true, and some plant varieties do not produce any seeds at all. New plants from these varieties must be produced from plant parts obtained from the parent plant.

Production of Plants by Separation and Division

One method of propagating plants asexually is to separate underground plant parts such as bulbs and corms. This plant propagation procedure is called **separation**, Figure 6–10.

A **bulb** consists of a short, starchy underground stem covered with layers or scales. Tulips and lilies are examples of plants having bulbs. Bulbs develop small structures called **bulblets** at the base of the bulb scales. They may be removed after the tops have dried up in the fall. Large bulbs are sorted out to be used for spring flowers, but some bulbs may require additional growing seasons to get big enough for flowering.

A **corm** is similar to a bulb, but it contains a longer stem and fewer scale leaves. Gladiolus plants are reproduced from corms using the separation procedure.

Propagation of plants by **division** is done by cutting underground stems and roots into smaller pieces and planting them to produce additional new plants.

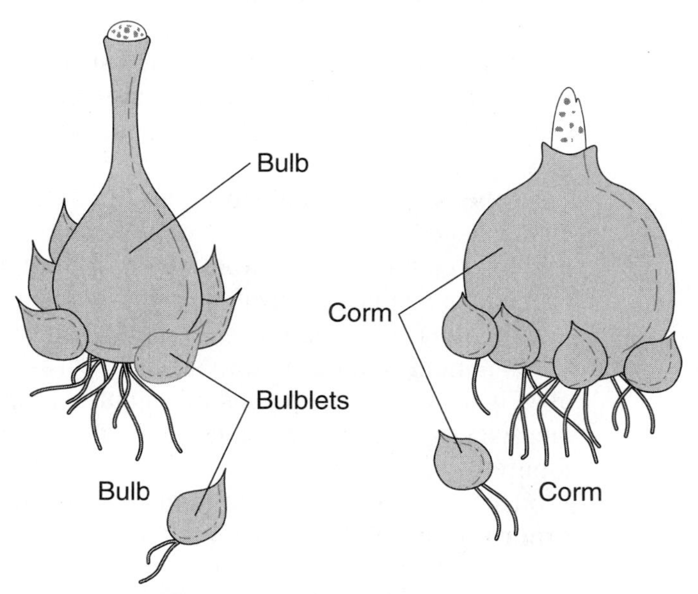

FIGURE 6–10 Plants that grow from plant parts such as bulbs can be propagated by separating bulblets from the parent material. Each bulblet is capable of growing into a new plant.

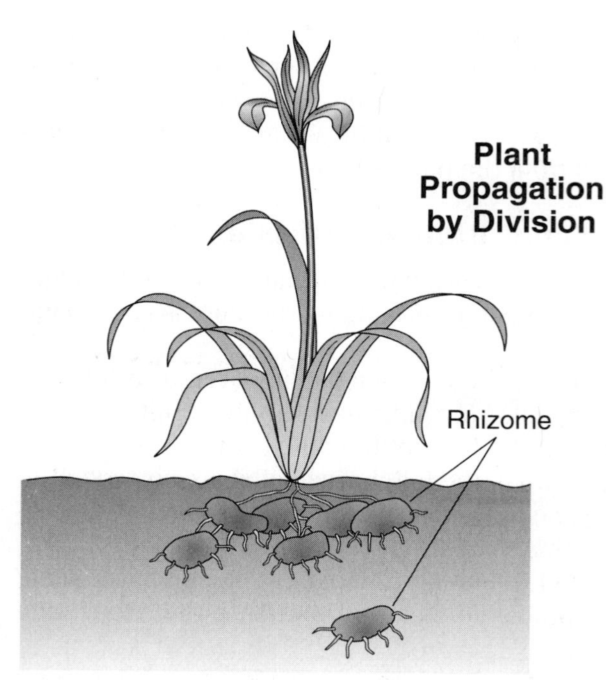

FIGURE 6–11 Rhizomes are underground stems. Dividing rhizomes into smaller pieces makes it possible for several plants to be propagated.

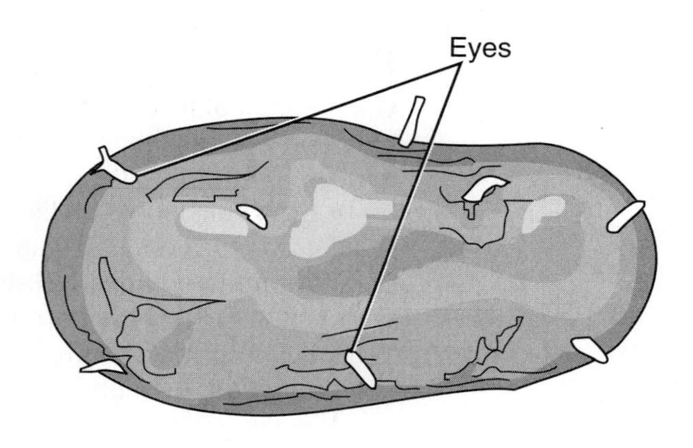

FIGURE 6–12 A tuber is an underground stem or shoot with "eyes" from which new growth of leaves, stems, and roots can occur. A tuber such as a potato can be divided and planted.

The procedure is effective with plants such as the iris, which grows from a horizontal underground stem called a **rhizome**, Figure 6–11. During the growing season it grows in length. The rhizome is cut into shorter lengths to produce new plants.

Tubers are specialized underground stems or shoots. They are easily distinguished by the eyes that are located on the fleshy stem. The **eye** is a structure located on the tuber from which a sprout develops. The potato plant is an example of a plant propagated by division. The process is accomplished by slicing the tuber into small pieces, each of which contains an eye. The cut tubers are commonly called "potato seed," even though they are not really seeds at all, Figure 6–12. The pieces are planted in much the same manner as other large seeds, and they produce roots and stems that develop into new plants, Figure 6–13.

The division process may also be used to propagate plants that have **tuberous roots**. Tuberous roots are different from tubers in that they are roots having root growth on one end and buds on the other end. The sweet potato is a plant with tuberous roots. Shoots that grow from the main tuberous root are pulled off and planted. They root easily and grow into new plants.

FIGURE 6–13 An entire new potato plant can be generated from a potato piece that contains an eye.

FIGURE 6–14 Cuttings are plant parts that are treated with rooting hormones and placed in potting soil to grow roots. *(Photo courtesy of Gaylen Smyer)*

Plant Production From Cuttings

Propagation of plants from **cuttings** is accomplished by cutting plant parts off living plants. Cuttings may be mature or growing plant parts including leaves, stems, roots, or buds. Some plant species such as dogwood or lilac may root better if the cutting is immature growth or **softwood**. Others, such as juniper or evergreen, may root better if the cutting is mature wood or **hardwood**.

Rooting is accomplished by treating the cutting with a rooting hormone and placing it in a rooting medium, Figure 6–14. Water is added to the medium and the atmosphere around the cuttings is controlled by covering the materials with plastic or placing them under a mist system.

Cuttings require an atmosphere similar to those required to germinate seeds except that light is needed to stimulate production of plant food. The process of photosynthesis occurs in the living cells of the cuttings and provides nourishment for the new plant during the rooting process.

Plant propagation using cuttings is practiced with many of the plants used by horticulturists.

Propagation Of Plants By Grafting

The structure of most woody stems is similar in that they are covered with an outer layer of bark that seals in moisture and protects the inner layer of tissue called **cambium**. The cambium is composed of a porous material that allows water and plant nutrients to be transported from the roots to the leaves. The inner section of the stem is composed of wood, Figure 6–15.

Grafting is an important procedure for propagating fruit trees. It is done by uniting two plants in such a way that their tissues grow together. The prac-

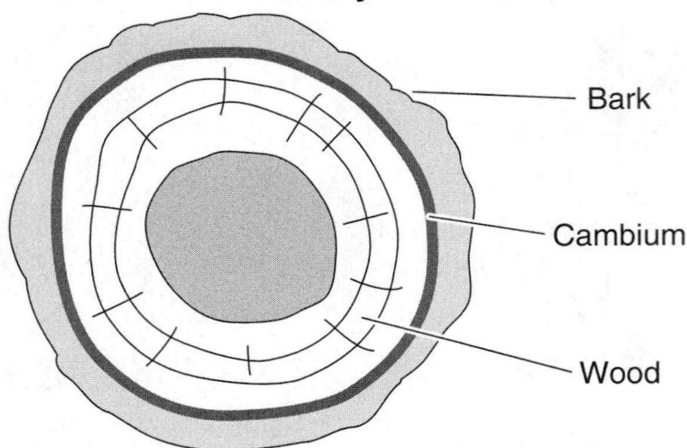

FIGURE 6–15 Water and nutrients flow through the cambium layer in woody plants. A successful graft requires that the cambium of both the scion and the stock is aligned to allow materials to flow freely.

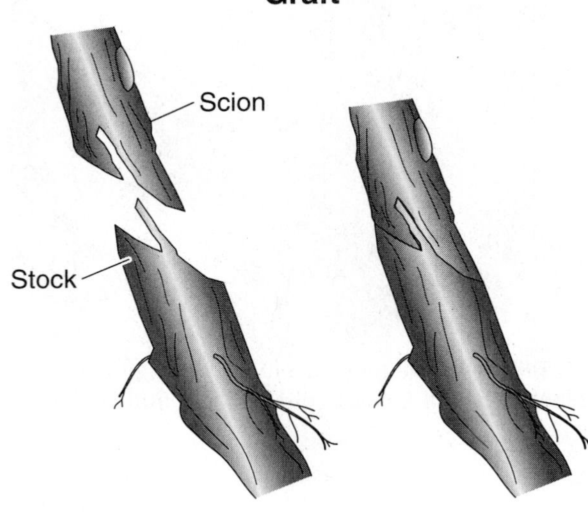

FIGURE 6–16 Whip graft

tice is used because it allows plants with superior qualities to be duplicated. A plant produced by grafting is genetically identical to the plant from which the scion (top of the plant) was obtained, Figure 6–16.

The **rootstock**, or root, that is used should be selected for the climate in which the plant will be grown to be sure that it is hardy enough to avoid winterkill. The kind of rootstock that is used will also determine the size of a tree. A rootstock that reduces plant vigor will result in the tree being a dwarf. Most fruit trees are grafted to rootstock that is more durable than the natural roots.

A successful graft requires careful attention to details. The plants from which the stock and the scion are obtained must be related closely enough for the plant tissues to grow together, and the scion wood should be healthy one-year-old wood. Grafting is most successful when the rootstock and the scion are grafted while they are dormant.

The cambium layers of the scion and the rootstock must be matched carefully. When this is done, water and nutrients can flow between the upper and lower sections of the plant. When a graft is completed, it must be carefully sealed to prevent the sap of the plant from leaking out and water from entering the graft. This can be done using grafting wax or waterproof wrapping materials made of plastic or rubber.

Grafting allows different parts of a fruit tree to produce related varieties of the fruit on the same tree. Apples of the red delicious and golden delicious varieties may be produced on separate grafts to the same tree.

Producing New Plants by Budding

Budding is a form of grafting in which a bud is used instead of a scion. A leaf bud is carefully removed from the desired plant and inserted into the tissue of the rootstock. This may be done in different ways depending on the type of plant to which the budding process is applied and the growth stage of the plant.

T-Budding is a process in which a leaf bud is implanted in a T-shaped slot in the bark of the rootstock. This process is used when the plant is in a stage of active growth, Figures 6–17, 6–18, 6–19, and 6–20.

Chip budding is similar to T-budding except that a chip of mature wood is removed from the rootstock, and a matching chip of wood that includes a bud is grafted into the rootstock. Chip budding is used when the rootstock is not growing actively. It is also used with grapes.

FIGURE 6–17 The first step in budding is to remove a healthy bud from the parent stock. *(Photo courtesy of Gaylen Smyer)*

FIGURE 6–18 A T-shaped slit is cut in the bark of the rootstock. *(Photo courtesy of Gaylen Smyer)*

FIGURE 6–19 The bud is carefully inserted into the slit that was prepared in the bark of the rootstock. *(Photo courtesy of Gaylen Smyer)*

FIGURE 6–20 The bud is tightly sealed in the graft area. *(Photo courtesy of Gaylen Smyer)*

CELL AND TISSUE CULTURE

Plant propagation using tissue culture methods is a new tool for plant scientists. Plants are grown from tissue obtained from the parent plant such as buds, leaf parts, or terminal shoots. It allows many new plants, that are genetically identical to the parent stock, to be propagated rapidly.

The tissue is sterilized and placed on a sterile nutrient **agar** jell or similar growing material along with growth regulating hormones, and it is sealed to prevent contamination, Figures 6–21 and 6–22. Tiny plant sprouts soon begin to grow from the plant tissue. The sprouts are carefully removed with tweezers and placed in a new medium to stimulate root growth, Figure 6–23.

Great care must be taken at every step of this procedure to maintain a sterile environment and avoid contamination of the materials, Figure 6–24.

This new procedure makes it possible to produce exact duplicates of valuable plants in large numbers. Plant tissue culture is one of the most valuable plant technologies to be discovered in recent times, Figure 6–25.

Artificial Seeds

A new technology known as **somatic embryogenesis** uses a tissue culture process to regenerate tiny plants from plant cells. Each plant embryo is allowed to mature to a selected stage of growth where its development is stopped. Each miniature plant is then coated with a synthetic polymer similar

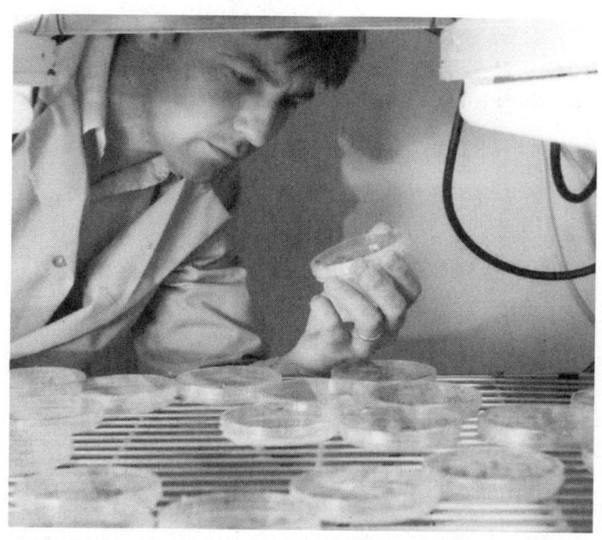

FIGURE 6–21 Tissue culture requires a carefully controlled environment to develop callus tissue. *(Photo courtesy of Utah Agricultural Experiment Station)*

FIGURE 6–22 Special hormones are applied to callus tissue to stimulate the growth of leaves and stems. *(Photo courtesy of Utah Agricultural Experiment Station)*

FIGURE 6–23 Individual plants are carefully protected while they develop roots, stems, and leaves. *(Photo courtesy of Utah Agricultural Experiment Station)*

FIGURE 6–24 Special growth chambers provide a proper atmosphere in which developing plants can survive and grow. *(Photo courtesy of Utah Agricultural Experiment Station)*

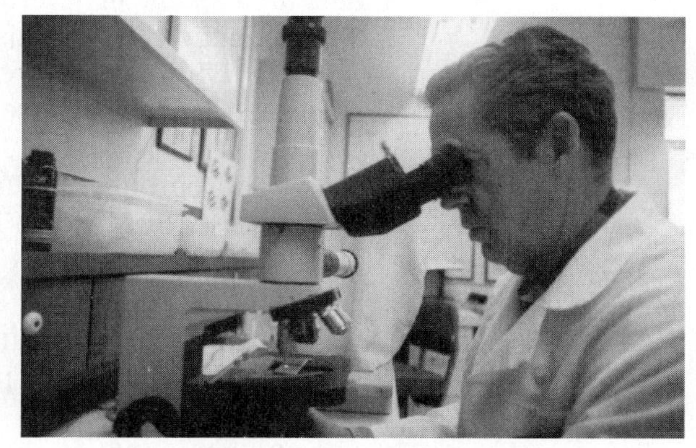

FIGURE 6–25 Scientists who do research in biotechnology are highly educated in biochemistry and life processes. *(Photo courtesy of USDA and Utah Agricultural Experiment Station)*

to the gelatin capsules that are used with powdered medicines. The coating may also contain plant nutrients, plant hormones, or other substances that promote plant growth. Some companies are even experimenting with adding other materials to the "seedcoat" such as fertilizers, insecticides, and herbicides.

The product that results from this process looks like a seed, but it is an "artificial seed." It is really an embryonic plant with a protective coating. When these artificial seeds are planted, they germinate uniformly within a day or two, and they mature at the same time. This process is being tested on such plants as lettuce, celery, alfalfa, and carrots. This new process will make it easier to get the crop to germinate and grow uniformly in the field.

Somatic Hybrid Plants

Plant scientists have learned to create new hybrid plants using a new form of cell and tissue culture. Using enzymes, the walls of two plant cells from different plants are dissolved. This allows the protoplasm from both cells to fuse together to form a new cell. This new hybrid cell contains the genetic material from both parent plants. This hybrid plant cell is then regenerated into a complete plant that has some of the characteristics of each parent. The process has been used to produce potatoes that are resistant to specific herbicides. The procedure has also been successfully used with other crops such as tobacco and rice. This procedure helps scientists to overcome problems associated with incompatible plant cells.

CHAPTER SUMMARY

Plants are capable of several different forms of reproduction. They include sexual reproduction from seeds, and asexual reproduction from vegetative parts of plants. Hybrid seeds are produced by crossing two distinct parent lines. Hybrid seed is used because the plants that are produced are capable of higher yields than either parent.

Asexual reproduction in plants is accomplished by regeneration of missing plant parts from pieces of live plant materials. Separation and division is a process in which underground plant parts are cut apart, and each part develops into a new plant.

Propagation of plants from cuttings is done by cutting plant parts off living plants and rooting the parts in a special rooting medium in the presence of light.

Grafting is another form of asexual propagation in which the scions or buds from one plant are united with a rootstock provided by another plant.

Tissue culture is a method of plant reproduction in which tiny plants are generated from plant buds, leaf parts, and terminal shoots under sterile laboratory conditions.

Asexual forms of plant reproduction are important because they make it possible to produce plants that are genetically identical to the plants from which they were derived.

CHAPTER REVIEW

Discussion and Essay Questions

1. Describe distinguishing characteristics of sexual and asexual reproduction in plants. Contrast the differences between the two processes.
2. Illustrate the parts of a seed, and explain the role each part plays in the germination process.

3. How has hybrid seed affected crop production in modern agriculture?
4. What procedures should be followed when separation and division are used as methods of plant reproduction?
5. Describe how plants are reproduced from cuttings.
6. How is the grafting process used to reproduce plants asexually?
7. List the conditions under which sexual propagation of plants is desirable.
8. Under what conditions is asexual reproduction of plants likely to be an advantage?
9. What is the role and importance of tissue culture as a plant reproduction tool in developing improved varieties of plants?

Multiple Choice Questions

1. The process that occurs when an organism reproduces itself is called:
   A. germination.
   B. fertilization.
   C. propagation.
   D. fermentation.

2. The plant organ in which pollen grains develop is the:
   A. ovary.
   B. pistil.
   C. tetrad.
   D. anther.

3. A megaspore mother cell is a diploid cell that is produced inside the:
   A. ovule.
   B. micropyle.
   C. pollen sac.
   D. stamen.

4. Which of the following organs is not part of the pistil or flower?
   A. stamen
   B. ovary
   C. stigma
   D. style

5. Pollination is the process in which:
   A. pollen is transferred from the anther to the stigma.
   B. pollen is produced in the anther.
   C. pollen is gathered by worker bees.
   D. people become allergic to plants.

6. A fertilized egg cell is called a:
   A. seed.
   B. ovule.
   C. zygote.
   D. polar nucleus.

7. As a seed germinates and begins to grow, the plant embryo is nourished by the:
   A. root.
   B. pollen grain.
   C. yolk.
   D. endosperm.

8. A seed with two cotyledons is called a:
   A. dicot.
   B. triplicate.
   C. diploid.
   D. monocot.

9. The fertilization of the polar nuclei in the ovule of a flower produces the part of a seed called the:
   A. pollen grain.
   B. embryo.
   C. endosperm.
   D. seed coat.

10. The asexual plant reproduction process that is used with bulbs and corms is called:
    A. division.
    B. separation.
    C. propagation.
    D. grafting.

11. A plant tissue that must be aligned or matched up during the process of grafting a scion to a rootstock is the:
    A. bark.
    B. cambium.
    C. stem.
    D. callus.

12. The formation of roots, buds, leaf parts, and terminal shoots from callus tissues is caused by substances called:
    A. herbicides.
    B. stimulants.
    C. cryoprotectants.
    D. hormones.

LEARNING ACTIVITIES

1. Examine and test a container of certified seed to determine whether the seed meets seed law standards for purity and germination qualities.

2. Conduct a laboratory exercise in which each of the following plant propagation skills is demonstrated and practiced:
   a. separation and division using rhizome materials from iris plants
   b. cuttings using leaves from African violets
   c. grafting sucker wood from an apple tree to a rootstock approved for local growing conditions

Technology: Food and Fiber

- Plant Management Technologies
- Agricultural Production Technologies
- Processing Agricultural Products
- Agricultural Marketing Technologies

CHAPTER 7

Plant Management Technologies

Plants and the soils they grow in are among the most valuable resources available to the human race. We depend upon plants for the most basic human needs. They are sources for food, clothing, shelter, and heat, Figure 7–1. They nourish and shelter our animals. They provide beauty to our environment. In a world where many natural resources are in danger of being depleted, plants are a renewable resource.

OBJECTIVES

After completing this chapter, you should be able to:

- contrast differences among mechanical, chemical, and biological forms of weed and pest control.
- explain why integrated pest management is considered to be a valuable form of pest control.

FIGURE 7–1 Abundant supplies of fruits and vegetables depend upon proper use of agricultural technologies. *(Photo courtesy of Utah Agricultural Experiment Station)*

- describe the importance of tissue analysis in determining fertility requirements for crop production.
- define the role of irrigation in producing food for a growing world population.
- describe the major forms of irrigation in use in modern agriculture.
- identify different types of irrigation equipment.
- appraise the use of remote sensing technologies to analyze climatic conditions, crop stress, and field conditions.

TERMS FOR UNDERSTANDING

The following vocabulary terms should be studied carefully as you read:

mechanical pest control
organic farming
chemical control
withdrawal period
hydroponics
aeroponics
petioles
tissue analysis
petiole testing
pH
irrigation
aquifers
flood irrigation
hydrologist
headgates
irrigation furrows
sprinkler irrigation
setting
main lines
handlines
wheel lines
solid set
center pivot sprinkler
lateral move sprinkler
risers
low-pressure sprinkler
drip irrigation
chemigation
fertigation
meteorologists
aerial mapping
infrared radiation
infrared thermometry
multispectral imagery

WEED AND PEST CONTROL

Weeds and pests have existed for as long as people have attempted to cultivate crops. Every species of living organism has checks and balances in the environment that limit its numbers. Many forms of weed and pest control have been tried, ranging from removing the pest by hand to applying soil sterilants to the soil that eliminates most forms of life.

Mechanical Control

Mechanical pest control is accomplished by removing or destroying the pest by hand or through the use of machines. Examples of this kind of control include weed removal by farm laborers who use hoes to cut weeds out of the growing

FIGURE 7–2 In many parts of the world, the most common form of mechanical control for weeds is a hoe. In mechanized nations, a tractor-mounted cultivator provides mechanical control.

FIGURE 7–3 Many types of cultivator tools are available. They are all designed to mechanically destroy weeds.

crops. Mechanical control of weeds is usually accomplished by cultivating or removing weeds from crops with cutting tools on tractor-mounted tool bars, Figure 7–2. Many different kinds of cultivating tools have been developed for specific uses, but all of them remove weeds mechanically, Figure 7–3. Many satisfactory methods of mechanical weed control have been developed.

Mechanized control of harmful insects is difficult to accomplish. **Organic farming** and gardening is practiced by people who depend on mechanical controls or sprays containing natural ingredients. These people refrain from the use of synthetic chemicals on their crops. Sometimes human labor is used to gather harmful insects from the fields in efforts to reduce the insect populations and to prevent crop damage. The greatest weakness in this control method is that many insects are capable of producing several generations in a single growing season. Even when they are physically removed from the crop, a new generation of insects soon hatches and crop damage from insect activity continues to occur.

Insects are sometimes lured to traps using baits such as pheromones that give off odors attractive to insects. In some situations, lights have been used to attract insects to electrical grids. Controlling insects using mechanical means is not generally very practical or cost effective, especially with insects that are relatively small.

Chemical Control

A widely-used method of managing weeds and pests is known as **chemical control**, Figure 7–4. Chemicals are available that are quite specific to plants, i.e., they kill certain plants and may have little or no effect on other plants. Chemicals can be obtained that destroy broadleaf plants but do not damage grasses. Such chemicals provide good weed control on crops like corn.

FIGURE 7–4 Chemicals that control weeds and pests are applied to many crops by dusting or spraying measured amounts on the plants. *(Photo courtesy of Utah Agricultural Experiment Station)*

FIGURE 7–5 Healthy plants that are free from insect damage are needed if we are to have an adequate and high-quality food supply. *(Photo courtesy of Utah Agricultural Experiment Station)*

Farmers must use caution in marketing crops that have been treated with chemicals. Agricultural products intended for animal or human food may be safely used when the **withdrawal period** or recommended time between chemical treatment and processing has been observed, Figure 7–5.

Extensive testing of chemicals is required before they are released on the market. In most circumstances, the chemicals used to control weeds and pests may be safely used when the directions for use are followed.

One problem associated with chemical control of weeds and pests is that over an extended period of time some individual organisms from the harmful species build up a chemical resistance. Offspring from resistant pests and weeds may become more difficult to control due to their abilities to tolerate some chemicals.

Chemical Safety

It is not the use of chemicals that threatens or damages the environment; it is the misuse of chemicals that causes these problems. Chemicals that are used to produce crops have been carefully tested before they are approved for general use, and the conditions are prescribed under which they may be used.

The following practices will contribute to safer use of chemicals in crop production:

1. Test soils for nutrient deficiencies and apply only the amount of fertilizer that is needed.
2. Apply small amounts of fertilizers as they are needed, instead of making a single large application.
3. Apply pesticides to only those crops for which they are approved for use, and in the amounts that are recommended on the label.

4. Dispose of empty chemical containers and leftover chemical mixtures according to the directions that are given on the label. DO NOT POUR THEM OUT!
5. Apply herbicides and insecticides on calm days to avoid having the materials drift to areas where they may cause harm.
6. Use selective herbicides that kill only certain kinds of plants if they are available.
7. When it is possible to do so, use pesticides that are readily broken down to nontoxic materials.
8. Avoid chemical storage problems by purchasing only the amount of chemicals that you will use.
9. Wear protective gear and clothing, and follow suggested clean-up procedures as you apply chemicals.
10. Avoid long periods of exposure to toxic chemicals.

Biological Control

Biological control of weeds and pests is a method of controlling undesirable species by introducing their natural enemies into the environment. In most cases where this has been successful, natural enemies have been obtained from an area of the world from which the pest or weed originated. To successfully accomplish biological control of a harmful species of plant or insect, the natural enemy must be able to adapt to the new environment. When pests and their natural enemies live in compatible environments, the success rate for biological control is good. Care must be taken to assure that introductions of new organisms into an environment do not result in new pests when the population of new organisms expands.

The performance of natural enemies of pests and weeds may be improved by creating a friendly environment for the natural enemy. This is done by providing food sources and plants that are attractive to the natural enemy. Many natural enemies are attracted to the pest by chemicals that they detect in the environment. Chemicals that attract the natural enemies of a pest are sometimes added to the environment to entice the natural enemies of a particular weed or insect to congregate and remain in an area.

Biological Pesticides

One method for controlling insects is to use organisms to kill them. This form of biological control makes use of several different kinds of natural enemies to restrict insect populations below the economic threshold. For example, gypsy moths are controlled by infecting them with a bacterium known as *Bacillus thuringensis*. It causes a protein crystal to form in the digestive tract that causes the death of the moth.

Nematodes are tiny roundworms that act as parasites toward the larvae of several species of insects. When nematodes that are otherwise harmless are applied on a field, they kill such pests as weevils, termites, maggots, and cutworms. Some kinds of viruses are now available to control caterpillars. Another kind of organism, a protozoan, reduces grasshopper populations by destroying their eggs. These organisms are so small that they can be applied by spraying the field. They offer effective alternatives to chemical insecticides.

Integrated Pest Management

Integrated pest management is a method of controlling harmful insects and providing protection to useful insects. The concept is briefly discussed in Chapter 5 of this text. It appears to be a more sensible approach to controlling harmful insects than mechanical, chemical, or biological control methods that are used alone.

An integrated pest management program uses limited applications of chemicals to control insects, but it depends mainly on the use of natural insect enemies and other forms of control to reduce harmful insect populations, Figure 7–6. The objective of this kind of control is to keep some of the pests alive as a food source for the natural enemies. It strives to establish a natural balance between harmful insects and their natural enemies.

HYDROPONICS

A plant production system in which plants are grown without soil is known as **hydroponics**, Figures 7–7 and 7–8. Plants are usually planted in an inert material such as sand or vermiculite that provides support for plant roots. The nutrients required for growth of the plants are dissolved in water and circulated over the plant roots at regular intervals.

Hydroponics makes it possible to completely control the nutrient intake of plants. It is even possible to control certain characteristics of fruits and vegetables such as acid levels in tomatoes. Tomatoes raised using the hydroponic system are often much lower in acid than tomatoes grown in soil. This makes them desirable as a fresh table product, and they bring a premium market price.

Most hydroponic plant production systems are operated in controlled environments such as greenhouses. **Aeroponics** is similar to hydroponics because plants are not rooted in soil. A nutrient solution is sprayed on the roots at regular intervals, Figure 7–9. The nutrient solution is carefully mixed to contain exact amounts of plant nutrients. Every aspect of production is carefully

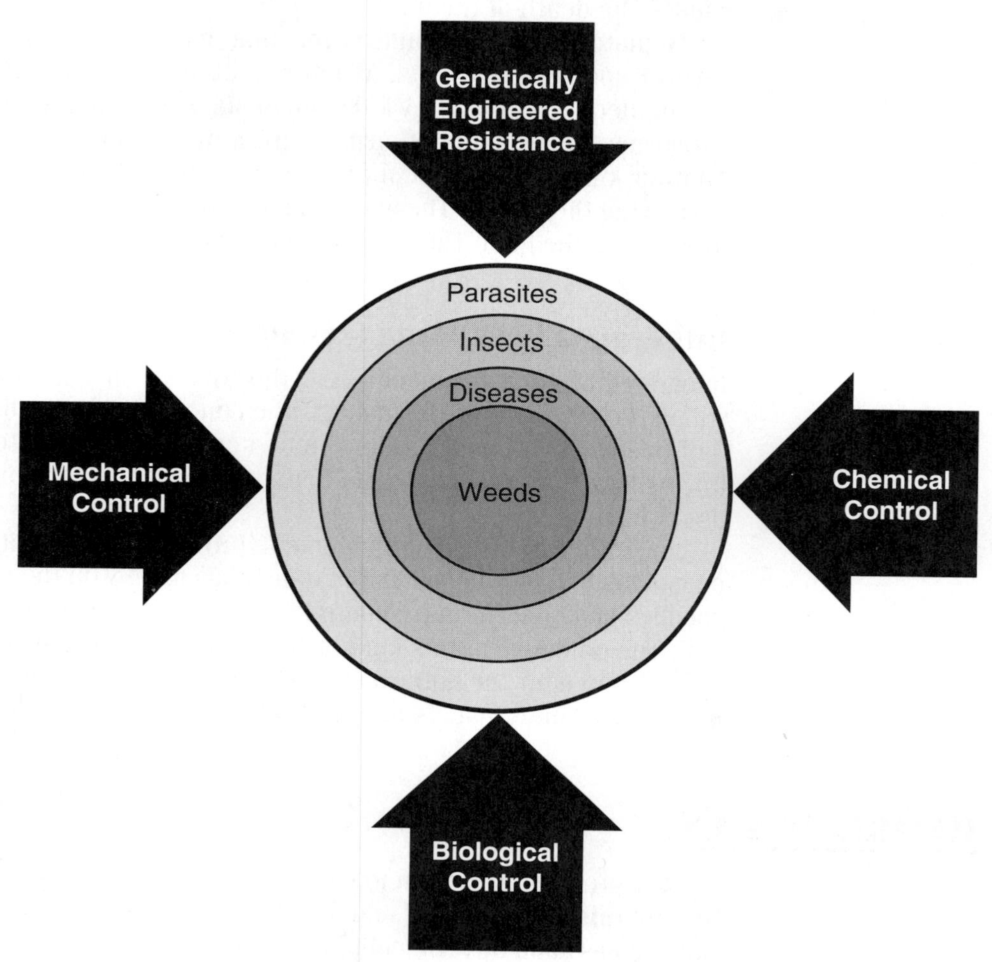

FIGURE 7–6 A multipronged attack using a variety of control methods is the most acceptable form of pest control.

controlled, and the production of good hydroponics systems is high in comparison with conventional methods of production.

TISSUE ANALYSIS

Accurate measurements of the nutrient needs of many plants can be obtained by testing the **petioles**, which consist of leaves and stems of actively growing plants. The last few leaves on a growing stem are considered to be the best plant parts to collect, Figure 7–10. The amounts of nitrogen, phosphorous, potassium, and other important plant nutrients are carefully determined for

FIGURE 7–7 Hydroponics is a system in which plant nutrients are supplied to plants in a water mixture. This system uses a drip irrigation to deliver water and nutrients.

FIGURE 7–8 Vertical tubes make effective use of limited space when hydroponics is practiced.

FIGURE 7–9 Aeroponics is a method of growing plants without soil. *(Photo courtesy of Utah Agricultural Experiment Station)*

FIGURE 7–10 Tissue analysis is a test for nutrient deficiencies in growing plants. Growing plant parts are collected and analyzed in laboratories.

each sample of plant tissue. This testing procedure is called **tissue analysis** or **petiole testing**.

A comparison of the actual nutrient content of the plant tissue is made with the ideal nutrient content of the plant tissue. Research data on nutrient requirements is available for a variety of crops. This information is used to determine how much fertilizer is needed by the crop to achieve the most profitable yields.

The accuracy of tissue analysis depends to a large extent on the quality of the plant tissue samples that are collected for testing. The comparisons must use plants in the same stages of maturity and grown under similar conditions. This is because plants require different amounts of each nutrient during the rapid growth stage than they require during the more mature stages of plant growth.

Fertilizers that are added to the plant root zone in the soil before planting do not always remain in the root zone as the growing season advances. Fertilizers are water soluble and can be leached away if too much water is available to the crop or if the soil texture is too coarse. Weeds also remove large amounts of crop nutrients from the soil, Figure 7–11.

Tissue analysis makes it possible for farmers to discover nutrient problems before the plants become stressed and damaged. Plants that become stressed at nearly any stage of growth tend to suffer a reduction in harvest yields. Stress due to nutrition can usually be prevented when tissue analysis technology is practiced during the critical stages of plant growth.

When a nutrient shortage is identified, fertilizers that contain the needed nutrients can be added to the soil or applied to the surfaces of the plant leaves, Figure 7–12. Some forms of plant nutrients can be absorbed directly through

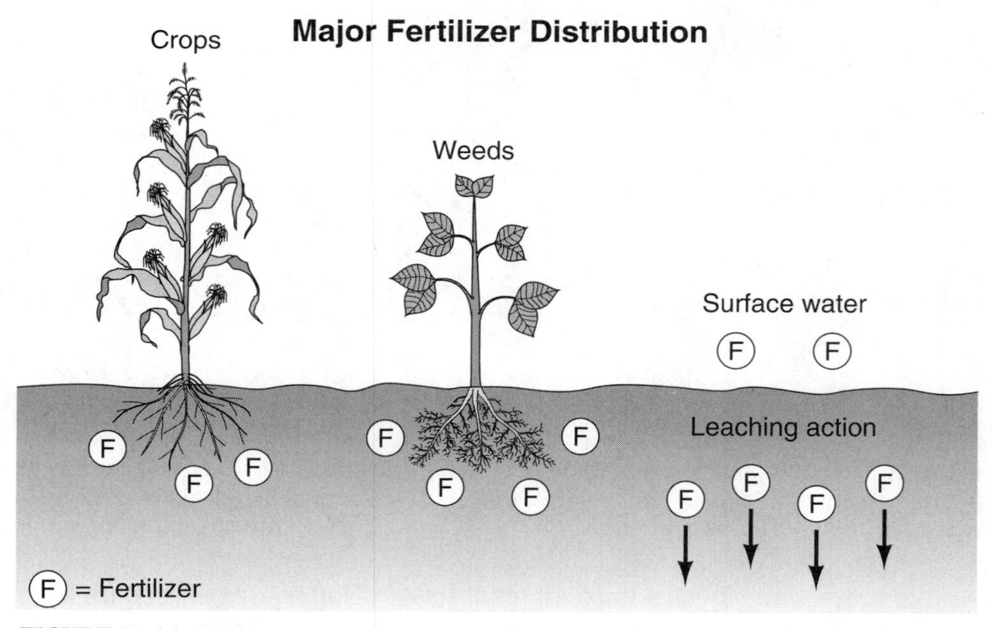

FIGURE 7–11 Fertilizer materials are sometimes lost to crops because they move out of the plant root zone due to uptake by weeds or leaching action caused by excess water movement down through the soil.

FIGURE 7–12 Fertilizers are available in many forms. It is possible to match fertilizer applications to plant deficiencies by using proper tissue testing procedures and properly selecting fertilizers.

the leaf surfaces. These fertilizers are often more costly than those added to the soil, but they can be used when it is difficult or impossible to get into fields without causing crop damage. Some forms of fertilizer can be added through irrigation water or applied from airplanes or helicopters to avoid crop damage during the fertilizer application process.

The most important benefit gained by using tissue analysis for field crops is that stress due to poor plant nutrition can be reduced or avoided. Other benefits to be expected from this technology include more efficient use of fertilizers and increased crop yields.

Soil pH

Soil **pH** is a measurement of the degree of acidity or alkalinity of the soil. Plants grow in soils that range in pH from acid (pH=3–6) to alkaline (pH=8–11). Most plants grow best in soil that is somewhat neutral (pH=7), but some plants are well adapted to acid or alkaline soils.

The pH of soils can be changed by adding soil amendments to the fields. Acid soils may be treated with ground limestone to neutralize the acid in the soil. Alkaline soils are treated with a soil amendment called gypsum. This material neutralizes the alkaline conditions found in some soils.

IRRIGATION MANAGEMENT

Water is the most important plant nutrient. In areas of the world where water is abundant, plant life flourishes. Where water is in short supply, few

FIGURE 7–13 Severe droughts affect even irrigated regions when reservoirs fail to fill with water from winter and spring stream flows. *(Photo courtesy of Utah Agricultural Experiment Station)*

crops can be grown, Figure 7–13. **Irrigation** is the practice of applying water to crops on land where natural rainfall is insufficient or where sufficient amounts of rain cannot be depended upon from year to year.

As the human population has increased, more and more productive crop land has been required to supply the needs of the people. In those areas of the world where water supplies are available, large tracts of desert land have been converted into productive farms. Rivers and streams have been diverted to provide water for land that was too dry for raising crops before irrigation. Wells have been drilled and water is pumped from **aquifers**, which are large bodies of water located underground.

The earliest form of irrigation probably consisted of using human labor to carry water to plants. This method has been abandoned except in the most primitive farming areas of the world because it is too labor intensive. It has been replaced by **flood irrigation**. Controlled flooding is the most widely-practiced form of irrigation. It is accomplished by bringing the water to the crops, Figure 7–14. Early settlers in the deserts of western America diverted rivers

FIGURE 7–14 When water is plentiful, it is often delivered to arid land in concrete ditches and canals. *(Photos courtesy of Utah Agricultural Experiment Station)*

FIGURE 7–15 Much of the western region of the United States is a desert. Large dams have been built there to store water for irrigation, hydroelectricity generation, and recreation.

and streams into man-made ditches and canals that carried water to the arid croplands. Today the need for fresh water is so great that large dams have been constructed on many rivers to store water during periods of low use until it is needed by crops, Figure 7–15.

CAREER OPTION

Hydrologist

A person who chooses a career as a **hydrologist** will work as a scientist who studies water, Figure 7–16. Hydrologists need a strong background in science, and a university or college degree.

The natural water cycle involves the movements of water between the earth and the atmosphere. It also involves the reactions of water with living organisms and other subtances. A hydrologist works on problems associated with water such as flood or drought forecasting, soil and water conservation, drainage, water quality, and irrigation.

FIGURE 7–16 A hydrologist is a person who studies water problems and issues and seeks ways to resolve them. *(Photo courtesy of Utah Agricultural Experiment Station)*

FIGURE 7–17 Large canals lined with concrete deliver water to the fields for irrigation.

FIGURE 7–18 Irrigation water is often applied to crops by furrow irrigation. *(Photo courtesy of Utah Agricultural Experiment Station)*

Flood irrigation is still widely practiced today, but water is usually precisely controlled. Much of the water for flood irrigation is transported through canals lined with concrete, Figure 7–17. Structures called **headgates** release water to canals and ditches. Once the water arrives at the fields, it is often siphoned from the ditch to the **irrigation furrows**, Figure 7–18. Metal tubes of several different diameters are available to control the amount of water delivered from the ditch to each crop furrow.

Gated pipes are sometimes used instead of ditches to deliver water to the fields, Figure 7–19. These special pipes are fitted with valves that can be opened to any desired setting to precisely control the amount of water delivered to each crop row.

In comparison with flood irrigation methods, the use of **sprinkler irrigation** has greatly reduced the amount of water required to raise a crop. When

FIGURE 7–19 Precise control over the amount of water that is delivered can be achieved using gated pipes. *(Photo courtesy of Utah Agricultural Experiment Station)*

sprinkler irrigation is used, water is delivered to the field in pipes. The water pressure in the pipeline is usually provided by pumps. Some systems operate under gravity pressure when the crop is located at lower elevations than the source of irrigation water, Figure 7–20.

The amount of water delivered to the crop can be controlled by changing the size of the sprinkler nozzles to match the correct amount of water to the desired time interval between settings. It can also be controlled by increasing or restricting the length of time that the sprinkler is operated on each setting. The **setting** is defined as the length of time the irrigation system is operated at a single location. Once the correct amount of water has been applied, the sprinkler line should be moved to a new setting.

Many different kinds of sprinkler irrigation equipment are available. Some of the systems require large amounts of labor to operate them properly while other systems are completely automated. Some of the most commonly used types of irrigation equipment are described below:

Main lines consist of large pipes through which water is delivered to the sprinkler system from its source at a well, canal, ditch, or pond, Figure 7–21. Large valves are located at appropriate intervals along the main line to which smaller sprinkler pipes can be attached. The smaller sprinkler pipes are moved across the field from one setting to another.

Handlines consist of sprinkler pipes constructed in lengths that are short enough to be moved using hand labor, Figure 7–22. Each pipe length is carried to a new location at the end of each setting. For most types of soils and crops, this procedure is repeated every eight to twelve hours. The system is less expensive to buy than most other systems, but labor costs are high.

Wheel lines are similar to handlines except that each sprinkler pipe is mounted in the center of a large lightweight wheel, Figure 7–23. The entire

FIGURE 7–20 Water pressure is maintained in sprinkler irrigation pipes by gravity flow or by using large water pumps powered by internal combustion engines or electrical power.

FIGURE 7–21 Main line pipe is used to deliver irrigation water from its source to the point of use. Valves to which lateral sprinkler lines can be attached are located at strategic intervals.

FIGURE 7–22 Handlines are located on the ground and must be hand-carried to new locations two or more times each day.

FIGURE 7–23 Wheel lines are used to mechanically move sprinkler irrigation systems to different locations as needed. *(Photo courtesy of Utah Agricultural Experiment Station)*

sprinkler line can be rolled between settings using power from hydraulics or small gasoline engines.

Solid set irrigation systems use small diameter irrigation pipe that is put in place at the beginning of the irrigation season, Figure 7–24. It is never moved until harvest time. The cost for this system is quite high in comparison to handlines and wheel lines, but most of the labor costs are eliminated. This system is most often used with high value crops.

Center pivot sprinkler systems take water from a source located in the center of the field and rotate continuously around it in a large circular pattern, Figure 7–25. This system is completely automated and requires little labor except for regular maintenance procedures. The speed at which the sprinkler travels through the field may be adjusted according to the needs of the crop. This system is also available as a low pressure sprinkler system to reduce energy consumption and soil compaction.

FIGURE 7–24 Solid set sprinkler systems consist of pipes that are close enough together that they do not require moving during the growing season.

FIGURE 7–25 A center pivot irrigation system is completely automated. It moves continuously around the field on wheels.

FIGURE 7–26 Lateral or linear move sprinkler systems are self-propelled units that are used to irrigate rectangular fields.

Lateral move sprinkler systems are completely automated systems designed for rectangular or square fields, Figure 7–26. Instead of a single water source in the center of the field, the lateral move system uses a series of mainline **risers**, or vertical pipes, to bring water from the buried mainline to the surface. A computer control system is used to locate and link up with the main line risers as it advances across the field.

Low-pressure sprinkler irrigation is a form of irrigation designed to operate using lower water pressures than standard sprinkler irrigation systems, Figure 7–27. The system is used with equipment such as center pivot and lateral move systems that continually move across the field while in operation. The systems are designed to reduce soil erosion and compaction caused by high-pressure irrigation systems.

Drip irrigation is a form of irrigation that uses small flexible tubes to apply water to the root zone of a crop, Figure 7–28. It makes very efficient use of

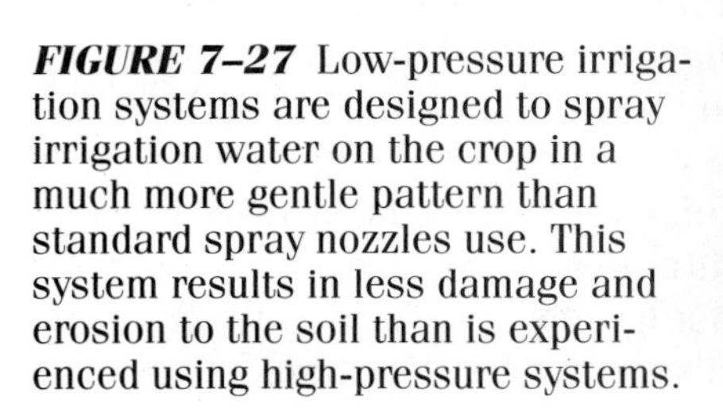

FIGURE 7–27 Low-pressure irrigation systems are designed to spray irrigation water on the crop in a much more gentle pattern than standard spray nozzles use. This system results in less damage and erosion to the soil than is experienced using high-pressure systems.

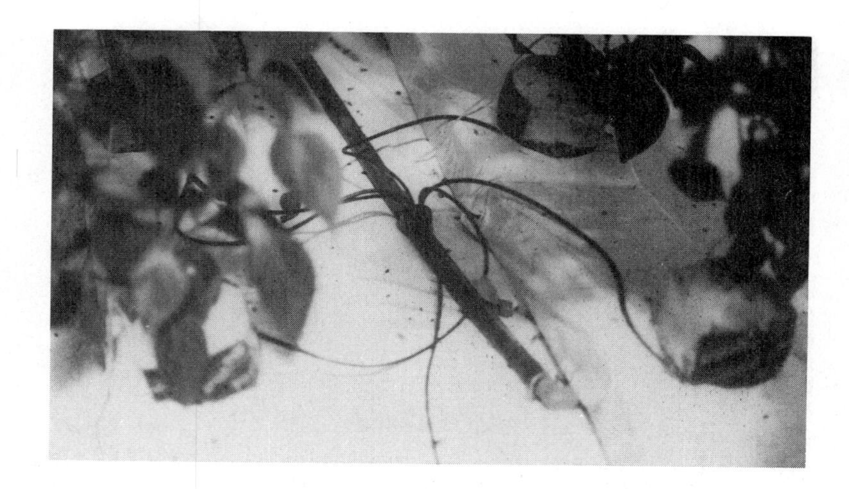

FIGURE 7–28 Drip irrigation is a permanent installation. It makes very efficient use of water because it only delivers water to the root zones of plants.

water because water is applied only to the crop and not to the areas between plants. It is an ideal system for use with horticultural crops and other long-term plantings that require frequent irrigations.

Chemigation and Fertigation

Chemigation is the application of agricultural chemicals through irrigation water. **Fertigation** is the application of water soluble fertilizers through irrigation water. Both practices are common in areas where irrigation is common. Precise amounts of chemicals or fertilizers can be delivered to crops by dispersing them in the irrigation water through a precise injector system. These methods are very efficient ways to apply some chemicals and fertilizers. Chemicals and fertilizers can also be distributed through irrigation water that is applied through furrows or by flooding, but it is difficult to be precise in the amounts that are applied using this irrigation practice.

REMOTE SENSING TECHNOLOGY

Remote sensing technology is a method of recording existing field and crop conditions as they appear from a vantage point high above the ground. This technology is most often practiced by taking aerial photographs using aircraft or satellite systems. Special photographic films are used that are capable of measuring radiant heat and other plant and soil characteristics from great distances.

Satellite imagery is a technology that has been used to a great extent by **meteorologists** to describe and predict conditions of the weather and the climate, Figure 7–29. No other industry depends more upon the climate than does agriculture. The ability to accurately predict long periods of drought and other climatic conditions is enhanced by satellite imagery.

Aerial mapping is a procedure using photographs transmitted to the earth from satellites, or taken from high flying aircraft to develop maps for use by

FIGURE 7–29 Satellite pictures have made it possible to predict weather patterns with improved accuracy in comparison with older methods. *(Photo courtesy of Utah Agricultural Experiment Station)*

farmers and agricultural agencies. Such maps can be used to calculate acreages in farms and fields. Soil types, plant stress patterns, disease problems, and chemical damage can all be mapped using special films.

Infrared radiation is a form of energy that is detected as heat and is invisible to the human eye. It can be detected using special photographic film. Infrared photography is a useful tool in measuring the amount of heat radiated from the leaf surfaces of plants. Problems associated with insect damage, diseases, irrigation practices, and other stresses suffered by plants can be detected using this technology.

Infrared thermometry is a system that uses a special thermometer to measure infrared radiation from a field of growing crops. Measurements from the field are analyzed using a microprocessor to identify the needs of the plants. The device is portable and provides instant interpretations of the information it gathers.

Multispectral imagery is a system that uses computers to enhance special aerial photographs, Figure 7–30. The procedure is used to measure the environment surrounding plants and includes mapping of soil types and problem areas. Mapping of plant vigor, growth, and temperature is used to identify potential crop problems. Experienced technicians using the enhanced photographs are able to verify the existence of problems before they begin to cause serious stress to the plants.

CHAPTER SUMMARY

Plants are renewable resources that are of vital importance to other forms of living things. Soil is a limited resource upon which plants are dependent. As additional land for crop production becomes scarce, it is important that we con-

FIGURE 7–30 Multispectral imagery is a crop management tool that is used to identify problems with a crop before they become serious. The white and checkered patterns indicate areas of the field where plants are stressed from lack of water. Other areas appear to have adequate water in the soil. These differences are probably due to differences in soil types.

serve our soil. Chemical fertilizers have been used extensively to maintain soil fertility. New technologies are now emerging that contribute to soil fertility while requiring fewer chemicals.

Irrigation technologies have advanced rapidly in recent years. Sprinkler irrigation conserves large amounts of water in comparison with flood irrigation, and drip irrigation makes very efficient use of water in comparison with sprinkler irrigation.

Remote sensing devices are able to measure plant stress and map fields showing the locations of stressed plants. In some cases the devices are even able to diagnose the problems. Multispectral imagery combines aerial mapping with computer analysis to measure environmental conditions that affect crops. These new technologies are often able to detect problems early enough to correct them before crop yields begin to decline.

CHAPTER REVIEW

Discussion and Essay Questions

1. Explain how weeds are controlled using mechanical methods.
2. Describe the beneficial and/or negative effects of prolonged use of chemical insecticides.
3. Discuss the importance of tissue analysis in determining nutrient needs of plants.
4. Explain the importance of irrigation in producing crops for the growing world population.

5. Describe the major irrigation systems now in use for crop production.
6. What types of irrigation systems are likely to be used ten years from now?
7. Identify the major types of remote sensing technologies that are in use for crop management purposes.
8. How can the use of multispectral imagery contribute to better management of crops?

Multiple Choice Questions

1. The use of a cultivator to remove weeds from a crop is an example of which of the following types of weed control?

   A. chemical
   B. mechanical
   C. biological
   D. philosophical

2. Introduction of natural enemies of harmful insects is an example of which of the following types of pest control?

   A. chemical
   B. mechanical
   C. biological
   D. philosophical

3. Integrated pest management uses which of the following methods of control?

   A. biological
   B. chemical
   C. mechanical
   D. all of these

4. Hydroponics is a system for raising crops in the absence of:

   A. water.
   B. nutrients.
   C. soil.
   D. chemicals.

5. Tissue analysis is a crop production practice that is used to:

   A. identify nutrient needs of a crop.
   B. determine the quality of the foliage.
   C. identify a plant.
   D. determine whether a scion and a rootstock are compatible for grafting.

6. Which of the following problems leading to removal of fertilizers from the soil has been demonstrated to pollute the ground water?

   A. fertilizer uptake by crops
   B. leaching action due to excess water or coarse soil texture
   C. fertilizer uptake by weeds
   D. evaporation of liquid fertilizers

7. The name of a large natural underground storage area for water from which irrigation water is obtained is:
   A. aquifer.
   B. well.
   C. reservoir.
   D. canal.

8. Which of the following types of irrigation equipment *is not* used in sprinkler irrigation?
   A. center pivot system
   B. gated pipe
   C. wheel line system
   D. lateral move system

9. An advantage of a low pressure sprinkler irrigation system is that it:
   A. does not require a pump.
   B. removes weed seeds from the water.
   C. reduces soil compaction and erosion.
   D. may be used to apply insecticides to crops.

10. Remote sensing technologies are used to create maps of environmental conditions in fields of crops. They do this by measuring which of the following kinds of energy?
    A. electrical
    B. radiant
    C. geothermal
    D. atomic

LEARNING ACTIVITIES

1. Divide the class into learning groups and assign each group to research a specific irrigation system. Provide names and addresses of major irrigation equipment companies. Have each group report the results of its investigation to the class. (Send the letters out 2–3 weeks before you begin to teach the irrigation unit.)

2. Invite local officials from three or four agricultural agencies to participate in a panel discussion. The topic: Chemical fertilizers—their uses and abuses.

CHAPTER 8

Agricultural Production Technologies

The technologies involved in production of agricultural goods on the farms and ranches are in a state of constant change. Machines are capable of performing almost all of the work that was done by hand only a generation ago. Modern machinery is a technological miracle. Machinery has reduced the amount of labor that is required on farms to produce agricultural products. It is now possible for much of the labor force to pursue other careers that improve the quality of our lives.

As long as people take care of the soil in which plants grow, it will continue to yield a harvest season after season. Sustained yields cannot continue, however, without conservation of the soil and the plant nutrients that are stored there, Figure 8–1.

FIGURE 8–1 Soil conservation on rolling farm land is practiced using strip cropping farm practices.

OBJECTIVES

After completing this chapter, you should be able to:

- describe cultural practices that are used to smooth and level fields.
- account for increased crop production in fields from which excess water has been removed by installing field drainage systems.
- explain the beneficial effects gained by construction of field terraces on hilly farm ground.
- analyze the benefits and problems associated with the use of chemical fertilizers.
- compare traditional tillage practices with minimum tillage and no-till farming practices.
- relate the importance of modern tillage, planting, and harvesting equipment to the world food supply.

TERMS FOR UNDERSTANDING

The following vocabulary terms should be studied carefully as you read:

laser
field drains
contour farming
terraces
soil fertility
essential elements
deficiency
macronutrients
primary nutrients
micronutrients
chemical fertilizers
alternative agriculture
nitrates
Rhizobium
site-specific farming
precision farming
geographic information system
global positioning system
LISA
tillage
implements
minimum tillage
no-till farming
seeding rate
seed plates
aerial seeding
reaper
dehydration

FIELD PREPARATION

The practice of smoothing fields before planting has been going on for many years. In areas where high value fruit or vegetable crops are produced, or where flood irrigation is practiced the fields must be level to allow irrigation water to flow evenly, Figure 8–2. Failure to prepare fields by smoothing and leveling often results in serious water control problems when it comes time to irrigate the crop.

In some areas farm land is unsuitable for cultivation of certain crops due to the uneven terrain. Farmers have attempted to solve this problem using

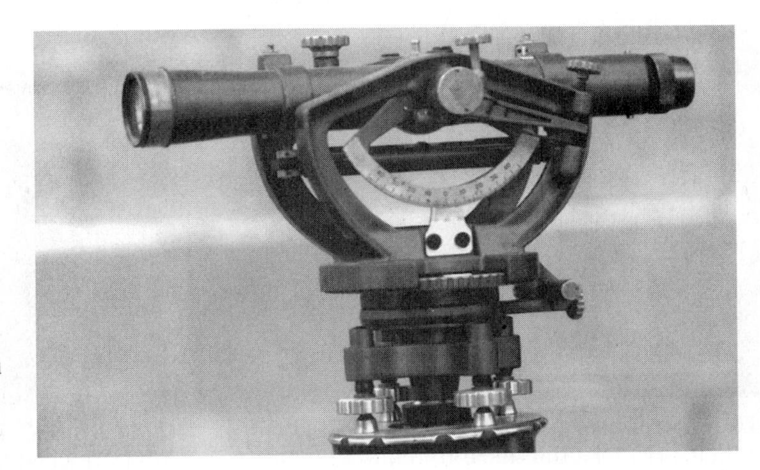

FIGURE 8–2 A transit is used to establish elevations in a field. Once elevations are known, land can be leveled, contours can be established, and irrigation structures can be constructed. *(Photo courtesy of Utah Agricultural Experiment Station)*

FIGURE 8–3 Large earthmoving machines have replaced teams and fresnos in preparing level fields. *(Photo courtesy of Utah Agricultural Experiment Station)*

everything from shovels to large construction equipment, Figure 8–3. In areas of the world where labor is inexpensive, many fields have been smoothed using human labor.

In the days before modern earthmoving equipment became available, teamsters with teams of horses and mules changed the terrain of many fields. The fresno was a machine shaped to scoop and carry loads of soil. Once the load of soil was in the desired location, the driver could easily dump the load. The work was slow and tedious in comparison with similar projects today.

During the years when land leveling equipment was pulled by horses, fields tended to be rather small. As land leveling equipment improved and greater amounts of soil could be moved, the land-leveling projects became more sophisticated and the size of irrigated fields increased.

Land-leveling contractors today frequently use large earthmoving machines controlled by laser beams, Figure 8–4. A **laser** is a narrow and intense beam

FIGURE 8–4 Laser beams are capable of performing many tasks that used to require human labor. Lasers activate electrical switches, control land leveling operations, sort materials, and even burn code numbers into packages.

FIGURE 8–5 Field drains are underground pipes that are used to collect and carry excess water out of fields. They aid in lowering the water table allowing greater plant root development and they stimulate increases in production.

of light that travels in a straight line from its source to a target. This device is used in the land-leveling process by emitting a laser beam at a constant elevation in a field. The adjustable blades of earthmoving machines such as carryalls and graders are equipped with controls that respond to the fixed elevation of the laser beam. This causes the machine to cut to a prescribed depth on the high areas of the field and dump the soil at an appropriate rate to fill the low areas of the field.

Due to plowing and other tillage practices, fields do not remain level over long periods of time. A land plane is an implement that is often used to prepare a firm, level field surface.

An important part of field preparation on some farms is the design and installation of **field drains** to remove excess water from the soil, Figure 8–5. In areas where the water table is high, soil conditions can be improved by digging drainage canals and installing underground tile drain pipe in the fields.

The drainage lines are placed at appropriate intervals in the field, and they carry the excess water from the fields to the drainage canals, Figure 8–6. Farm drainage systems are frequently designed by teams of soil specialists who work for the United States Department of Agriculture.

Many of the soils in the United States have been mapped, and areas that can benefit from drainage systems are easily identified. Soils that are waterlogged remain cold, and plant root development is restricted by the level of the water table, Figure 8–7. Well-drained soils are generally productive and can be used to raise a greater variety of crops than wet soils.

In farming areas where fields are located on uneven terrain, it is often necessary to adopt contour farming practices. **Contour farming** is a system of

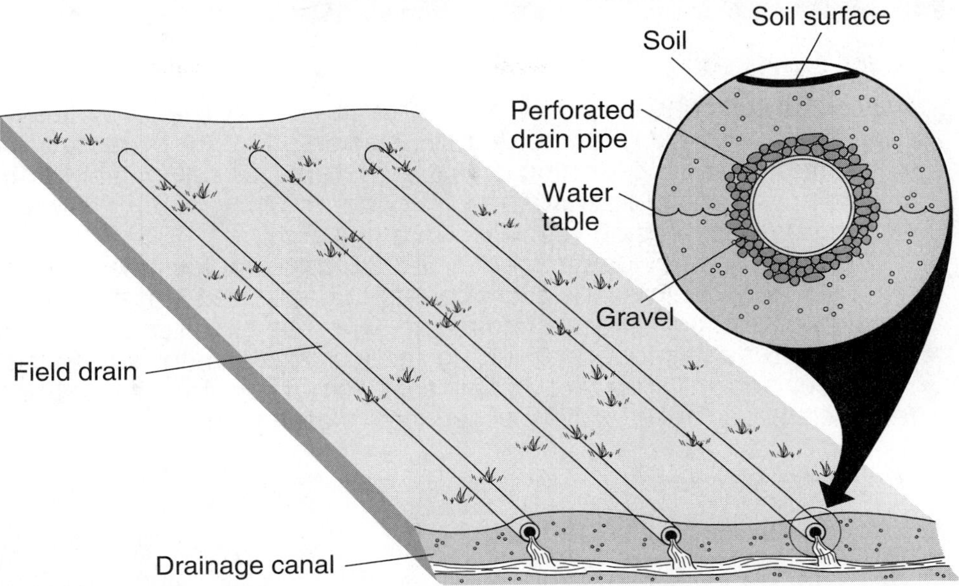

FIGURE 8–6 Field drains must be spaced close enough together to draw down the water level between the pipes. Correct spacing of drains is dependent on the soil type and the level of the water table.

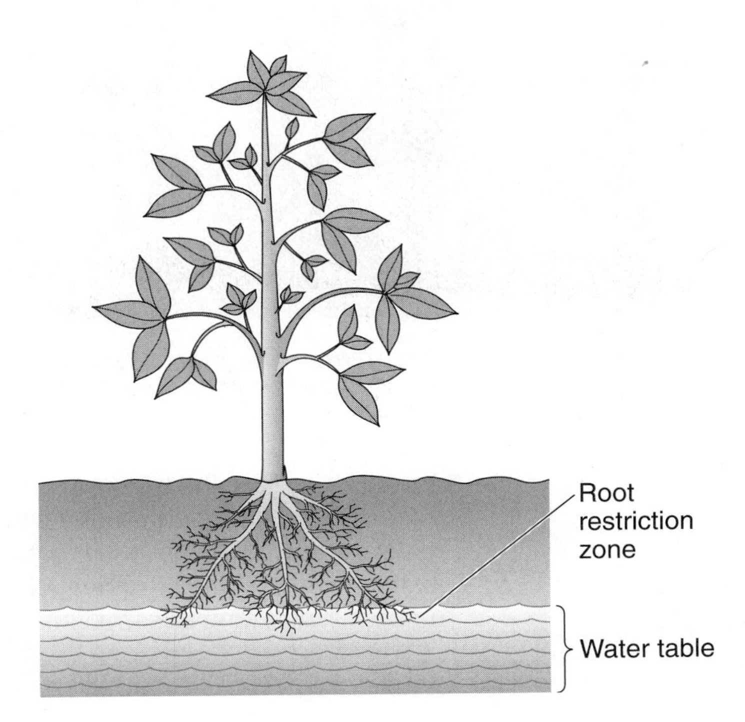

FIGURE 8–7 A shallow water table is as effective as bedrock at restricting the depth of plant roots. Field drains are effective at lowering the water table.

CAREER OPTION

Soil Conservation

A career in soil conservation is an interesting career that provides opportunities for a variety of work experiences. A qualified person might expect to spend time in the field mapping soils, designing drainage systems, or solving soil problems on farms. A person with an appropriate university degree may qualify to work as a teacher, a researcher, or extension specialist for a university.

The United States Department of Agriculture employs a large staff of conservation officers who work in nearly every county in the United States, Figure 8–8. State departments of agriculture require the expertise of soil specialists, and private laboratories offer a variety of consulting services that employ soil specialists.

Career options are also available that do not require a baccalaureate degree. Technicians perform a variety of services related to soil fertility and conservation.

FIGURE 8–8 Soils and nutrient management specialists do cooperative research to determine the best fertilizer rates to optimize corn growth. *(Courtesy USDA/ARS #K–3694–5)*

farming in which field boundaries are laid out along elevation lines. The practice is used on hilly ground and all of the field operations are performed by moving around the hill at a constant elevation instead of moving up and down the hill, Figure 8–9. This system of farming is used to prevent soil erosion due to water runoff on excessive slopes, Figure 8–10.

Construction of **terraces** is often necessary in addition to farming on the contour. A terrace is constructed by throwing up a bank of soil along the con-

FIGURE 8–9 Contour farming is a practice which reduces soil erosion by farming around hills instead of tilling up and down the slopes.

FIGURE 8–10 Moving water is a highly destructive force that causes severe soil erosion on unprotected land.

tour of a hill with a basinlike holding area for water above the bank, Figure 8–11. During periods of heavy water runoff, the excess water is held on the hillside by the terrace until it has been absorbed into the soil, Figure 8–12. The terrace prevents water erosion and promotes absorption of water into the soil.

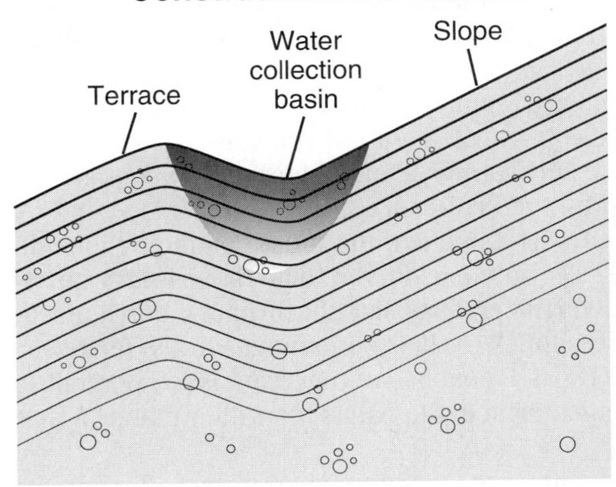

FIGURE 8–11 A properly-constructed terrace maintains the same elevation around the contour of a hill. It consists of a bank of soil that creates a basinlike depression that collects and holds runoff water from the slope above until it seeps into the soil.

FIGURE 8–12 Terraces trap surface water and create a site where it can be absorbed into the soil instead of moving down across the slopes.

SOIL FERTILITY

Since the earliest days of recorded history people have searched for ways to increase crop yields. When virgin land was plentiful, farmers simply moved to new land when their soil "wore out." Entire cultures depended upon flood waters to deposit new silt on the bottom lands beside the rivers. The silt was rich in nutrients and renewed the fertility of the soil. Ancient farmers also discovered that animal manures and decaying materials were good sources of plant nutrients. They fertilized their soils by adding these materials to their fields.

Farmers in the modern world can no longer move to new lands or wait for new silt to cover their farms. They must conserve and improve the soil they have. To accomplish this they must replace what they take from the soil. Soil is a renewable resource only to the extent that it can be improved through proper care.

Soil fertility is the amount and availability of plant nutrients in the soil. Sixteen elements are known to be required for plant growth. They are known as **essential elements** because normal plant growth is impossible when any of them is not present in the soil. The absence of an essential element is referred to as a **deficiency**. Distinct symptoms appear in plants when one or more of these elements is missing.

Essential Elements of Soil

The plant nutrients that are used by plants in the greatest abundance are carbon (C), hydrogen (H), and oxygen (O). Plants obtain them from water that is absorbed from the soil through the plant roots, and from the carbon dioxide (CO_2) that is abundant in the air.

Six of the essential elements are required in moderate amounts, and they are obtained by plants from the soil. These plant nutrients are also known as **macronutrients**. They include nitrogen (N), phosphorous (P), potassium (K), sulphur (S), magnesium (Mg), and calcium (Ca).

Nitrogen, phosphorous, and potassium are the most common elements found in commercial fertilizers. They are considered to be the **primary nutrients** of plants. The labels on fertilizer bags list the primary nutrients in order as N-P-K. This order corresponds to their percentages in the fertilizer. A fertilizer described as grade 8–16–8 consists of 8 percent N, 16 percent P, and 8 percent K. The rest of the material in the bag is inert filler material that is of little value to plants.

Seven other essential plant nutrients are obtained from the soil. They are iron (Fe), boron (B), manganese (Mn), chlorine (Cl), copper (Cu), zinc (Zn), and molybdenum (Mo). These plant nutrients are used only in small amounts, and they are called **micronutrients**. Soils that are deficient in any of these elements are considered to be infertile, because plants cannot grow well in such soils until the missing nutrients are provided.

In addition to the sixteen essential elements, plants require environments in which water, light, and air are available. The role of light in the key plant process known as photosynthesis was discussed in Chapter 5. Soil is another factor affecting plant growth, but it is not essential, as we have learned from the discussion of hydroponics in Chapter 7. Essential elements can be provided as dissolved nutrients in the water supply.

Memorizing the Essential Elements

The essential elements and their chemical symbols can be easily remembered by associating the chemical symbols with the words of this phrase: "See Hopkin's cafe managed by mine classy cousin Mo." Now rewrite the phrase using the chemical symbols of the 16 essential elements: "C HOPKNS CaFe Mg B Mn Cl CuZn Mo." The essential elements are carbon (C), hydrogen (H), oxygen (O), phosphorous (P), potassium (K), nitrogen (N), sulphur (S), calcium (Ca), iron (Fe), magnesium (Mg), boron (B), manganese (MN), chlorine (Cl), copper (Cu), zinc (Zn) and molybdenum (Mo).

Chemical Fertilizers

The fertility of soil is measured by taking samples of soil from the field and testing them for the presence of plant nutrients. When a soil sample is found to be lacking a particular plant nutrient, the problem can be corrected by adding materials to the soil that contain concentrated forms of the nutrient. Common sources of plant nutrients used to replace missing soil elements are industrial by-products such as ammonia, Figure 8–13. Plant nutrients obtained from sources other than plants and animals are called **chemical fertilizers**.

Chemical fertilizers play important roles in sustaining the fertility of the land, Figure 8–14. Farmers are able to fertilize the soil to meet the fertility

FIGURE 8–13 Ammonia is a nitrogen-rich by-product of the industrial age. It becomes bound on the surfaces of soil particles when it is injected into the topsoil. It is stored under pressure in large metal storage tanks.

FIGURE 8–14 Modern agriculture depends upon chemical fertilizers to provide adequate nutrition for crops. Fertilizers are often applied to fields as part of the seed bed preparation process. *(Photo courtesy of Utah Agricultural Experiment Station)*

needs of the crops that are grown, but a trend toward reduced dependency on chemical fertilizers is occurring. Current research in soil fertility is looking for ways to sustain crop production with fewer chemicals.

Modern agriculture depends heavily on chemical fertilizers to sustain soil fertility. Concerns over chemical pollution of the water in streams and aquifers has created interest in alternative systems of crop production. High production costs have added to this interest. Resistance by some consumers to chemical treatments of food crops has stimulated the development of **alternative agriculture**. This type of agricultural production attempts to maintain soil fertility by using organic fertilizers from plant or animal sources.

The use of crop rotations to enrich the soil has proven beneficial when legume plants are included in the rotation. The legume class of plants takes nitrogen from the air and puts it in the soil in a form that other plants can use. This process is called nitrogen fixation. It is accomplished by a form of bacteria that lives in nodules on the roots of legume plants. A nodule is a structure where colonies of nitrogen-fixing bacteria live. A mutually beneficial relationship known as symbiosis exists between the bacteria and the host plant.

Nitrogen Fixation

The air in the earth's atmosphere consists of 78 percent nitrogen, but plants are unable to use nitrogen in its elemental form. The nitrogen in the air must first be converted to compounds called **nitrates**. This is done by bacteria that are found living in the root nodules of legume plants such as alfalfa, clover, beans, and peas, Figure 8–15. The most important of the nitrogen-fixing bacteria is the ***Rhizobium***.

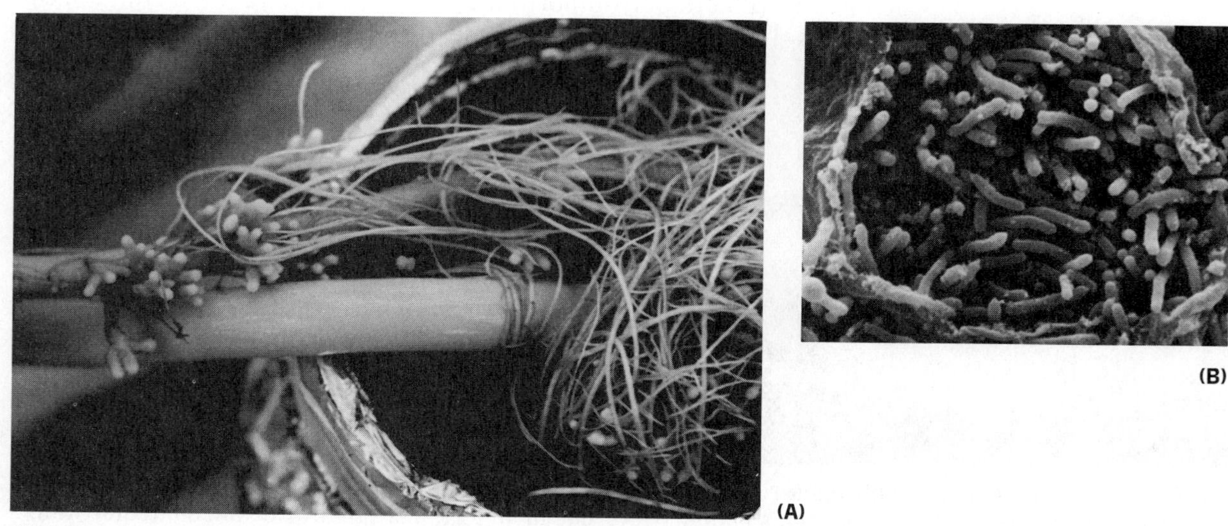

FIGURE 8–15 (A) Nitrogen-fixing nodules on the roots of an alfalfa plant. *(Courtesy Dr. J. Burton, The Nitragin Company)*. (B) Detail of *Rhizobium* bacteria (magnification 5000X). *(Courtesy Dr. Carroll Vance, Research Plant Physiologist, Department of Agronomy and Plant Genetics, University of Minnesota)*

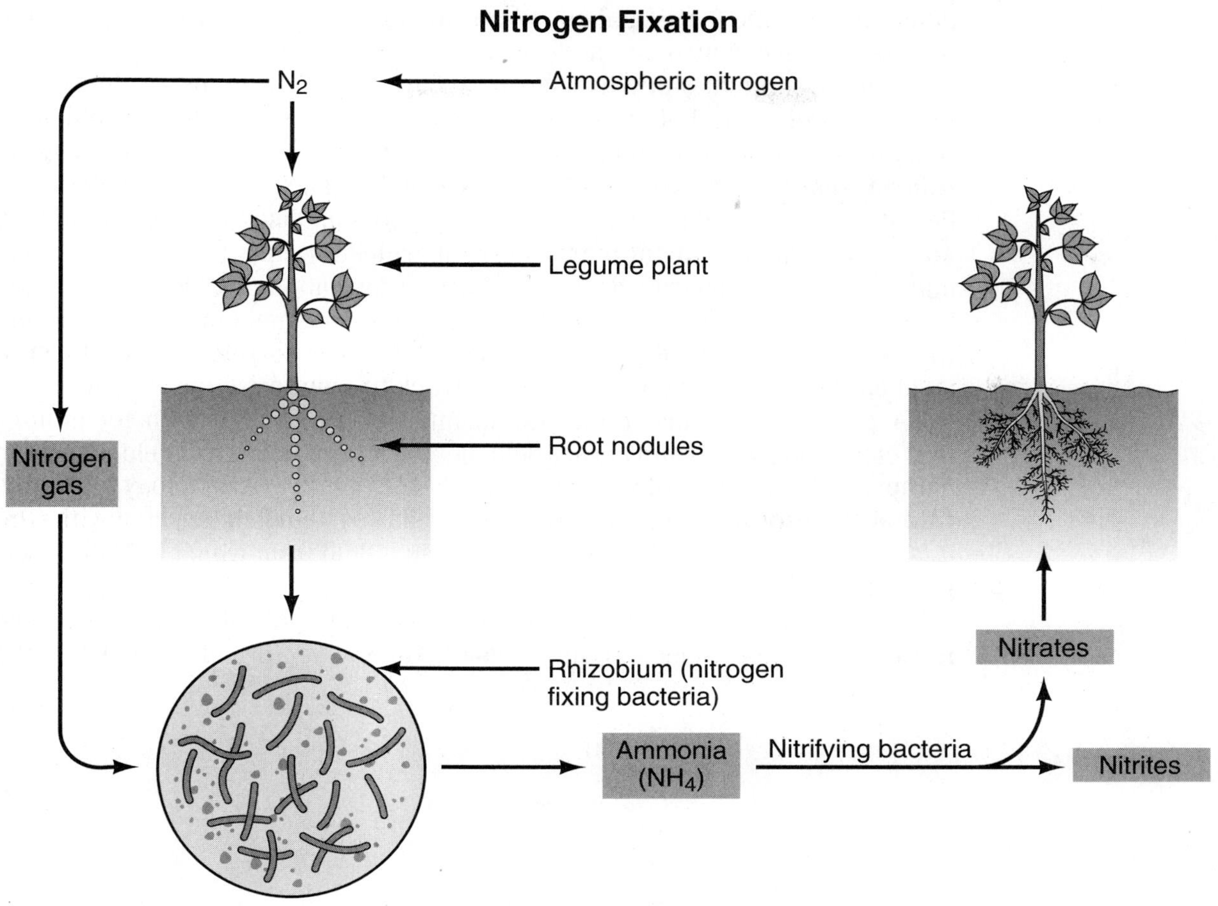

Nitrogen Fixation

N_2
Atmospheric nitrogen
Legume plant
Root nodules
Nitrogen gas
Rhizobium (nitrogen fixing bacteria)
Ammonia (NH_4)
Nitrifying bacteria
Nitrates
Nitrites

FIGURE 8–16 *Rhizobium* bacteria obtain sugars and minerals from their host plants. They convert nitrogen from the air to nitrate compounds that provide nutrients to the plants.

Nitrogen-fixing bacteria use sugars produced by plants as their source of energy, and they convert elemental nitrogen to ammonia. Nitrifying bacteria change the ammonia to nitrate compounds that can be used by plants. A healthy population of *Rhizobium* is capable of producing up to 300 pounds of nitrogen per acre each year. This process is called nitrogen fixation, Figure 8–16.

SITE-SPECIFIC FARMING

Site-specific farming, also known as **precision farming** is a crop management system that takes into account the differences that exist in different areas of a field. Precise soil maps of soil nutrients and crop yields are used to apply different fertilizer, pesticide, and seeding rates to different locations in the same field. Site-specific farming also involves locating pest populations in the field to allow precise control methods to be used where they are needed.

Computer technology is used to create accurate maps of field data. Computer software that is used for mapping purposes establishes field locations on a grid, Figure 8–17. A system of this kind is called a **Geographic Information System** (GIS). Field data is added for each sector of the field. The field data is obtained by taking soil samples from precise locations in the field and testing them for plant nutrients. Crop yields for each location are determined by using measuring devices on harvesting equipment. Yield information is fed into the computer data system as a harvesting machine moves through the field. The exact field location of the harvester is correlated with the crop yield data using the satellite system to track the harvester.

A geographic information system must be integrated with technology that can accurately identify exact field locations from which the field data were gathered. An example of such a system is the satellite technology known as **Global Positioning System** or GPS. When this system is integrated with GIS, it is capable of locating the exact location in a field from which field data was gathered.

Satellite technology makes it possible to adjust application rates of seeds, fertilizers, and pesticides to the needs of different sectors of a field while the

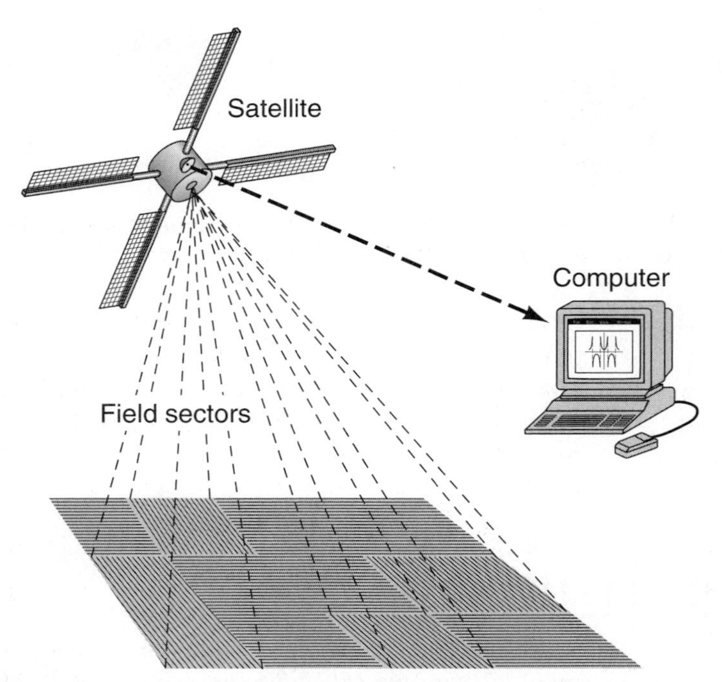

FIGURE 8–17 Site-specific farming is made possible by a satellite-based Geographic Information System that accurately identifies crop needs in individual sectors of a field.

machinery is moving. Electronic equipment located on planters, sprayers, and fertilizer equipment uses satellite transmissions to increase or decrease the flow rates of crop inputs according to the crop needs that are identified on the soil map. This farming system has great potential in combination with a sustainable agriculture system.

LOW INPUT SUSTAINABLE AGRICULTURE (LISA)

Low input sustainable agriculture, commonly known as **LISA**, is a farming strategy in which an attempt is made to reduce the use of agricultural chemicals while maintaining a consistent level of production. It is an attempt to use alternative methods of sustaining soil fertility and controlling pests that are less damaging to the environment than the chemicals that are now used. It does not promote the elimination of chemical inputs, but it does promote a reduction in chemical use on farms.

Soil nutrients must be carefully managed if crop production is to be sustained while inputs such as commercial fertilizers and pesticides are being reduced. One management tool involves planting legumes such as peas, alfalfa, and clovers in rotation with grains and other nitrogen-consuming crops. Clover or alfalfa crops that are planted at the same time as the grain crop mature soon after the grain is harvested, providing continuous ground cover and adding nitrogen to the soil.

When it is practical, legume crops are used as cover crops during the fall and winter. They are plowed into the soil to provide organic fertilizer when the fields are prepared for planting higher value crops the next spring. This practice adds nitrogen and humus to the soil, and it protects the soil against wind erosion. It also helps reduce competition from weeds that use soil nutrients.

Effective weed control in low input farming requires the implementation of several weed control strategies. One of these strategies is to cultivate row crops more frequently during the growing season. Another is to control weeds that grow around the perimeter of the field to keep them from producing new seeds.

Crop rotations are important in low input farming. They help to control insects by providing a different plant community in the field each year. This makes it difficult for insects that prefer specific plants to become established in the field environment. Crop rotations help prevent the gradual buildup of harmful insect populations that occurs when the same kind of crop is grown in a field for more than one season.

Crop varieties that are bred to resist insects, diseases, and pests play important roles in LISA farming systems. The need for pesticides can be substantially reduced or eliminated when resistant crop varieties are used. Biotechnology and crop breeding programs account for most of the resistant crop varieties.

Integrated pest management (IPM) was discussed in Chapter 5. This approach to managing pests is important in low input sustainable agriculture systems. The IPM management approach is effective because it uses a variety of strategies to control insects, but it also allows for the use of chemical pesticides when the economic threshold is reached.

Low input sustainable agriculture practices contribute to reduced soil erosion, higher water quality in some domestic wells and in surface water, and a safer environment for rural families. Another important benefit of reducing crop inputs while maintaining yields is that the costs of production may be reduced. This is because chemical fertilizers and pesticides are expensive, and if cost effective alternative controls for pests can be found, then they should be used.

SEEDBED PREPARATION

Preparation of the seedbed is done using **tillage** machines such as plows, disks, chisels, and other assorted implements designed to dig and mix the soil and reduce the size of soil aggregates, Figure 8–18. A smooth, even soil surface is prepared prior to planting the seeds.

Seedbed preparation to primitive farmers consisted of poking a sharp stick in the ground to create a hole in which the seed was planted. Wooden plows eventually replaced pointed sticks as farmers attempted to control the growth of weeds by tilling the soil. Wooden plows have now been replaced by steel plows in many parts of the world. Steel plows have been supplemented by many other tillage machines designed to break down soil aggregates to create a smooth, firm seedbed.

Most farms today feature a variety of different tillage **implements**, or machines that are used for seedbed preparation. It is a common practice on many farms to plow fields followed by disking and harrowing operations

FIGURE 8–18 Tillage implements are used to prepare a smooth, firm seedbed in which to plant crops. This disk is used to work fields and to reduce soil aggregate sizes, allowing better germination of the seeds.

before the seed is planted. Some crops require additional field preparation such as forming corrugates or raised beds.

The amount of seedbed preparation that is required depends upon the crop to be planted and the soil type found in the field. Small vegetable seeds, grass seeds, and legume seeds such as clover and alfalfa may require a greater amount of seedbed preparation than is needed by larger seeds such as corn and other grains.

Heavy clay soils require different tillage practices than are used on sandy soils. Clay soils are usually plowed in the fall season because the freezing and thawing action that occurs on soil particles during the winter months tends to break down large clods resulting in improved soil texture.

Sandy soils often require protection from wind erosion during the winter. This can usually be provided by leaving the stubble of the previous crop on the field and delaying tillage operations until spring.

Minimum Tillage Farming

Minimum tillage is a form of seedbed preparation in which only limited preparation of the soil precedes the planting of the crop. Field operations are reduced or eliminated. Instead of several trips across a field in preparation for planting, several operations may be combined.

One advantage of the minimum till system is the reduction of soil exposure to erosion forces such as wind and moving water. With fewer trips across the field, costs are reduced and less soil compaction occurs. Less labor is needed when farming programs adopt minimum tillage practices.

No-till Farming

No-till farming is a system in which farm ground is not tilled prior to planting, Figure 8–19. The new crop is planted in the stubble of the old crop. One

FIGURE 8–19 No-till farming is practiced by planting seed through the straw and plant residue that protects the surface of the soil. *(Photo courtesy of Utah Agricultural Experiment Station)*

trip across the field is usually sufficient to inject fertilizer and plant the seed in the stubble of the last year's crop.

No-till farming has the advantage of reducing farm operating costs by eliminating field operations. Traditional tillage implements are not needed, and a single trip across the field applies the fertilizer and gets the crop planted.

One disadvantage of the system is the high cost of a no-till planter. Because it often operates in hard soil, it must be constructed of stronger material than is required in conventional machinery.

Another disadvantage associated with no-till farming is that crop yields can be severely reduced by drought conditions. This is because moist soil is not brought up into the seeding area. Conventional tillage practices tend to carry moisture upward in the soil. This often results in higher seed germination rates for crops planted in tilled areas in comparison with crops seeded using no-till or minimum tillage practices.

PLANTING

Many clever devices have been used on farms to plant seeds. The modern planters of today are much advanced when compared to planting seeds by hand. Planters are designed to control the **seeding rate**, or the amount of seed planted per acre. They also control the depth at which the seed is planted, and they pack soil firmly around the seed to assure close contact of the seed with moist soil.

Planters of many kinds are available for planting a variety of crops, Figures 8–20 and 8–21. Small grains and most small legume seeds are planted with a

FIGURE 8–20 Row crops are usually planted with planters mounted on tool bars to allow adjustment of the spacing between rows. Seed is loaded into individual canisters from which it is metered to the planting mechanism.

FIGURE 8–21 A grain drill is designed with one or two separate seed boxes to allow seeding of small seeds at the same time that large seeds are being planted. Row spacings are permanent.

FIGURE 8–22 Seed plates are specially designed parts in planters that control the amount of seed planted in each row.

machine called a grain drill. The drill can be set for different seeding rates, and it must be adjusted for each plant variety that is planted.

Row crops are usually planted by mounting several planters on a tool bar at the desired row spacings. Most planters of this type use exchangeable rotating **seed plates**, each designed for different seed sizes to control seeding rates, Figure 8–22. This type of planter is usually used for such crops as corn, beans, sugar beets, and some vegetable crops.

Some planters are designed to broadcast or scatter the seed on the soil surface. These kinds of planters are often used to plant grasses and other plants in untilled areas such as rangelands or areas where the terrain is too steep for safe operation of standard planting equipment. Some of these planters are hand-operated while others are mounted on aircraft for **aerial seeding** in rough terrain.

Some types of self-propelled sprinkler irrigation systems have been equipped with special sprinkler heads for planting purposes. Seeds may be metered into the water and applied through the irrigation system for even distribution on the field.

HARVESTING

The development of machines for harvesting crops is one of the most important accomplishments of the twenty-first century. One farmer with a modern grain combine and a fleet of trucks and drivers can harvest more grain in a day than a hundred men and their teams could harvest in the era of the threshing machine.

The invention of the mechanical reaper by Cyrus McCormick in 1831 was one of the most significant events in the evolution of agriculture. It was the beginning of mechanized agriculture. The **reaper** was able to cut the stalks of ripened grain and tie them into bundles to be hauled in for threshing, Figures 8–23 and 8–24. This work had been done by hand prior to this time. In com-

FIGURE 8–23 The McCormick reaper was one of the most important technologies to be used on a farm because it proved that complicated tasks could be performed by machines.

FIGURE 8–24 Shocks or bundles of grain used to be a common sight during the harvest season. These shocks were hauled to a stationary threshing machine where the grain was separated from the straw. *(Photo courtesy of Utah Agricultural Experiment Station)*

bination with the threshing machine, it revolutionized the process for harvesting grain, Figures 8–25 and 8–26.

The development of new technologies has resulted in modern harvesting machines for many crops. Most major crops can now be harvested mechanically. Hay is cut, baled, hauled, and stacked entirely without hand labor, Figures 8–27, 8–28, 8–29, and 8–30. It can also be chopped and blown into a storage bag or a silo from which it is delivered to the feed bunks through a

FIGURE 8–25 Early grain combines that moved across the field were much more efficient than hauling the shocks of grain to the threshing machine. *(Photo courtesy of Utah Agricultural Experiment Station)*

FIGURE 8–26 The modern grain harvest often involves fleets of combines and trucks that harvest vast acreages of grain in a single day. *(Photo courtesy of Utah Agricultural Experiment Station)*

FIGURE 8–27 Hay harvesting has changed greatly since the days of the scythe and the pitchfork. The derrick that was used to unload and stack loose hay was once considered to be a great technological advancement. *(Photo courtesy of Utah Agricultural Experiment Station)*

FIGURE 8–28 Hay balers were the newest thing in hay harvesting in the 1940s. Much hand labor was required to load, haul, and stack baled hay and straw. *(Photo courtesy of Utah Agricultural Experiment Station)*

FIGURE 8–29 The need to transport hay for long distances led to the development of hay balers that produce rectangular-shaped bales weighing up to a ton. *(Photo courtesy of Utah Agricultural Experiment Station)*

FIGURE 8–30 Big bales are moved using loading equipment powered by hydraulics. *(Photo courtesy of Utah Agricultural Experiment Station)*

mechanical delivery system. In some instances, it is cubed or pelleted, Figure 8–31.

Root crops such as potatoes and sugar beets are dug mechanically and loaded on trucks with minimal hand labor, Figure 8–32. Forages, fruits, nuts, vegetables, and a variety of other crops are gathered by mechanical harvesters, Figure 8–33.

FIGURE 8–31 Hay is sometimes prepared for export and long-distance hauling by chopping the forage and compressing it into dense hay cubes. *(Photo courtesy of Utah Agricultural Experiment Station)*

FIGURE 8–32 Mechanical diggers are capable of harvesting several rows of root crops such as sugar beets in a single pass through the field. *(Photo courtesy of Utah Agricultural Experiment Station)*

FIGURE 8–33 Many crops such as forages, grains, vegetables, nuts, cotton, and some fruits are now harvested mechanically. *(Photo courtesy of Utah Agricultural Experiment Station)*

Most crops grown today are harvested at the end of the growing season and put into storage or sold. Before the development of mechanized harvesting machines, many crops were destroyed by winter storms before the produce could be harvested. Mechanized harvesting methods have played major roles in providing abundant supplies of high-quality foods.

FARM STORAGE

The construction of crop storage facilities on the farms has been very beneficial to both producers and consumers of agricultural products. The ability to store some of the current year's crop has allowed farmers to market crops throughout the year instead of selling everything at harvest time when prices for agricultural goods are traditionally low. Farm storage gives the producer a marketing option. If prices appear to be unfavorable at harvest time, crops need not be sold.

Storage structures of many kinds are found on farms. Barns have always been part of the farm scene. They provide dry storage areas for forage crops and shelter for livestock. Silos and silage pits are storage structures for forage crops that are preserved for cattle feed. They are typically found on many livestock farms and ranches, Figures 8–34, 8–35, and 8–36.

The feed that is preserved in a silo or silage pit is called silage or ensilage. These structures preserve forage crops that have relatively high water content by sealing off the oxygen supply from the plant materials. This prevents the

FIGURE 8–34 Airtight silos provide storage for high-quality livestock feeds. *(Photo courtesy of Utah Agricultural Experiment Station)*

FIGURE 8–35 Plastic bags provide storage space and an airtight storage environment for chopped forages. *(Photo courtesy of Utah Agricultural Experiment Station)*

FIGURE 8–36 Balage is a forage product that is baled with high moisture content (65 percent moisture), and the bales are sealed in heavy plastic bags to form a product similar to silage.

Anaerobic Respiration

Source

Starch (stored plant tissue)

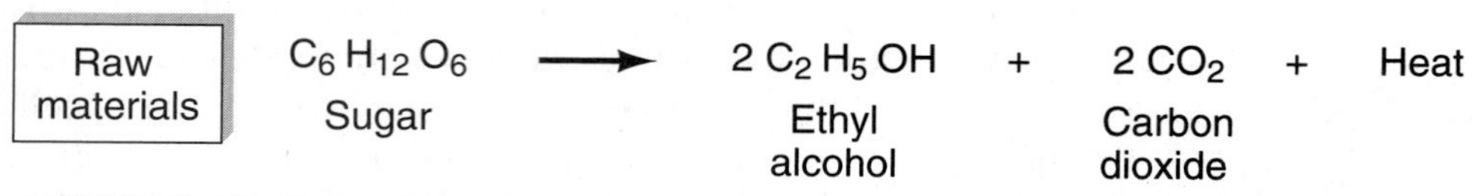

FIGURE 8–37 Anaerobic respiration is a fermentation process that can be used to preserve forage crops that have high moisture content.

process of respiration from occurring. As we learned in Chapter 5, oxygen is required for respiration to take place.

In the absence of oxygen, energy is released from plant sugars through a process called fermentation. The end products of the fermentation process are alcohol or lactic acid, carbon dioxide (CO_2), and a small amount of heat energy (2 ATP), Figure 8–37. This is the same process that is used to make the ethyl alcohol that is added to gasoline to make gasohol. The fermentation process is similar to respiration except that respiration produces water and CO_2 plus a large amount of energy (38 ATP) in the form of heat.

Graineries, warehouses, and cellars provide storage areas for grain, cotton, and other fiber crops, fruits, and vegetable crops. Grains and fiber crops require dry storage areas to protect them from molds and mildews. They also need protection from some insect pests. Fruits and vegetables require cool storage areas with enough moisture in the air to prevent them from drying out during storage. The cool temperatures slow the metabolism rates in these crops, making it possible for them to endure storage stresses.

One factor that causes stress to stored fruits and vegetables is the loss of water from the living tissues of plant materials. This is called **dehydration**. When it occurs in fruits and vegetables, they become shriveled. They lose their value and appeal as foods, because they are no longer fresh and crisp.

Controlled environment structures provide storage for perishable crops, Figure 8–38. These structures may be cooled, and sometimes relatively inert gases such as nitrogen are added to the atmosphere in the storage area. This procedure has the same effect as protecting these crops from atmospheric oxygen. Respiration and metabolism rates are reduced substantially. Vegetables and fruits that are harvested in the summer and fall seasons can now be marketed the following summer and throughout the year. As farmers learn to market their produce more efficiently, farm storage facilities are likely to increase in importance.

FIGURE 8–38 Large storage facilities in which the environment is controlled to preserve perishable crops are used on fruit and vegetable farms. They often control temperature and humidity, and sometimes the gases in the air are controlled.

CHAPTER SUMMARY

Modern production methods in agriculture have resulted in much more efficient production of agricultural goods than was possible in the past. Machines have been designed to perform much of the manual labor that was part of farming for many centuries. New farming practices based on science are preserving and improving the fields, and methods have been devised for reducing erosion of the soil.

Tillage, planting, and harvesting equipment has made it possible for farmers to operate larger farms and harvest greater crop yields than ever before. Improved storage of agricultural products has reduced spoilage losses and made it possible to enjoy many agricultural products throughout the year.

CHAPTER REVIEW

Discussion and Essay Questions

1. Describe the benefits of land leveling as a part of field preparation in the production of irrigated crops.
2. List some reasons why crop yields are likely to increase when drainage systems are installed in fields where the water table is near the surface.
3. Explain why contour farming and construction of terraces on hilly farming areas are beneficial farming practices.
4. How do fertilizer applications to the soil contribute to higher crop yields?
5. Name some problems associated with the use of chemical fertilizers, and propose solutions to the problems.
6. Compare traditional tillage practices with minimum tillage and no-till farming practices.

7. Describe the problems and benefits associated with minimum till and no-till farming methods.
8. Explain the influence of modern farm machinery on the availability of the world food supply.

Multiple Choice Questions

1. The practice of mixing topsoil and smoothing fields before they are planted is performed with which of the following kinds of machinery?

   A. forage implements
   B. harvesting equipment
   C. tillage equipment
   D. cultivating implements

2. The purpose for which field drains are installed is to:

   A. lower the water table.
   B. flush agricultural chemicals out of fields.
   C. provide sources of water for livestock.
   D. maintain the water level in drainage ditches.

3. A shallow water table is undesirable because:

   A. it restricts the downward growth of plant roots.
   B. it increases the risk of flooding.
   C. crops often require no supplemental water.
   D. it cannot be depended on for artesian water supplies.

4. The soil conservation practice of laying out field boundaries along elevation lines on uneven terrain is called:

   A. aerial mapping.
   B. terracing.
   C. conservation tillage.
   D. contour farming.

5. Construction of terraces along the contour of a hill is done for the purpose of:

   A. removing excess water from the field.
   B. creating artificial boundaries to control the migration of harmful insects.
   C. creating reference lines for making topographical field maps.
   D. holding excess water in place on the field until it is absorbed by the soil.

6. Each of the following elements is considered to be an essential element for plant nutrition except:

   A. phosphorous.
   B. sodium.
   C. boron.
   D. chlorine.

7. Which of the following essential elements is also a primary nutrient for plants?
   A. potassium
   B. carbon
   C. copper
   D. molybdenum

8. Which of the following essential elements of plants is also a micronutrient?
   A. magnesium
   B. hydrogen
   C. manganese
   D. nitrogen

9. The name of a nitrogen compound that is used by plants and which is the final product of the nitrogen fixation process is:
   A. nitrite.
   B. ammonia.
   C. nitrous oxide.
   D. nitrate.

10. A farming strategy that utilizes crop rotations, green manure crops, and other crop management practices to maintain soil fertility and control pests with limited dependence on chemical fertilizers and pesticides is called:
    A. LISA.
    B. Turbo Farm.
    C. TQM.
    D. IPM.

11. A crop management practice that promotes planting seeds in the stubble fields without first preparing a seedbed is called:
    A. minimum tillage farming.
    B. no-till farming.
    C. low input sustainable agriculture.
    D. integrated pest management.

12. Controlled environment storage facilities are able to preserve fresh fruits and vegetables by:
    A. increasing their metabolic rate.
    B. slowing the rate at which respiration occurs.
    C. sterilizing the environment.
    D. freezing them.

LEARNING ACTIVITIES

1. Invite some retired farmers to share their memories of farm work with the class. The experiences should relate to the farming methods in common use when they were young, and progress to the farming practices in use at the time they retired. (This activity might be done in the form of a panel discussion or as a story telling activity.)

2. Assign each class member to prepare a written report on the development of a specific type of farm machinery. Students could also be assigned to give the report orally to the class.

CHAPTER 9

Processing Agricultural Products

Agriculture is more than farming. As production farming has evolved to larger and more efficient business units, the agricultural processing industry has kept pace. The industry has become a source of pride, and it provides consumers with food and textile products that are the envy of the world.

OBJECTIVES

After completing this chapter, you should be able to:

- equate the growth of the agricultural processing industry to the use of technology on farms and ranches in agricultural production.
- associate the need of the processing industry for clean water with the obligation of the industry to return clean water to the environment.
- explore the role of high technology in the operation of a modern processing plant.
- explain the need for the quality control division in the agricultural processing industry.
- describe several methods of preserving perishable products to increase shelf life.
- justify the use of food additives by food processors.
- appraise the importance of packaging materials in agricultural processing.

TERMS FOR UNDERSTANDING

The following vocabulary terms should be studied carefully as you read:

raw products
consumer goods
robotics
laser technology
quantum theory
photon

quality control
antibiotic
quality
uniformity
perishable
shelf life
blanching
dehydration
freeze-drying
lyophilization
sublimation
immersion freezing
freezing point liquid
cryogenic liquid
irradiation
fermentation
hydrogenation
pasteurization
food additives
packaging
retortable pouch
biodegradable

Clean, pure water is an essential ingredient of most types of agricultural processing. A reliable source must be available. It is used to clean the raw product, the machinery, and the processing plant, Figure 9–1. It is used to make steam for cooking and providing heat and power. After use, the water must be treated to remove impurities before it can be released into the environment. Processing plants are required to process waste water or have access to a water treatment facility.

CONTROL SYSTEMS

Modern agricultural processing plants feature some of the finest examples of modern technology. Hundreds of mechanical devices have been designed to perform a large number of tasks. **Raw products** are the agricultural products that are produced on farms and ranches. The role of the agricultural processing

FIGURE 9–1 Cleanliness is very important in food processing facilities. Much of the processing equipment is made of stainless steel, and plenty of clean water is used to keep equipment clean. *(Photo courtesy of Utah Agricultural Experiment Station)*

FIGURE 9–2 Processing activities range from grading eggs for quality to preserving perishable fruits and vegetables. *(Photo courtesy of Utah Agricultural Experiment Station)*

FIGURE 9–3 Modern agricultural processing plants make use of computers and other electronic devices to control processing activities. *(Photo courtesy of Utah Agricultural Experiment Station)*

industry is to prepare raw products for consumers, Figure 9–2. After they have been processed to the satisfaction of the people who buy and use them, they are known as **consumer goods**.

Much of the machinery used in agricultural processing is powered by electric motors. Large amounts of electrical energy is consumed, and well-trained electrical technicians are needed to keep equipment operating.

A processing plant is like a big machine with many systems and moving parts—it is highly mechanized. The function of a processing employee is to understand the functions of processing equipment and keep it operating. Maintenance technicians require special training in the operation, maintenance, and repair of complex machines.

Computers can be used to monitor product lines and maintain control over processing activities that are occurring simultaneously. This is done by linking electronic devices to computers, Figure 9–3. Computer technicians, programmers, and engineers provide technical support to keep the systems operating and to design improvements.

Robotics is the study of the design and use of robots, Figure 9–4. Modern processing plants are using this new technology to improve the efficiency of agricultural processing by designing robots that can perform the same movements workers use to perform simple tasks. Once such a machine has been constructed, it can be linked to a computer that is programmed to repeat the same movements required to complete a specific task. When a particular function is no longer needed, a robot can be reprogrammed to perform a new task.

Robots are used to perform simple tasks that require the same procedure to be performed over and over again. A robot is well-suited to such work. It can be programmed to perform a task the same way every time. As long as nothing in the manufacturing or processing procedure changes, a robot can perform its job continuously without becoming tired, bored, or inefficient.

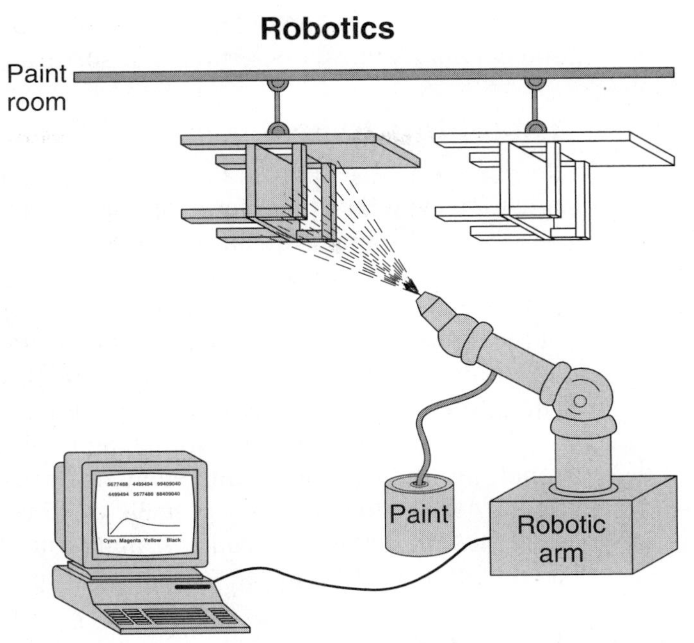

FIGURE 9–4 Most applications of robotics in agricultural processing involve the use of a mechanical arm linked to a computer that controls its functions.

A robot is especially useful in performing work on assembly lines or in performing simple tasks that are dangerous to humans. It can perform a task more consistently than a human worker, which adds to product quality, Figure 9–5. Robots that are equipped with an electric eye can be programmed to respond to visual images such as colors or shapes. Some robots are equipped with special sensors, making it possible for them to be programmed to

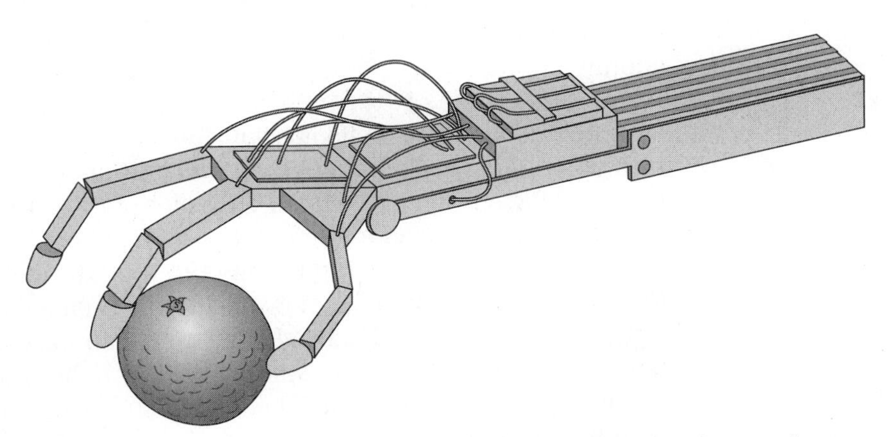

FIGURE 9–5 Robotics technology is advanced enough that a robot can be programmed to grasp an object as sensitive as a piece of fruit without damaging it. Researchers are developing robots to harvest citrus fruit.

respond to the way an object feels. Robotic research is underway to develop the capability of robots to pick and sort fruits and vegetables.

Laboratory Robots

Research scientists at the Monsanto research center in St. Louis, Missouri have programmed robots to perform laboratory tests on plant and animal tissue samples. Special sample jars have been designed in which are placed the materials that are to be tested. The filled jars are positioned in trays from which the robotic hand retrieves them. The robot removes the lid, takes a measured amount of the material in the jar, replaces the jar lid, places the measured sample in a container with a test reagent, performs the laboratory procedure, and measures, records, and prints the test results.

Laboratory technicians are needed only to program the functions of the robot, put the equipment and samples in place, start the process, and clean up after the testing process is completed. The laboratory staff is able to go home at night with the process running, and return the following morning to retrieve a computer printout of the test results.

Laser Technology

Laser technology is used in a wide variety of processing applications. Some types of lasers are used to brand codes on packages. Lasers can be used as electric "eyes" to watch product lines and control machine operations by activating on/off switches. They are also used in robotics to locate, recognize, and sort materials.

The first operational lasers were developed in 1960. The principle of science that is the basis for laser technology was described by Albert Einstein as the **quantum theory**. The theory deals with the way that atoms absorb and radiate light. Energy such as electricity, radio waves, or light that is applied to certain kinds of atoms causes the atoms to become "excited" by absorbing energy.

An atom that is in the excited state gives off a unit of light energy called a **photon** as it returns to its normal state. When this occurs in a tube with mirrors on both ends, the photons excite additional atoms as they are reflected back and forth through the tube. The color of a laser is determined by the kind of atoms that are used in the energized chamber.

Some light is allowed to escape the tube through the mirror at one end, Figure 9–6. This continuous flow of photons is called a laser beam. It is a narrow and very intense beam of light that can stay focused over great distances. This allows great amounts of energy to be transferred from one place to another. This focused beam of energy is sufficient to operate switches or even melt materials.

Laser technology in combination with robotics is rapidly taking over many of the tedious processing tasks that require many repetitions of the same

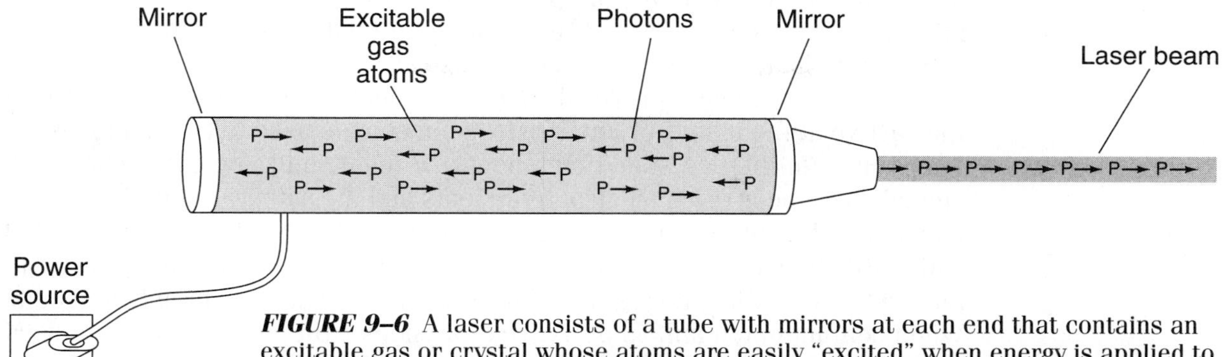

FIGURE 9–6 A laser consists of a tube with mirrors at each end that contains an excitable gas or crystal whose atoms are easily "excited" when energy is applied to them. Reflected photons excite other atoms causing a continuous flow of light to be emitted from one end of the tube. The laser beam transfers power from the energy source to a different site where it can be used to operate electrical switching devices, melt materials, or carry information.

activity. Tasks of this nature often prove to be boring and unfulfilling to people, but they can be performed flawlessly by specially designed robots.

QUALITY CONTROL

Quality control is a processing function that is responsible for the quality of finished products. High quality finished products can be produced only from high quality raw materials, Figure 9–7. Quality control technicians inspect and test incoming raw materials to be sure that they meet quality and purity standards before the raw product is used for processing, Figure 9–8.

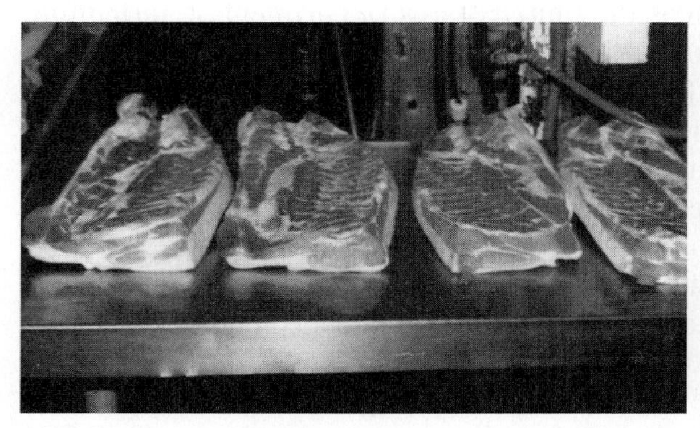

FIGURE 9–7 High quality food at the supermarket begins with high quality products at the farm and high quality processing methods. *(Photo courtesy of Utah Agricultural Experiment Station)*

FIGURE 9–8 Milk is stored on the farm in stainless steel tanks until it is transported to the processing plant. High quality dairy products begin with high quality raw milk at the farm.

Product contamination can be observed in several forms in food products, and great care is taken by processors to assure that food is free from contamination. The most common contaminants of food products are antibiotics, chemical residues, bacteria, and dirt.

Serious problems can result if raw products are contaminated before they arrive at the plant. An **antibiotic** is a medicine used to treat animals for infections. Antibiotics sometimes occur in meat, milk, or eggs when sick animals have been given medical treatments and the withdrawal period for the medication has not been observed. Milk or meat that has been contaminated with antibiotics can be a serious health hazard for people who are allergic to the medication. Milk processors are very careful to test for contaminated milk. When contaminated milk is processed for cheese, the bacteria that are used in the cheese-making process may be killed by the medication. The result is a poor cheese product.

Chemical residues may sometimes be found on the surfaces of fruits and vegetables, but this is infrequent because food safety regulations require that chemical use is suspended well ahead of harvest time. Most residues that remain are removed when the produce is washed. Farmers who market products that are contaminated with illegal chemical residues are liable for heavy legal penalties.

Most bacterial contamination of food occurs when animal or poultry carcasses become contaminated with fecal matter during processing. Dirty processing or storage equipment also contributes to bacterial contamination. Meat and dairy product inspectors have the authority to condemn milk or meat that is produced or processed under unsanitary conditions.

Dirt contamination of food products refers to any contaminant that is filthy or unclean that comes in contact with a product. Meat that is processed or stored in an unclean location, or milk that is contaminated by fecal matter from the udder of the cow are examples of this kind of contamination.

Government agencies such as the United States Department of Agriculture and the state departments of agriculture regularly inspect both the food products and food processing facilities and equipment. Rigorous standards for quality are in place, and violations of the standards invoke penalties. For example, a beef carcass that is found to contain antibiotics or illegal drugs will be condemned and cannot be used for human food. The person who sold the animal to the processor will also be liable for penalties.

Fruits and vegetables must be free from contamination from herbicides and pesticides when they are marketed to the public. Chemical companies test their products to determine the amount of time required for chemical residues to be degraded and reduced to safe levels for humans, Figure 9–9.

Raw products for the textile industry such as wool and cotton must be free of foreign fibers that contaminate and destroy the value of textile products such as cloth. When plastic twine is used on bales of hay or straw that are used for feeding or bedding sheep, plastic fibers sometimes contaminate the wool fibers

FIGURE 9–9 Laboratory testing of products is necessary to assure consumers that chemical residues have been reduced to safe levels and product quality is maintained at high levels. *(Photo courtesy USDA/ARS #K–3512–3)*

FIGURE 9–10 Domestic wool has been adversely affected by the use of plastic twine on hay bales. Care must be taken at the farm to maintain clean wool that is free of contaminants.

in the fleece. Such materials are very difficult to remove from the wool, and cloth that contains fibers from the twine is of little value, Figure 9–10.

Consumers are quick to boycott a product they think may be contaminated, Figure 9–11. The duty of the quality control arm of the processing industry is to assure consumers that the products offered for sale are free from contaminants and safe to use.

FIGURE 9–11 Textile buyers are charged with the responsibility of selecting clean, high quality raw products. *(Photo courtesy of The American Sheep Industry Association)*

CAREER OPTION

Quality Control Technician

A person who chooses a career as a quality control technician is responsible for testing raw products as they arrive at the processing plant, and testing finished products to ensure that the quality is acceptable, Figure 9–12. Products are tested to determine the content of salt, fat, water, and contaminants.

A quality control technician tests and adjusts equipment used for packaging to make sure that packages contain the proper amount of product. He or she adjusts scales and other equipment as necessary to ensure that a uniform product is produced and packaged.

Technical training is required in order to learn dependable testing procedures. Sophisticated equipment is used for many of the tests that are conducted, and quality control technicians must be proficient in the use of scientific instruments. Technicians are also expected to accurately record product data and write reports. Most quality control technicians have technical or baccalaureate college degrees.

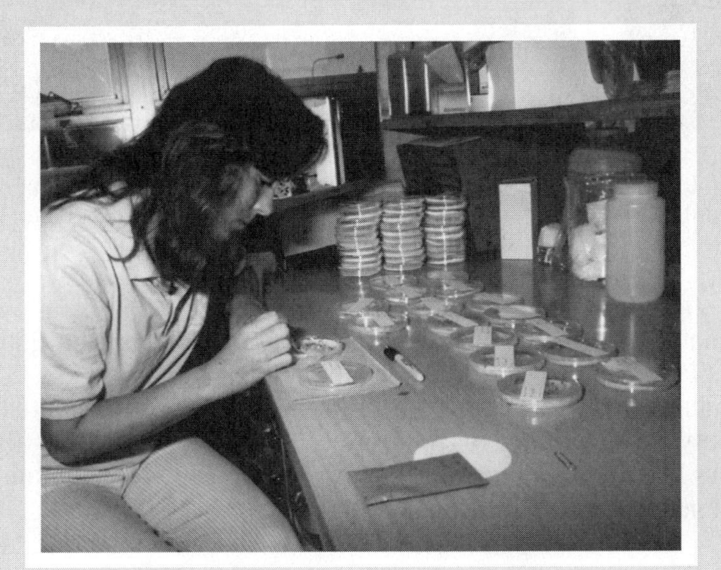

FIGURE 9–12 A quality control technician is responsible to make sure that products are consistent in quality. *(Photo courtesy of Utah Agricultural Experiment Station)*

Quality is a measurement of how good a product is. The purpose of quality control is to measure those characteristics of a product that will determine whether or not product quality is being maintained. Products are tested at critical stages of processing to detect processing errors before they become serious enough to affect the quality of the finished products, Figure 9–13.

Uniformity is a measurement of sameness. Uniform products do not vary significantly in the characteristic being measured. Consumers expect the products they buy to be uniform in quality, size, shape, texture, color, etc. Adjustments are made in the process as often as necessary to maintain product uniformity and quality.

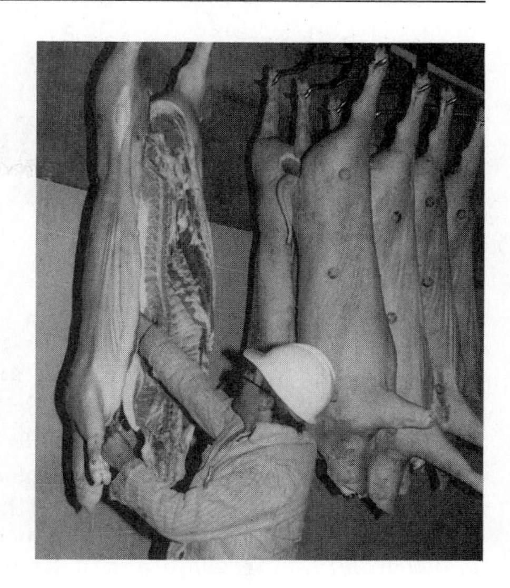

FIGURE 9–13 Meat is carefully inspected and graded for quality by federal meat inspectors. *(Photo courtesy of Utah Agricultural Experiment Station)*

PROCESSING

Processing includes all of the activities that occur as raw products are prepared for consumers. Modern facilities and equipment are designed to ensure that processing is done efficiently and that high quality consumer products are produced, Figure 9–14. Processing is done to preserve perishable products beyond the time when they would spoil under natural conditions or to convert products to forms that consumers can use.

Most raw vegetable products must be cleaned, trimmed, and cut into acceptable sizes before they are preserved and packaged. Many vegetable products must be separated from waste materials using hand labor. Processing includes the disposal of waste products. It also includes the treatment of water

FIGURE 9–14 Modern processing facilities for agricultural products are often referred to as plants, factories, and mills. This modern sawmill processes logs into a variety of wood products.

FIGURE 9–15 Modern technologies such as this deboning machine are able to reduce the labor required to process agricultural products. *(Photo courtesy of Utah Agricultural Experiment Station)*

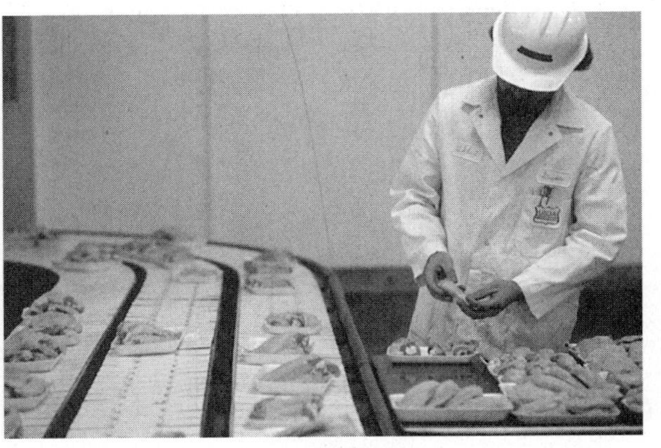

FIGURE 9–16 Sensitive conveyor systems move products from one location to another during processing activities. *(Photo courtesy of Utah Agricultural Experiment Station)*

that has been used in the processing activity before it is discharged from the processing facility.

Meat processing includes killing the animals, removing the skin, hair, or feathers, and removing the offal or waste materials such as internal organs from the carcass. Carcasses are cut into wholesale and retail cuts, and specialty meats are prepared, Figure 9–15. All of these processes require workers who have been taught to perform specific tasks.

Some of the most sophisticated machines ever invented are used to process agricultural products. Conveyor systems that move products are found in nearly every kind of processing plant, Figure 9–16. They carry products from one processing area to another as different procedures are used to modify the products. Many processing plants rely on automatic packaging devices that place the product in wrappers, cans, boxes, or bags.

PRESERVING

Most food products are **perishable** or susceptible to spoilage. Food spoilage occurs when conditions are favorable to the growth of the microorganisms that break down plant and animal tissues. Although microorganisms that cause spoilage are found all around us, spoilage of products occurs more readily when there is a nearby source of microorganisms such as contaminated equipment or dirty processing conditions. Bacteria and other microorganisms require food, so food particles that are present on equipment and table tops make great sites for microorganisms to multiply. Microorganisms require moist conditions if they are to thrive in an environment. They also benefit from warm temperatures. When these conditions are present over an extended period of time, bacteria

FIGURE 9–17 Food preservatives are added to some food products to extend the shelf life of perishable items. *(Photo courtesy of Utah Agricultural Experiment Station)*

and other microorganisms that cause spoilage are likely to be present in large numbers.

Quality declines rapidly once the product reaches maturity. Food processors have developed processes that will delay product spoilage. One process includes partial cooking of the product. Other preservation processes include dehydration, fermentation, cold processing, irradiation, hydrogenation, pasteurization, and smoke curing of products. In addition, food additives have been developed that are capable of greatly increasing the **shelf life** of most products. Shelf life is the length of time that a perishable product can be stocked in a store before it loses quality, Figure 9–17.

Partial cooking of food products is practiced to reduce the amount of bacteria present in the product, and to improve the convenience of the product for consumers. Partial cooking is usually followed by freezing or cooling the product for short term storage or canning it for long-term preservation. Many products are preserved in this manner. They include milk and dairy products, which are pasteurized by heat followed by storage at low temperatures. Many vegetable products are blanched in a scalding hot water bath before they are packaged and frozen. Potato products such as french fries are partially cooked at the time they are processed to reduce the cooking time required at fast food restaurants.

Blanching Vegetables and Fruits

Many of the vegetables and fruits that we eat have been preserved by freezing them in small meal-size containers. Before these foods are frozen, they are cut to acceptable sizes, thoroughly washed, inspected, and blanched. **Blanching** is a process by which fruits and vegetables are briefly scalded in boiling water or steam. The length of time that is required for blanching depends on the size of the food pieces and the amount of material that is in the batch.

Blanching is done to inactivate natural enzymes in the fresh produce. Some enzymes continue to function at freezing temperatures, and spoilage can occur in produce that is not properly blanched. This process extends the storage life of frozen fruit and vegetables.

Dehydration

Products can be preserved by drying or **dehydration**. This process removes water from the product making it possible to store it at room temperature. Food dryers are available for home use, but the most common commercial dehydration process is **freeze-drying**. Another term for freeze-drying is **lyophilization**. This process is accomplished by freezing materials in a vacuum to draw off 99 percent of the moisture. Low pressure in the freezing chamber causes ice crystals to vaporize more quickly than they would in a chamber at normal atmospheric pressure. The process by which a solid crystal (ice) forms directly into a gas (water vapor) without passing through its liquid form (water) is known as **sublimation**.

Freeze-dried foods can be stored for long periods of time without spoiling, and these products do not take up very much storage space. They are especially useful to backpackers because they are lightweight foods that can easily be carried. They have also been used by astronauts during space flights because large amounts of these foods can be stored in a small storage area. Freeze-dried foods can be easily prepared for consumption by adding water.

Cold Processing

One of the most common methods of preserving a food product is to freeze the fresh product. This method of food preparation is called cold processing. Many food products are prepared for market in this manner. Examples of these products include meats, fruits, vegetables, and convenience foods. Freezing fresh foods extends the shelf life, and makes it possible to ship perishable products without spoilage or loss of quality.

Several factors should be considered in order to preserve high quality in frozen foods. Air must be removed from packages to prevent freezer burn. This is usually done by removing air from the food packages with a vacuum tube. The temperature at which foods are frozen is important. Frozen foods should be maintained at zero degrees Fahrenheit or colder to maintain quality.

Foods should be frozen quickly for best results. This can be done by blowing very cold air across the packages. Air temperature should be 20–30 degrees below zero during the initial freezing process. Packaged products can also be subjected to **immersion freezing**. Two kinds of liquids are used for this purpose: freezing point liquids and cryogenic liquids.

A **freezing point liquid** contains sugar, salt, or alcohol such as glycerol, and it does not form a solid at 32 degrees. These liquids freeze at much colder temperatures. They are effective at drawing heat out of food products at a rapid rate. It is important that foods are not placed in direct contact with these liquids.

A **cryogenic liquid** is obtained from a gas that has been converted to a liquid by subjecting it to high pressure. Those most commonly used to freeze foods are liquid nitrogen (-196°C) and liquid carbon dioxide (-79°C). These liquids are nontoxic, and food can be immersed directly in them.

Irradiation

Irradiation is a controversial food processing method that subjects foods to radiant energy such as x-rays to kill microorganisms that cause food spoilage. It is opposed by some consumers because it uses radiation to kill harmful organisms. However, it has been demonstrated to be a safe method of food preservation that is used to extend the shelf life of such foods as shrimp, citrus fruits, strawberries, and other food products. The radiation passes through the foods and does not remain in treated food products. Foods that have been irradiated are safe to eat. This practice is also used to kill unseen insects on fresh produce before it is shipped.

Fermentation

The **fermentation** process is one of the oldest forms of food preservation. It was discussed in Chapter 8 as a method of preserving feeds. Bacteria, yeasts, and enzymes are used to modify the chemical makeup of products. It is a form of respiration that occurs when oxygen is not present. It has been used since ancient times to make cheese, Figure 9–18. Fermentation is used to convert plant sugars to alcohol in the wine and brewing industries. Cucumbers and cabbage are fermented to make pickles and sauerkraut. Milk is fermented to make

FIGURE 9–18 Fermentation is a food preservation process whereby bacteria converts sugars to acids or alcohol that protects the food from spoilage. *(Courtesy of USDA)*

yogurt. All of these products, plus many more, depend upon the fermentation process.

Hydrogenation

Hydrogenation is a form of food processing that is used with fats and oils. This process adds hydrogen to the unsaturated chemical bonds of liquid oils causing them to become more solid at room temperatures. This is the process that is used to make margarine and shortenings from vegetable oils. The hydrogenation process reduces the tendency of oil products to spoil or become rancid.

Pasteurization

Pasteurization is a processing method by which some foods such as milk and beverages are processed, Figure 9–19. This process involves heating the milk or beverage to 131–158 degrees F for 30 minutes. The product is then quickly cooled and refrigerated. The bacteria that cause spoilage to occur are killed by the heat, and the shelf life of the product is greatly increased.

Additives

Food additives are materials that are added to processed foods to reduce spoilage. Additives include common materials such as salt. This additive has long been used to cure meat. Salt pork was one of the main food items carried on ships during the long voyages of early explorers like Columbus and Magellan.

Salt is still used to preserve food, but many additional food additives have been identified to aid in food preservation. Food additives have greatly improved the shelf life of many foods, and they have also been used to improve the appearance of the food. Some additives are used to preserve color

FIGURE 9–19 Pasteurization is a processing method by which some foods, such as milk, are processed. *(Courtesy USDA)*

as well as freshness. Food additives are included with the other ingredients listed on the food container.

In the textile industry, chemical additives have been identified that improve some qualities of natural fibers. When they are applied to carpets or other materials, they prevent or reduce the tendency of these materials to become stained when household materials are spilled on them.

Some consumers resist the use of additives in any form. Fresh products are preferred by most consumers, but the use of additives in food products has improved the quality of most products that must be stored. During some seasons of the year and in some locations, fresh products are scarce or unavailable.

PACKAGING

Packaging is the process of wrapping or sealing a commodity or product in a container, Figure 9–20. High quality packaging materials must be nontoxic, and they must protect the product from contaminants such as inks and dyes. Good packaging materials do not detract from the flavor or quality of the food product. They provide sanitary protection, and they control the movement of moisture and gases into the package. Some containers even protect against light infiltration because many foods are sensitive to light. Eggs and similar perishable products require packaging that is resistant to impacts.

Many different kinds of materials are used to package foods. Some packaging materials are designed to withstand the extreme cold that is encountered during immersion freezing and to tolerate the heat of cooking. Other packaging materials include metal cans, glass bottles, paper bags and containers, plastic containers and wrapping materials, and laminated packages that contain two or more different materials layered together.

Some new plastics that contain corn starch are biodegradable. Another new advancement in packaging is the **retortable pouch**. It consists of polyester

FIGURE 9–20 Packaging is the process of sealing a commodity in a container like the juice cartons. *(Photo courtesy of FFA)*

material on the outside, aluminum foil in the middle layer, and an inner layer of plastic. An entire industry has developed to provide packaging materials for foods and other products.

Agricultural products are packaged as part of the processing activity for several reasons. Many products have longer shelf lives when they are sealed in clean packages than when they sit out in the open air. Packaging provides a sanitary way of maintaining purity in a product. It is a convenient way for persons who are involved in the marketing process and for the consumers who buy the product to handle it.

Marketing experts design packages that appeal to the buying habits of customers, and printed messages are often included on packages for advertising purposes, Figure 9–21. Packages have taken on nearly as much importance as the products they contain, Figure 9–22. Hesitant customers may make the decision to buy or not to buy based on how attractive the product container is.

The buying habits of consumers have changed a great deal in the past few years. Only a generation ago, milk was purchased in glass bottles, which were returned to the store to be used again. Fruits and vegetables were purchased in large amounts and bottled at home to sustain the family until the following year. Root cellars were filled with bins of raw fruits and vegetables stored for winter use.

Very little food is stored in homes today. Most families buy what they need several times each week. Families are smaller than they used to be, and food is purchased in small convenient containers. Many items are even packaged in individual packages. The cost of packaging material is sometimes greater than the cost of the product it contains.

FIGURE 9–21 Modern packaging for consumer-ready items is attractive to the customer and designed to promote the concept of product convenience. *(Photo courtesy of Utah Agricultural Experiment Station)*

FIGURE 9–22 Attractive labels on containers play an important part in advertising to promote product sales. *(Photo courtesy of Utah Agricultural Experiment Station)*

Some packaging material is recycled and used again, but much of it is discarded as trash and must be disposed of. Packaging materials have created a disposal problem because many of them are not **biodegradable**, meaning that the materials do not break down or rot away. Instead, they persist for many years and sometimes become sources of pollution to the environment.

STORAGE

Many foods and other agricultural products are not consumed very soon after they are processed. Storage is a strategy that allows us to keep products until they are needed. Storage of agricultural products is closely associated with agricultural processing. Facilities are needed to store raw products until they are delivered to the processing plant. Since most agricultural products are harvested in the fall of the year, storage facilities are usually needed for the majority of our raw products. Some of this storage is available on farms, but a substantial amount of storage capacity is provided by processors, Figure 9–23.

A different kind of storage is needed for processed products than was needed for the raw product. Many products are stored in large warehouses where packaged goods are stacked on pallets. Perishable products are usually stored in coolers or freezers, Figure 9–24. Products that are not highly perishable are often stored in less expensive storage facilities, Figures 9–25 and 9–26. As products are sold, they should be removed from the storage areas in the same order in which they were processed. This is done to ensure that the length of the storage period is as short as possible.

FIGURE 9–23 Storage facilities must be available to agricultural processors. This feed processing plant requires separate bins for each feed ingredient. *(Photo courtesy of Utah Agricultural Experiment Station)*

FIGURE 9–24 Huge freezer facilities such as this food product storage warehouse are equipped with facilities for loading trucks and railroad cars. Large amounts of finished products must be stored until they are needed.

FIGURE 9–26 Processed sugar is stored in huge silos to await shipment. Bulk sugar is shipped to companies that package the sugar in their own bags for resale.

FIGURE 9–25 Huge tanks provide storage for syrup that is extracted from sugar beets. The syrup is later processed into sugar.

CHAPTER SUMMARY

Agricultural processing has become a major industry. The abundance of raw products from production agriculture must be processed, stored, and marketed. Consumers prefer that products be neatly packaged and easily prepared. The processing industry has successfully met consumer needs.

Processing of agricultural products requires an abundant supply of clean water. After the water has been used, the processor must clean and purify it before it is returned to the environment.

Modern processing plants rely on technology to convert raw agricultural products to products that are ready for use by consumers. Computers, lasers, robotics, and complex machines are all linked together to accomplish the processing task.

Quality control is a processing division that makes sure products are safe for the consumers, and that the quality of the products is uniform. Perishable products are preserved by cooking, dehydrating, smoking, fermenting, cold processing, hydrogenating, pasteurizing, and adding food preservatives. Some products must be sealed in containers or frozen for storage. Storage facilities are necessary to preserve foods and other products until they are needed.

CHAPTER REVIEW

Discussion and Essay Questions

1. Describe how the use of new technologies on farms and ranches has affected the agricultural processing industry.
2. Discuss why fresh water is needed by the agricultural processing industry, and explain why the processing industry is obligated to return clean water to the environment.
3. Explain the role of high technology in the agricultural processing industry.
4. Describe some of the technologies used to process agricultural products.
5. Explain the role of the quality control component of agricultural processing, and describe why it is important.
6. List several methods of preserving perishable products, and explain how they are different.
7. What factors contribute to spoilage of food products?
8. What are some advantages and disadvantages of using food additives in processed food?
9. Why is product packaging important to agricultural processors and consumers?

Multiple Choice Questions

1. An essential natural resource for agricultural processing that is used as a solvent and cooking medium, and that can be used to transfer power and heat is:

   A. water.
   B. ethanol.
   C. diesel.
   D. propane.

2. A machine that can be used to monitor product lines and maintain control of integrated processing activities is a(n):

   A. emulsifier.
   B. robot.
   C. FAX machine.
   D. computer.

3. A laser is a device that emits a beam of light that is used in the processing industry to:

   A. provide lighting in critical work areas.
   B. thaw frozen products.
   C. activate electrical switches.
   D. generate steam.

4. A technical machine that is used by processors to simulate movements of workers in performing simple tasks is:

   A. photometer.
   B. robot.
   C. generator.
   D. articulator.

5. Which of the following *is not* considered to be a chemical contaminant of food products?

   A. antibiotics
   B. herbicides
   C. pesticides
   D. bacteria

6. Which of the following government agencies is least involved in regulating contaminants in food products?

   A. United States Department of Labor
   B. United States Department of Agriculture
   C. state departments of agriculture
   D. United States Food and Drug Administration

7. Which of the following terms is a reliable measurement of how good a product is?

   A. uniformity
   B. quality
   C. diversity
   D. retail price

8. Which of the following conditions is considered favorable to microorganisms that cause food products to spoil?

   A. cold temperatures
   B. contaminated equipment
   C. dry environment
   D. clean processing facility

9. A food preservation method that converts plant sugars to alcohol is:

   A. fermentation.
   B. dehydration.
   C. cold processing.
   D. blanching.

10. A liquified gas that is used to quickly remove heat from food products by immersion freezing is called:

    A. freezing point liquid.
    B. super-cooled water.
    C. cryogenic liquid.
    D. hydrogenated oil.

11. The food preservation process that uses radiant energy to kill microorganisms is:

    A. cryogenic freezing.
    B. irradiation.
    C. hydrogenation.
    D. pasteurization.

12. A new type of packaging known as retortable pouch is composed of which of the following materials?

    A. polyester material
    B. aluminum foil
    C. plastic film
    D. all of these materials

LEARNING ACTIVITIES

1. Take a field trip to a local agricultural processing plant and instruct students to observe the following:

    a. types of machines that are used
    b. kinds of control devices that are in use
    c. testing procedures used in quality control
    d. methods of preserving the product
    e. types of packaging materials that are used
    f. types of storage units that are required

2. Select a farm product that is produced locally, and study the different steps required to prepare this product for use by consumers. Track the product from the time it leaves the farm until it is consumed.

CHAPTER 10

Agricultural Marketing Technologies

Marketing includes all of the business activities involved in moving agricultural products from the producer to the consumer. It includes buying, processing, and selling of raw materials and consumer goods. Marketing also includes transportation, advertising, and promotion of products. This chapter explores marketing technologies that are available to managers of farms and agricultural businesses.

OBJECTIVES

After completing this chapter, you should be able to:

- explain the importance of an enterprise budget in developing a marketing plan for an agricultural commodity.
- describe the roles of publications and printed materials in the marketing process.
- define the importance of the telephone and related technologies in marketing agricultural commodities.
- evaluate the effectiveness of radio and television as marketing tools.
- appraise the importance of computers and related technologies to the marketing process.
- predict the role of satellite communications in developing international markets.
- distinguish the differences between forward contracting, trading in options, and trading in futures contracts.
- define the relationship between the transportation system and agricultural marketing.

TERMS FOR UNDERSTANDING

The following vocabulary terms should be studied carefully as you read:

- commodity
- information age
- teletype
- computer networks
- modem
- internet
- information highway
- FAX machine
- commodity groups
- AMS
- direct sales
- farmer's market
- cooperative
- forward contracting
- futures trading
- options trading
- Boards of Trade
- futures contract
- hedging
- basis
- video merchandising
- telemarketing
- domestic markets
- international markets

Agricultural marketing is a system through which the marketing needs of agricultural producers and processors are met. Agricultural marketing includes ideas, processes, and systems through which agricultural goods are sold. Marketing technologies consist of equipment and machines that are used in the marketing process.

Markets for agricultural products may be expanded or restricted by the availability of a product, the ability to communicate with potential customers, the ability and willingness of the customer to pay for the product, and the ability to transport the product to the buyer. When any of these conditions cannot be met, there can be no sale of a product.

It is the job of an agricultural marketer to sell products by solving problems that interfere with sales. This is done by locating products that are for sale, finding customers for the products, arranging payment terms and conditions, and making arrangements for the product to be transported to the buyer.

ENTERPRISE BUDGETS

An agricultural **commodity** is a product that is usually generated from plants or animals. Effective marketing of an agricultural commodity begins long before it is produced. It begins by preparing an enterprise budget for the product. The budget should include all known production costs and estimated expenses for materials or services that will be required during the production cycle. It should also include the estimated or contract value of product sales. The budget helps to determine whether the commodity can be profitable and it helps to establish the price for which a product or commodity must be sold to achieve a profit, Figure 10–1.

Sample Livestock Budget

Production Year______

A. Summary of Investments	Current Total	Replacement Per Unit	* % Charged To Enterprise
Breeding Animals			
Market Animals			
Livestock Equipment			
Tractors/Trucks			
Machinery			

*Some costs to be pro-rated to other enterprises

B. Production Income	Quantity	Weight	Price	Total
Feeder Animals				
Market Animals				
Breeding Animals				
Culls				
Products/Services				
Total Receipts				

C. Enterprise Expenses (variable costs)	Quantity	Price/Unit	Total Cost
Feed: Forages			
Grains			
Supplements			
Pasture/Range			
Health Care			
Fees			
Hauling			
Marketing			
Machinery			
Equipment			
Labor			
Maintenance (Mach./Equip)			
Interest—Operating Capital			
Total Enterprise Expenses			

D. Ownership Expenses: Insurance, replacement, taxes, interest
- Livestock
- Machinery
- Equipment
- Real Estate Taxes

Total Ownership Expenses

E. Other Costs:
- Land Costs
- Management Costs

Total Other Costs

F. Total All Costs
- Overhead Costs

G. Net Income (Total Receipts—Total Costs)

FIGURE 10–1 A sample enterprise budget

The availability of any commodity or product is determined by how profitable it is to those who supply it. In countries where free markets exist, a dependable supply of an agricultural product can usually be assured when a farmer or rancher can earn a profit by raising it. The profit motive is very powerful as a stimulant to produce, and the lack of profit is just as powerful in reducing product supplies. Enterprise budgets are tools that are used by producers to determine whether a commodity should be produced.

A computer is a very useful tool in the preparation of an enterprise budget. It is used to calculate production costs based on current information and past production records. It can be used to track commodity markets and to obtain current market prices. It is also used to consult with industry experts to obtain marketing information and advice.

COMMUNICATIONS TECHNOLOGIES

Agricultural technologies play important roles in providing the information needed to develop sound enterprise budgets, Figure 10–2. We are fortunate to live in the **information age** when many technological advances have been achieved in communications systems. The most significant communication advancement is the ability to communicate with people all over the world using personal computers. We are also able to access information from many

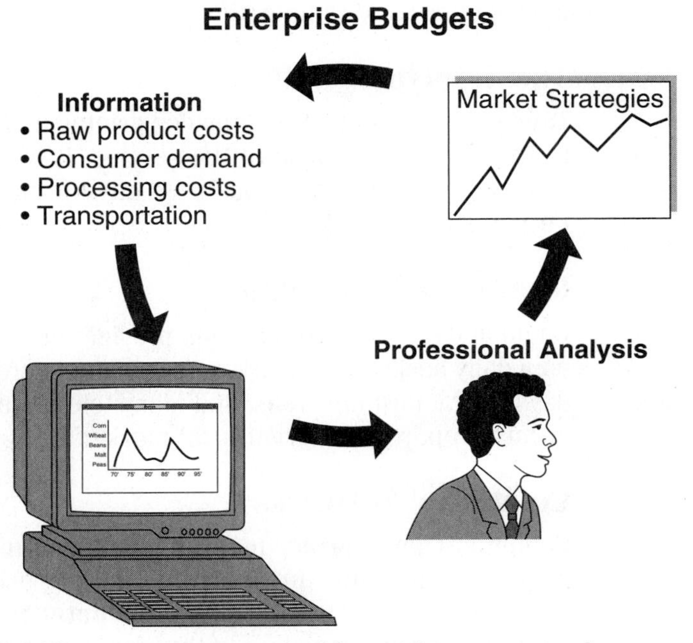

FIGURE 10–2 The computer is a valuable tool for processing large amounts of information and identifying marketing strategies.

sources without leaving our homes or offices. The information age has truly arrived.

New product and market information is available in many forms. Publications such as books, magazines, journals, and trade letters are widely distributed and read by agricultural managers. Modern printing technologies have increased the availability of printed materials, which makes marketing publications readily available to producers of farm commodities. Telephone, teletype, radio and television communications, computer networks, and FAX machines play important roles in providing agricultural marketing information to producers.

The Telephone

The telephone is one of our most important marketing tools because it makes it possible to communicate with potential customers in the domestic markets of North America and in international markets throughout the world. Many marketing transactions are completed by communicating using the telephone. It is a relatively inexpensive way to bring sellers and buyers together to do business.

The cellular telephone has become an important business tool to farm managers and their suppliers. It allows communications to continue as farmers are doing their work in the fields and on the road. This phone makes it much easier to stay in touch with parts suppliers and commodity buyers without interrupting the work schedule of the farm or ranch. It also makes it much easier for agricultural marketers and others to contact farm and ranch managers during regular working hours.

The Teletype

Teletype communication provides printed market information instantaneously to those who subscribe to the service. It is used to provide updated market prices throughout every business day for most commodities that are sold at large terminal markets.

Radio and Television

Radio and television broadcasts provide the latest farm marketing information on a daily basis. Farm market reports are heard on the radio several times each day in rural farming areas, and many television stations broadcast regular programs prepared by agricultural market analysts, Figure 10–3.

Computer Networks

Computers have become important tools for making marketing decisions. Many computer programs are available that are useful in compiling and analyzing information related to marketing. **Computer networks** provide sources of information that can be accessed through computers using telephone modems. A **modem** is an instrument that connects the computer to a telephone line.

FIGURE 10–3 A valuable service is provided in many rural communities by market analysts who report sales and prices for agricultural products through radio and television broadcasts.

Several networks are available to agricultural managers that provide complete information on markets, weather forecasts, economic forecasts, and other information that can be used to prepare budgets or to make marketing decisions.

The Internet. Many communication systems provide valuable information to agricultural marketers, but none compares with the amount and quality of information that is available through the computer **internet** system. This is a service to which customers may subscribe that connects them to the **information highway**, a world-wide computer network linking libraries, research data banks, government data banks, marketing reports, and electronic mail services to businesses and private individuals. Many other services are available to internet subscribers including consultation services with marketing experts for nearly every agricultural commodity.

The Fax Machine

A technology that has rapidly gained acceptance as a marketing tool is the **FAX machine**, Figure 10–4. This machine sends copies of documents, bills of

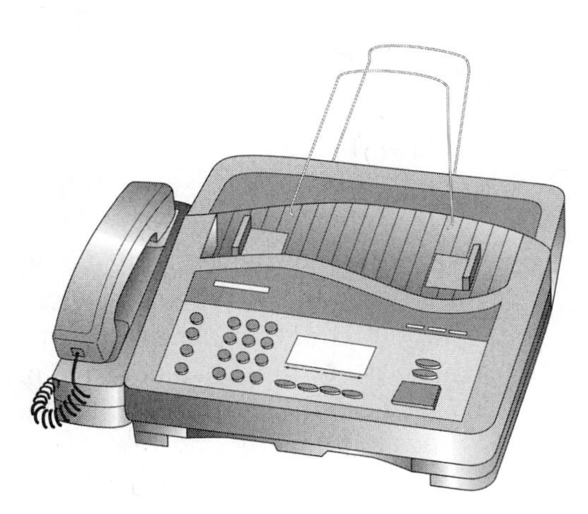

FIGURE 10–4 The FAX machine makes it possible to transmit documents and other printed material over a telephone line using electronic signals.

FIGURE 10–5 Satellite communications make it possible for agricultural marketers to have access to the latest market information whether they are located in population centers or on remote farms and ranches.

sale, contracts, and marketing information over telephone lines to any location that has a FAX machine installation. Printed materials can be supplied instantly in about the same amount of time it takes to make a telephone call. Computers are available with a "built-in" FAX and a modem to allow documents to be developed and FAXed from the computer.

Satellite Systems

Satellite communication systems are available that supply many kinds of information related to marketing of agricultural commodities, Figure 10–5. Managers are able to select the kind of information they need from a menu, and the information is viewed on a screen or printed out for later use. Agricultural producers pay a fee to gain access to the system, and the information on the system is updated frequently to reflect changes in markets.

ADVERTISING AND PROMOTION

Prices for many agricultural products are controlled by the industries that process the products, and not by the farmers who produce them. Advertising and promotion of these types of products by individual producers seldom results in higher prices for the products. Producers have learned, however, that advertising of consumer goods or processed commodities is often effective in creating higher demand for the raw product as the sales of processed commodities increase, Figure 10–6.

Massive advertising campaigns for consumer goods are financed through the combined funding sources of producers and processors of agricultural goods. **Commodity groups** have been organized in most areas to promote the merits of the products they produce. The members of these groups are the farmer/producers who raise the crops and livestock and the representatives of agricultural processors. The mission of these organizations is to protect the interests of members and promote their products.

FIGURE 10–6 Large amounts of money are invested each year in advertising products. Attractive charts, pictures, displays, TV ads, and sales brochures are only a few of the strategies that are used to attract customers to a product. This photo was part of an advertising piece for cheese. *(Photo courtesy Price Chopper Supermarkets)*

Commodity research and promotion boards have been organized by congress for the purpose of developing and promoting new agricultural products that are in tune with consumer needs and wants. These boards function under the direction of the **AMS** or the Agricultural Marketing Service of the United States Department of Agriculture. They are funded by growers and producers of agricultural commodities by a marketing assessment or checkoff. This fee is deducted from the checks of the growers each time a sale of the commodity is made.

States have organized commodity marketing commissions that promote the sales of products that are produced within each state. These commissions are mostly funded with checkoff assessments that are used for advertising and product promotion in domestic and international markets. Some state commissions, such as the Florida Citrus Commission, regulate the quality of the product that is sold by the industry, and they also regulate the companies and organizations that do business in the industry.

Technologies used by these groups to promote products often include radio, television, printed materials, logos, gimmicks, and signs. We are taught to believe that milk products that carry the quality checked logo or the real seal are better for us than those that do not, Figure 10–7. We have been conditioned to believe in the quality of California raisins, Idaho potatoes, Philadelphia cream cheese, and Texas pink grapefruit. Advertising and consistent high quality of the products stimulates sales at the consumer level, and eventually results in increased demand for the raw products at the farms.

The prices for some agricultural commodities are controlled by the producers. An example of farmers controlling prices is in the specialty food market. Some consumers are willing to pay premium prices for fruits and vegetables that have been raised without the use of chemicals. Because few farm-

FIGURE 10–7 Many marketing agencies use color, interesting sketches, and logos to help customers identify a product and associate it with high quality. *(Photo courtesy of Utah Agricultural Experiment Station)*

ers produce commodities in a chemical free environment, those who farm without synthetic chemicals are able to demand higher prices for their products. Advertising is important. The customer must be able to find the products. A good reputation will keep customers, but advertising is often needed to find customers.

Purebred livestock sold for breeding stock usually bring higher prices than they would when sold as slaughter animals. This is because the producer sets the price, and the buyer expects to benefit from the purchase by increasing the production of his livestock operation.

Advertising is important to the success of individual producers when they are able to set the price of the products they sell. Many livestock breeds sponsor livestock journals or magazines in which much of the advertising is done by farmers who have purebred animals for sale. Some of these farmers also send printed materials through the mail to potential customers, and some have begun to use video as an advertising medium.

COMMODITY SALES

Sales of agricultural commodities are achieved using a variety of approaches. A common marketing approach for some farm commodities is through **direct sales**. This marketing method occurs when a product is sold by the producer to the final consumer of the product, Figure 10–8. It is sometimes used for such products as fresh fruits, nuts, vegetables, eggs, animal feeds such as hay and corn, and other farm products that do not require processing before they are used by consumers. This sales strategy brings the buyer into direct contact and communication with the producer.

Farmer's markets and roadside stands are both forms of the direct sales approach to marketing. A **farmer's market** is usually established in an urban area. It is often set up in a warehouse or outdoor covered area, and it consists

FIGURE 10–8 Many enterprising farm families market their products by selling them directly to consumers at their roadside stands or through farmer's markets.

of sales stalls from which individual farmers sell their fresh produce to customers. This is an ancient method of conducting business, but modern transportation methods have greatly improved the process.

Direct sales through roadside stands is an effective way to market fresh produce if the location is on a heavily traveled road. Many successful roadside stands sell products that have been purchased for resale from other growers in addition to marketing produce from their own farms. This provides more variety in the choice of products available to customers, and usually results in an increased volume of sales for the homegrown products.

Another form of marketing through direct sales is the "pick your own" approach, Figure 10–9. Customers provide the labor required to harvest the crop in return for a lower price for the commodity. This method works best in areas located near urban population centers.

One of the most important marketing tools for agricultural commodities is a business organization known as a **cooperative**. This kind of business organization is owned by the farmers who use it. It allows them to pool their com-

FIGURE 10–9 "Pick your own" commodity sales offer the benefit of a lower price to the buyer and reduced harvest costs to the seller.

modities for marketing purposes so that they are competitive with large businesses in negotiating market prices. Examples of cooperatives that have been organized to market farm products include the United Dairymen, Cenex-Land O'Lakes, Blue Diamond, Sunkist, and Ocean Spray. Over 100 farm cooperatives representing over 350 brands of food products do business in the United States.

Many different agricultural commodities are processed and marketed by cooperatives. Agricultural cooperatives are involved in marketing 28 percent of all agricultural products as they enter the marketing system from the farms and ranches. Farmers use their cooperative businesses to conduct research, process raw commodities, manufacture goods, provide agricultural services, and to promote their brand names. Profits from farmer-owned cooperatives are distributed as dividends to the shareholders.

Innovative Marketing Approach: Partner/Family Leases

A modified approach to the "pick your own" marketing approach is to offer cash leases to a limited number of partners/families for the growing season. The partners are invited to participate in a planning activity to determine which crops will be produced each year. Partners buy the right to help with the work and to purchase products at a reduced cost. Partner families are invited to bring the children and participate in planting, harvesting, and even social activities during the growing season. The concept of ownership in the farm operation is the key to making this marketing approach work. Partner families are welcome at the farm at any time, and are made to feel that their participation in the farming activities is needed if the farm is to operate successfully.

Forward Contracting

Forward contracting is a marketing strategy that greatly reduces the risk of being forced to sell a commodity for a price at or below the cost of producing it. It is a signed contract to deliver a commodity of a specific quality grade for a specific price. This approach to marketing allows the producer to know the selling price of a commodity before it is ready for sale. When the seller knows the cost of production, and an acceptable profit margin can be built into the contract, the risk associated with farming is minimized.

Forward contracting is also beneficial to the buyer of the contract. The buyer is guaranteed delivery of a commodity of a specified quality at an acceptable price. This type of marketing arrangement makes it possible for agricultural processors to keep adequate amounts of raw agricultural products available to avoid interruption of the processing operation.

Futures Trading

Futures trading is a marketing technology in which contracts for agricultural commodities such as live cattle, hogs, wheat, corn, and some other products

CAREER OPTION

Market Analyst

A person who enters the profession of market analyst is responsible for gathering and analyzing data to determine market trends, customer preferences and buying habits, marketing methods of competitors, and local, regional, and national market conditions, Figure 10–10. He/she will need a professional degree from a college or university with emphasis on economics, statistics, and business.

FIGURE 10–10 A market analyst researches market trends and passes the information on to buyers and sellers of agricultural products.

are bought and sold for delivery at a future date. When the price of a futures contract is adequate to yield a profit on a commodity, the producer sells a contract on the commodity exchange through a broker. Such a contract may be met either by delivering the commodity or by buying a contract for the same future month at any time prior to the delivery date. The contract to sell is canceled by a contract to buy and no obligation exists to deliver the commodity.

It is usually more profitable for producers who hold futures contracts for delivery of a commodity to cancel their obligation to deliver the commodity by buying back the contract and selling the commodity in the cash market. Cash prices and futures prices usually rise or fall together. For example, when cash market prices go up, futures contracts usually go up too.

When the cash market and the futures market for a commodity goes down after the producer has hedged by selling a contract, a profit will be made on the futures contract when the producer buys a contract that is less expensive than the contract he or she sold, but the value of the commodity on the cash market will be less than was expected. When the cash market and the futures market goes up after the producer has hedged, a loss will be experienced on the futures contracts, but the profit from the commodity on the cash market will be higher than was expected. In both cases, the actual marketing profit that a producer earns by trading in futures and selling on the cash market will usually be very close to the marketing profit that was expected when the commodity was hedged.

FIGURE 10–11 Commodity exchanges make possible the use of futures trading as valuable marketing and management tools. *(Courtesy of Chicago Board of Trade)*

Futures trading is a good tool for locking in a favorable price if the producer actually owns or expects to own enough of the commodity to cover the contract, Figure 10–11. Trading in this market is pure speculation if the trader does not have the commodity to back up the contract.

Options Trading

Options trading is a form of marketing in which a producer buys an option giving him/her the right to purchase a futures contract for a commodity at a specific price by a specific future date. If a favorable futures contract is available during the option period, the producer may exercise the option to buy or sell.

Boards of Trade have been established in Chicago and Kansas City for several different agricultural commodities. These markets create competition for products, and provide new ways to buy and sell commodities. The Chicago Board of Trade began trading in futures contracts in the mid–1800s. A **futures contract** is a legal obligation to deliver or accept delivery at a specified location of a certain amount and quality of a commodity at a specific price during a particular future month. The practice of buying or selling a futures contract is called **hedging**. The difference in price between the local cash price of a commodity and the futures price on a given day is called the **basis**.

Futures contracts are useful to a seller who owns or plans to produce a commodity because the sale price is often established before the commodity is produced. It allows a producer who knows the cost of production to lock in a price at which a profit can be realized, Figure 10–12. Futures contracts are useful to processors and other buyers because they can lock in a price for raw products that allows them to make a profit on processed goods.

Video Merchandising

Video merchandising is an innovative approach to selling livestock that allows buyers in distant locations to view the sale offering without actually vis-

Futures Transactions

When Prices Go Down:

10,000 bu. of wheat

Total production cost @ $2.10/bu	Date	Cash	Futures	Basis
Sale of December futures contract @ $2.20/bu	August	Expected hedge price in Dec. @ $2.10/bu	Sell 1 Dec. Wheat contract (10,000 bu) @ $2.20/bu	Expected –$0.10
Sale of wheat on cash market @ $2.05/bu	December	Cash sale price @ $2.05/bu	Buy 1 Dec. Wheat contract @ $2.12/bu	Actual –$0.07
Purchase December futures contract @ $2.12/bu			Gain +$0.08/bu	Improvement +$0.03
NET EFFECT OF FALLING PRICES:		Cash Sale ($2.05) +	Futures Profit ($0.08) =	Actual Hedge Price ($2.13/bu)

When Prices Go Up:

10,000 bu. of wheat

Total production cost @ $2.10/bu	Date	Cash	Futures	Basis
Sale of December futures contract @ $2.20/bu	August	Expected hedge price in Dec. @ $2.10/bu	Sell 1 Dec. Wheat contract (10,000 bu) @ $2.20/bu	Expected –$0.10
Sale of wheat on cash market @ $2.15/bu	December	Cash sale price @ $2.15/bu	Buy 1 Dec. Wheat contract @ $2.25/bu	Actual –$0.10
Purchase of futures contract @ $2.25/bu			Loss –$0.05/bu	No change +$0.00
NET EFFECT OF RISING PRICES:		Cash Sale ($2.15) –	Futures Loss ($0.05) =	Actual Hedge Price ($2.10/bu)

FIGURE 10–12 A sample futures trading transaction

iting the farm or ranch. A videotape is prepared that shows the live animals moving about a pen much as they would be seen if the buyer was inspecting them on site. This method has worked well as a method of marketing feeder cattle and other classes of livestock.

Telemarketing

Telemarketing is a type of auction sale conducted over the telephone. Animals or commodities that are offered for sale are weighed, graded for quality, and sorted into uniform groups prior to the sale. Potential buyers are assembled on a telephone conference call prior to the sale, and the sale offerings are described to the buyers. The offerings are then sold by auction over the telephone to the highest bidder for immediate delivery. This marketing strategy has proved to be particularly effective in the sale of market lambs and other slaughter animals. Video and telemarketing strategies are sometimes combined to improve the marketing process for agricultural commodities such as beef.

PRODUCT DISTRIBUTION

Every form of marketing includes the transportation of products from the farms and ranches to the sale delivery point, Figure 10–13. Even in the case of direct sales at the farm, most products must be hauled to central locations before they are purchased by customers. Sometimes products are sold with the understanding that the buyer will be responsible for transporting the products, but most farm products must be delivered to a central point by the producer.

Trucks

The use of trucks is the most common method of transportation for farm products, Figures 10–14, 10–15, and 10–16. Even when other forms of transportation are used, trucks are usually needed to deliver commodities to shipping terminals. The state and interstate highway system is a critical part of shipping, and the modern trucking industry is an efficient and effective form of transportation, Figure 10–17.

FIGURE 10–13 Getting the product to the customer has always been a problem. Trade has improved considerably since the days when camel caravans and pack horses were the chief means of transporting products. *(Photo courtesy of Utah Agricultural Experiment Station)*

FIGURE 10–14 Large trucks haul grain from the grain growing areas to population centers and export terminals.

FIGURE 10–15 Modern refrigerated trucks make it possible to transport perishable products to nearly any location. *(Photo courtesy of Utah Agricultural Experiment Station)*

FIGURE 10–16 Modern livestock trucks have brought major improvements to the task of getting animals to markets.

FIGURE 10–17 Modern roads play major roles in transporting raw products to markets and consumer-ready products to population centers.

Railroad

Large amounts of agricultural commodities are transported to markets each day by the railroad system. It is an excellent way to move many products because huge amounts of a product can be easily carried between major population centers. Modern rail cars are available with refrigeration systems to keep products cool or frozen during shipment. Railroads play a vital role in delivering agricultural products to processing and distribution points, Figures 10–18 and 10–19.

FIGURE 10–18 Railroads played a big part in developing the agricultural sector of our economy. They provided a much-needed transportation link between rural and urban areas.

FIGURE 10–19 The coming of the railroad to most areas made it possible to raise animals and other agricultural products in remote areas and to deliver them to populated areas for consumption. *(Photo courtesy of Utah Agricultural Experiment Station)*

Barges and Ships

Barges and ships are used to move many products to export markets in foreign countries. This is because huge amounts of these products can be transported in this manner at a minimal cost. Grain is a product that is usually shipped internationally by ship or barge, Figures 10–20 and 10–21. Other types of products are often loaded into large metal containers for shipping. The shipping containers that are used most often have been designed to stack easily for transporting by truck, railroad, or ship.

FIGURE 10–20 Large amounts of the domestic grain crop are shipped to international markets through huge grain terminals located at shipping ports.

FIGURE 10–21 Ships and barges are capable of carrying huge cargos of grain to nations that import grain to feed to their people and animals.

FIGURE 10–22 Modern airplanes provide transportation to nearly every area of the world. They are used to transport food and other supplies to people who require immediate disaster relief. Aircraft are even fitted to carry live farm animals to distant parts of the world. *(Photo courtesy of Utah Agricultural Experiment Station)*

Planes

Some agricultural products are transported by air to **domestic markets** located in the United States, Figure 10–22. The fresh flower industry frequently uses air transportation to speed its products to customers while the flowers are still fresh. Live baby chickens are shipped regularly by air from hatcheries to distant farms. Puppies are shipped nearly every day from dog kennels to pet stores in many parts of the country. Transport planes are also used to ship live animals to **international markets** located outside the boundaries of the United States.

The highly advanced transportation technology that is available for shipment of agricultural goods is responsible in part for the availability of agricultural products in all areas of the United States and in many parts of the world.

CHAPTER SUMMARY

Marketing includes all of the business activities that are required to move products from the producer to the consumer. It begins with the preparation of an enterprise budget, which estimates the cost of production, and it is completed when the product is purchased by the consumer.

Modern communication technologies such as radio, television, printed materials, telephones, teletype, and FAX machines have greatly improved the marketing process. Computers and computer networks are important marketing tools. Satellite communication systems are also used to deliver market information.

Advertising has become an important marketing strategy. Commodity groups have been able to increase demand for the products they produce by sponsoring massive advertising campaigns. Advertising plays an important role in setting the price for farm commodities, such as organic foods or purebred livestock, that are in short supply or that appeal to specific groups of consumers.

Several important selling technologies are available for use in marketing farm products. They include direct sales approaches, trading in options and futures contracts, forward contracting, telemarketing, and video-marketing.

Transportation systems such as the railroads and highways provide an efficient ground transportation system. The airlines, barges, and ships provide alternative forms of transportation for both domestic and international markets.

CHAPTER REVIEW

Discussion and Essay Questions

1. Explain the importance of an enterprise budget in developing marketing plans for agricultural products.
2. How have books, magazines, and trade letters affected agricultural marketing?
3. Why are telephones and related technologies such as teletype and FAX machines important in marketing agricultural products?
4. Describe how radio and television technologies are used to market agricultural products.
5. How effective are radio and television advertising as marketing tools?
6. How important are computers and related technologies to the marketing process?
7. Identify ways in which satellite communications are used or may be used in the marketing process.
8. Distinguish the differences between forward contracting, trading in options, and trading in futures contracts.
9. Explain how the quality of the transportation system affects the agricultural marketing system.

Multiple Choice Questions

1. An enterprise budget is a document that is:
   A. prepared before a commodity is produced to determine whether it has profit potential.
   B. developed during the production of a commodity to assure accuracy.
   C. used to establish legal ownership of a commodity.
   D. used to report taxable income.

2. Which of the following conditions is likely to restrict a market?
   A. a consistent supply of a product.
   B. the ability to communicate with potential customers.
   C. the inability of customers to pay for a product.
   D. the ability to transport the product to the buyer.

3. The availability of a commodity in a free market is most influenced by:
   A. profitability of the commodity.
   B. how expensive the commodity is.
   C. how cheap the commodity is.
   D. usefulness of a commodity.

4. A modern marketing technology that is used to transfer contracts and documents to distant places is the:
   A. teletype.
   B. television.
   C. cellular phone.
   D. FAX machine.

5. The information highway is:
   A. a shipping route to major markets.
   B. a teletype system for reporting commodity prices.
   C. an interstate highway with lots of road signs.
   D. a world-wide computer network that is used to access information from many sources.

6. An agricultural organization that is organized to promote and advertise a commodity and protect the interests of its members/producers is called a(n):
   A. cooperative.
   B. commodity group.
   C. internet.
   D. Agricultural Marketing Service.

7. A commodity research and promotion board is funded by:
   A. assessments or checkoff fees.
   B. tax revenues.
   C. government grants.
   D. gifts and donations.

8. A farmers' cooperative is a business that:
   A. conducts research.
   B. processes raw farm products.
   C. distributes its profits to its farmers/shareholders.
   D. performs all of these functions.

9. Options trading is a form of marketing in which an option is purchased that allows its owner to:
   A. collect the full amount of a commodity sale before the commodity is produced.
   B. exercise the right to purchase a favorable futures contract for a specific price by a future date.
   C. exchange a commodity for a different commodity that is marketed by the same board of trade.
   D. exercise the option of selling a commodity to the government at a price that is set by the producer.

10. All of the following are forms of product distribution with the exception of:
   A. air transport.
   B. FAX technology.
   C. rail cars.
   D. trucks.

LEARNING ACTIVITIES

1. Simulate an experience with the commodity exchange by allocating 10,000 bushels of corn and 40 head of beef cattle to be marketed in 50 days as 1000 lbs./animal. Chart the options and futures markets along with the cash market, allowing a short period of time each day for making decisions to buy or sell. Provide an award to the student who does the best job of managing his/her marketing program. Be sure to emphasize that unless ownership of the commodity exists, trading on the commodity exchange is pure speculation instead of marketing.

2. Assign class members to gather examples of marketing activities for agricultural goods. Favorite promotions should be illustrated or summarized in a short paragraph. Decorate a bulletin board using the materials contributed by the class.

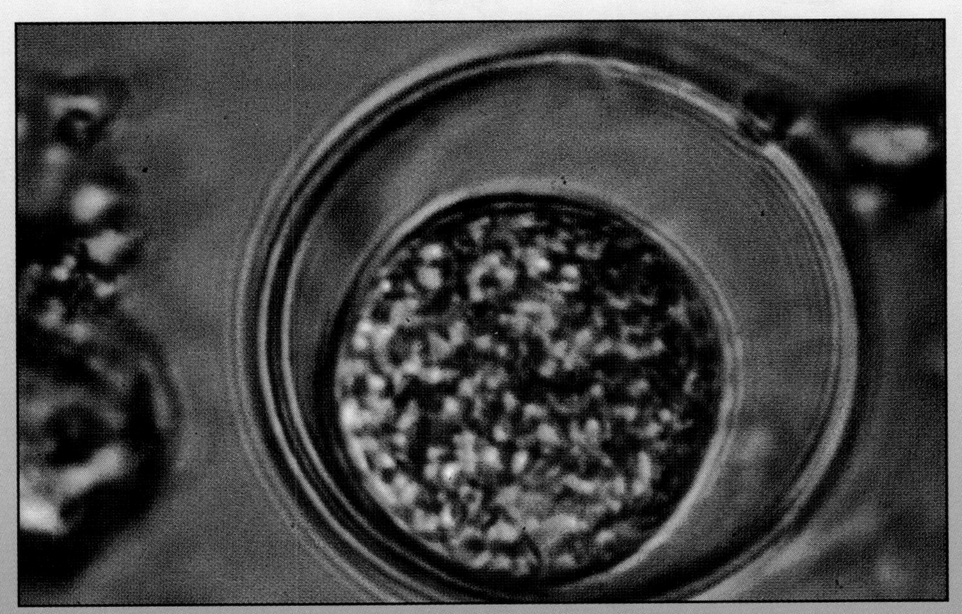

Life begins as a single cell, and the environment in which it exists determines whether it will continue to live and develop or die and be lost forever. (Photo courtesy of Utah Agricultural Experiment Station)

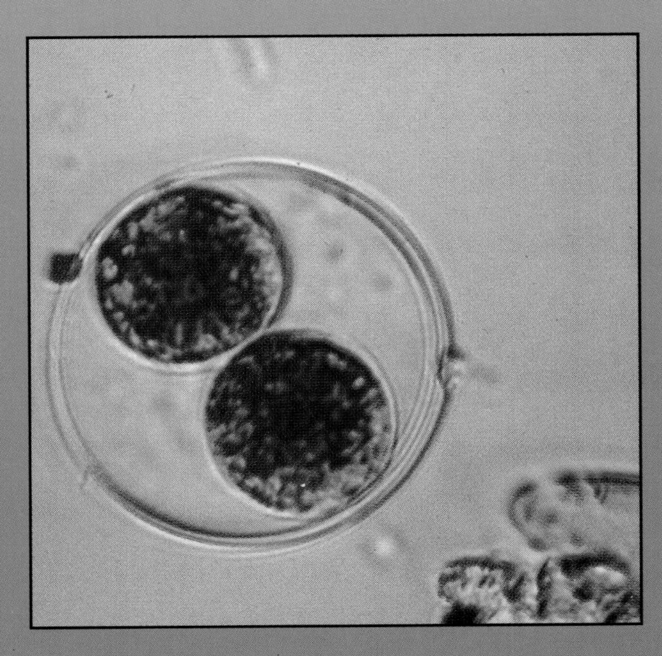

A living organism begins as a single cell which divides repeatedly until an entire organism is formed. (Photo courtesy of Utah Agricultural Experiment Station)

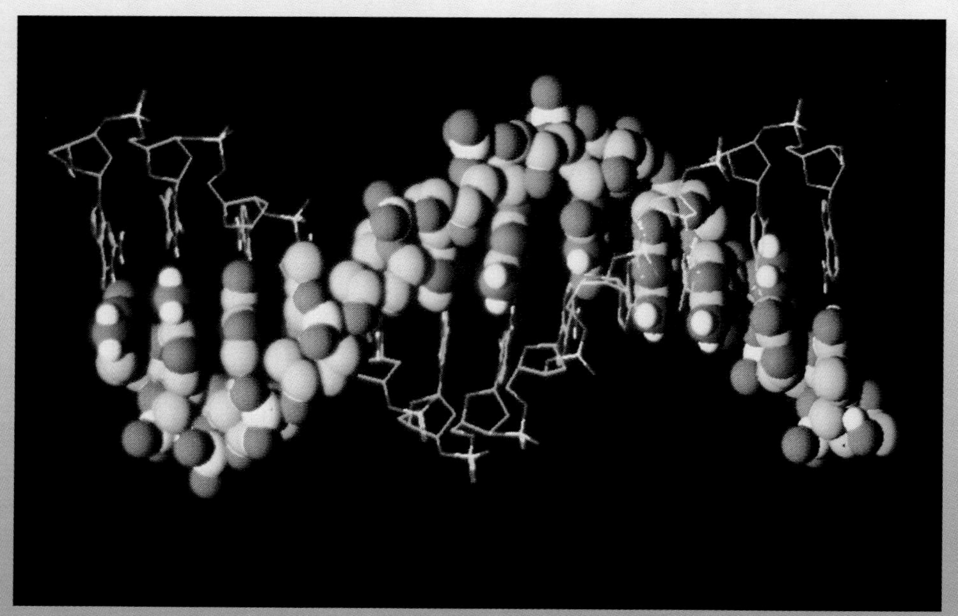

The chromosome contains the code of life for every living organism. (Photo courtesy of Utah Agricultural Experiment Station)

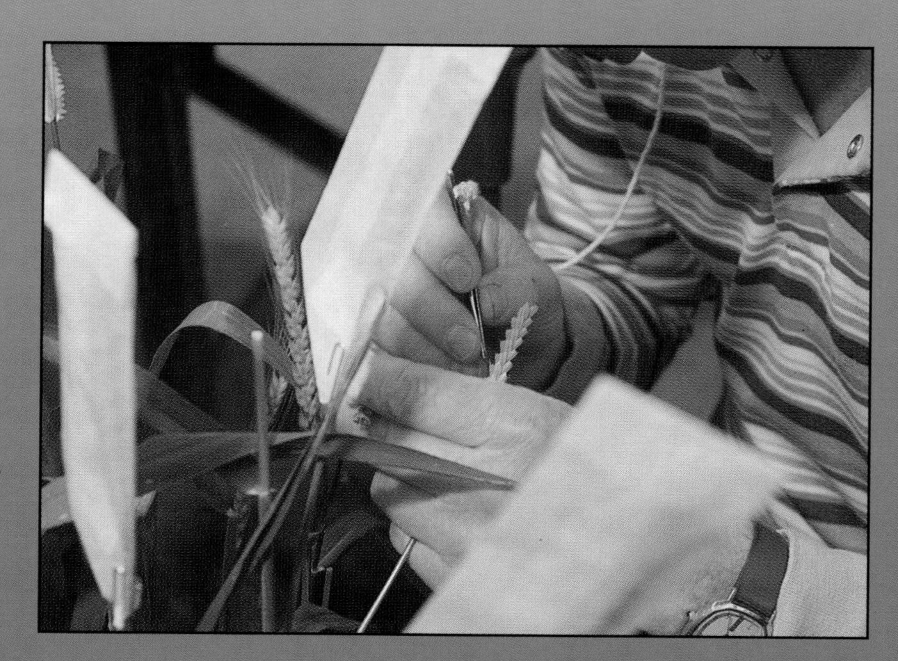

Many new plant varieties have been produced by controlling pollination. (Photo courtesy of Utah Agricultural Experiment Station)

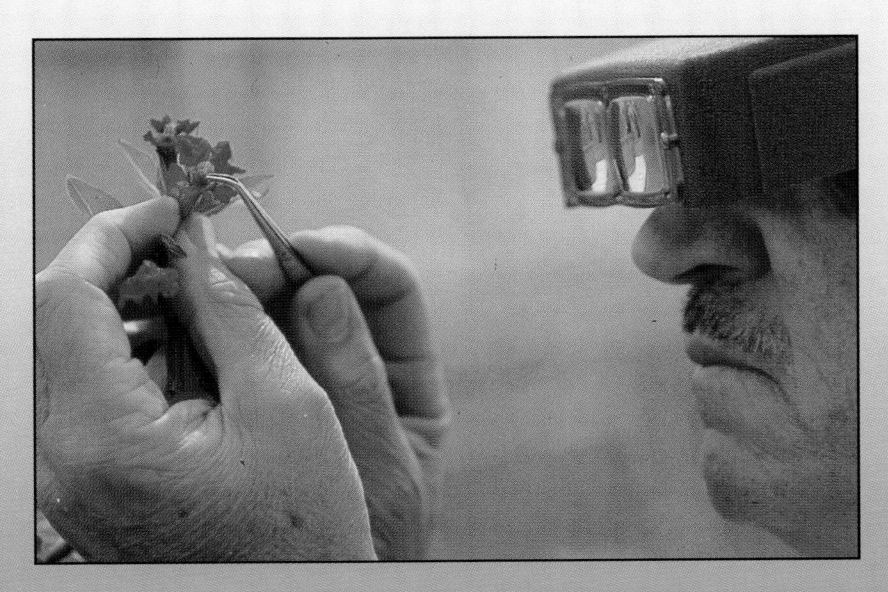

A research geneticist works with living organisms to discover ways to understand and manipulate the genetic code. (Photo courtesy of Utah Agricultural Experiment Station)

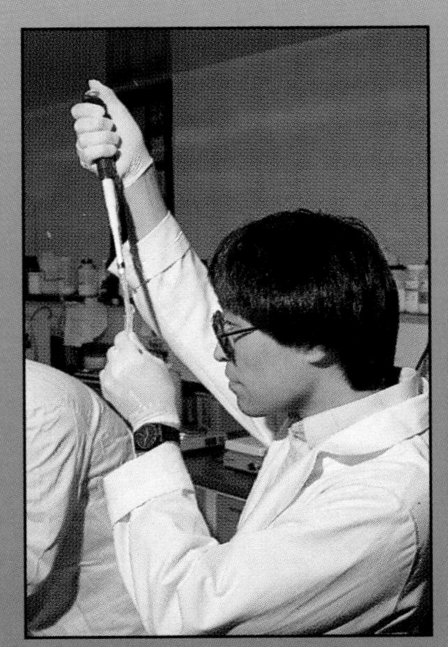

Science has done much in our lifetimes to modify environments, and to modify living organisms to be compatible with changing environments. (Photo courtesy of Utah Agricultural Experiment Station)

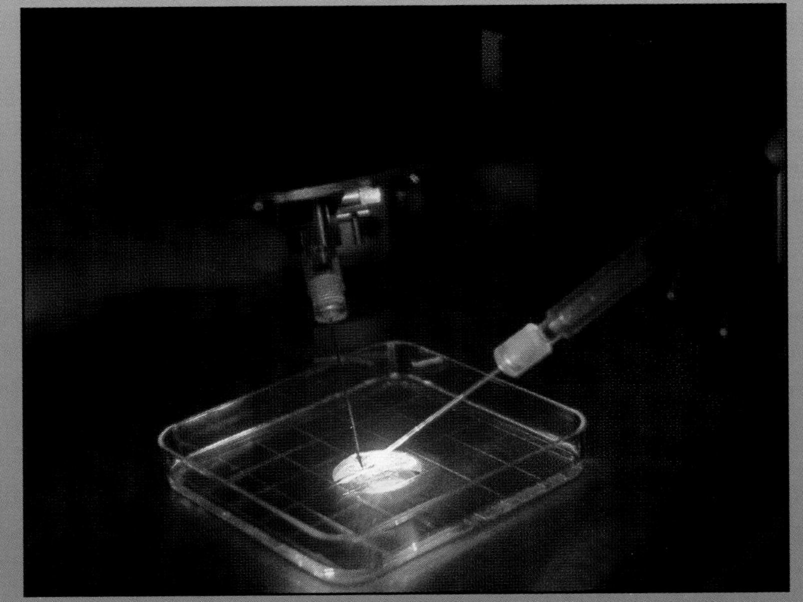

Modern science is exploring the secrets of life in the new science called biotechnology. Embryos are divided to form two or more identical individuals where only one existed before. (Photo courtesy of Utah Agricultural Experiment Station)

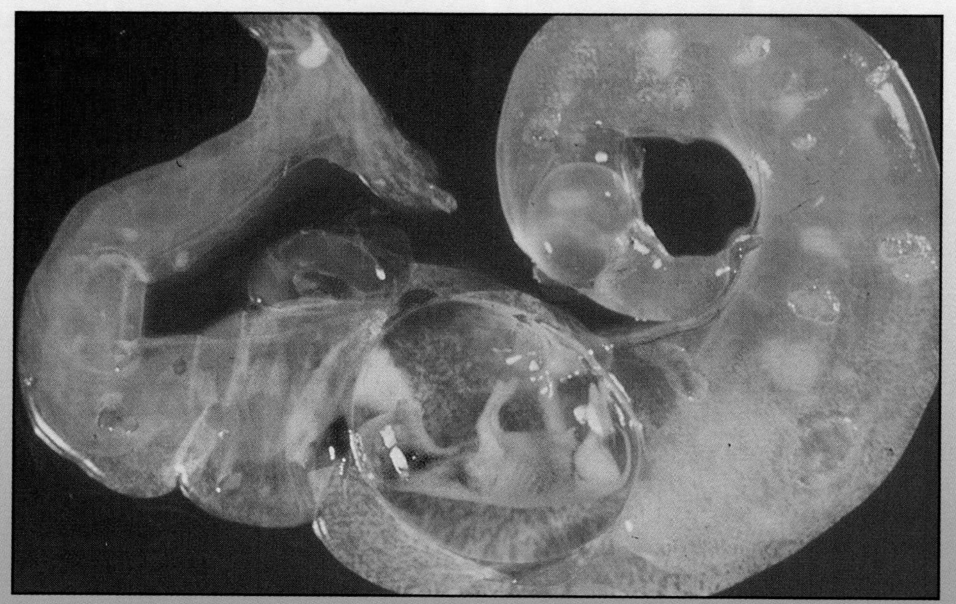

Reproductive management includes all of the events associated with producing and nurturing a living fetus. (Photo courtesy of Utah Agricultural Experiment Station)

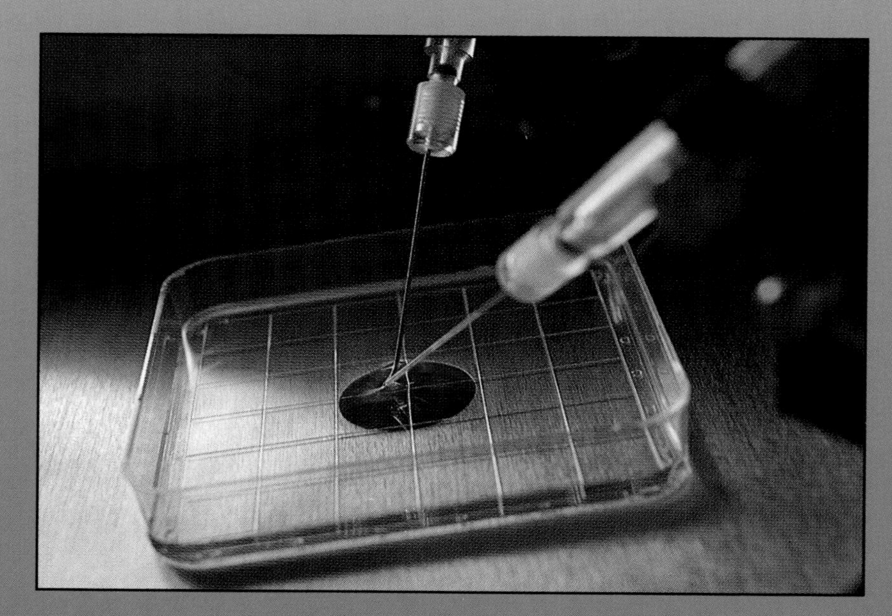

Animals obtained from splitting embryos are identical in the traits which they inherit from their parents.

Hybrid plant seeds are produced by removing male flower parts to prevent self fertilization. (Photo courtesy of Utah Agricultural Experiment Station)

Special hormones are applied to callus tissue to stimulate the growth of leaves and stems. (Photo courtesy of Utah Agricultural Experiment Station)

Once plant cells have developed into tiny plants, they are placed in a special growth medium. (Photo courtesy of R. Zemetra, University of Idaho)

The nature of soils and the effects of erosion continue to be important topics of agricultural research. (Photo courtesy of Utah Agricultural Experiment Station)

Protection of our water supply from chemical fertilizers and pesticides is an important issue which will be researched thoroughly in the next few years. (Photo courtesy of Utah Agricultural Experiment Station)

Conservation tillage practices are likely to increase as efforts accelerate to clean up and protect our environment. (Photo courtesy of Utah Agricultural Experiment Station)

Improvement of our public and private rangelands is becoming an important issue as environmentalists attempt to discredit agriculturists as conservationists. (Photo courtesy of Utah Agricultural Experiment Station)

Confinement housing for animals must provide adequate ventilation and light. Artificial lighting helps to increase egg production of laying hens. (Photo courtesy of Utah Agricultural Experiment Station)

A space-age growth chamber for calves protects them from harsh environmental conditions, provides a supply of fresh air and protects calves from exposure to diseases by isolating them from each other. (Photo courtesy of Utah Agricultural Experiment Station)

Multiple use as a management tool on public lands will become a hotly contested issue as grazing rights are reduced and wilderness issues surface in a society that is no longer agriculturally oriented. (Photo courtesy of Utah Agricultural Experiment Station)

With undesirable plant species removed from the range, productive grasses are seeded. Rangelands which are improved in this manner produce increased yields of forage for both domestic and wild animals. (Photo courtesy of Utah Agricultural Experiment Station)

Rape is a crop which shows potential to supplement or replace petroleum as a source of fuel for engines. The seed is rich in oil, and may prove to be a renewable source of fuel.

The Whooping Crane is an endangered species. Conservation officers have attempted to establish new breeding colonies of Whooping Cranes by placing their eggs in the nests of Sandhill Cranes. (Photo courtesy of Utah Agricultural Experiment Station)

Many species of living creatures have become extinct or endangered when humans or natural disasters have destroyed their habitats. The Trumpeter swan is threatened by loss of wetlands. (Photo courtesy of Utah Agricultural Experiment Station)

The Wolf and the Grizzly Bear are large predatory animals which are protected in some areas of the world. Some conservationists would like to establish populations of both species in wilderness areas. Many sportsmen and livestock ranchers who use these areas are opposed to the plan. (Photo courtesy of Utah Agricultural Experiment Station)

Modern technologies for producing fruit crops include this large wind machine which mixes warm air from upper layers with freezing air on the ground. It is possible to prevent frost damage to fruit blossoms using this machine.

Timber harvests are an important part of the multiple use concept for forest management. Well managed forests can support wildlife, livestock, logging, mining, and recreational activities.

Many specialty crops are raised which require unusual cultural practices. One of these is "hops". Many acres of land are covered with poles, cables and string to accommodate the needs of this crop.

High quality forage can be harvested whether it rains or shines by chopping forage crops in the high moisture stage and sealing them in plastic bags. The feed retains nutrients which may have been lost if the crop had been harvested as dry hay.

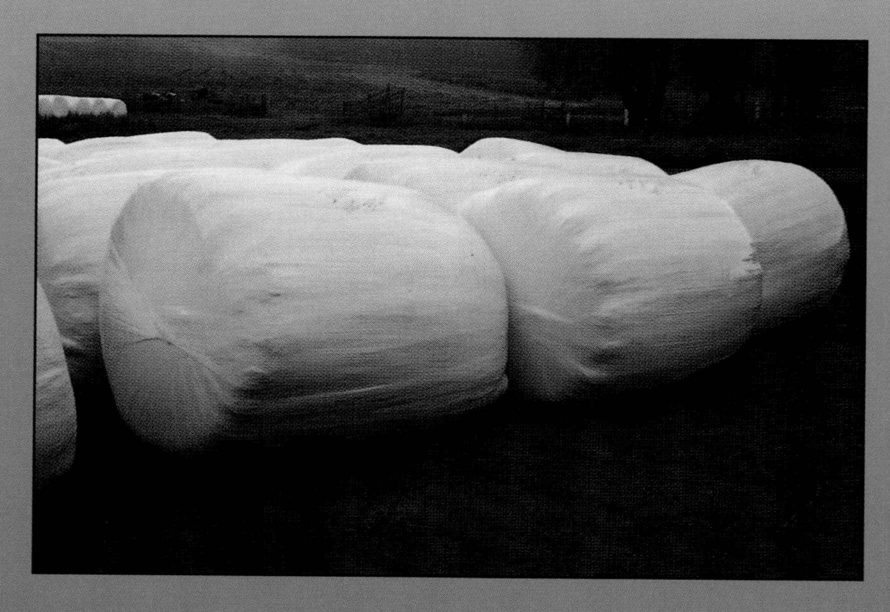

Strip farming is a practice which helps to reduce and eliminate soil losses. Conservation of soil is of high priority to serious agriculturists.

It is possible to harvest our rich forest resources and still maintain a healthy forest environment. Removal of mature trees allows sunlight to penetrate to the forest floor where new vegetative growth is stimulated. In a short time young trees begin to grow beside the stumps that mark the site of the harvest.

The mule, a hybrid cross between a donkey and a horse was an important source of power for performing farm work for many years. The mule was hardy and strong and could outwork other draft animals.

Mechanical power sources such as these early model tractors gradually gained popularity and eventually replaced draft animals as sources of agricultural power.

Modern farm tractors make it possible to perform farm work in complete comfort and with a high degree of efficiency. This modern miracle of technology has made it possible for agricultural producers to perform field work in a timely manner assuring that an abundant and high quality food supply is available to our people.

A healthy watershed is important in maintaining dependable flows of high quality water in streams and rivers.

One of the greatest tragedies of modern agriculture is erosion of the soil. Steps must be taken to reduce soil losses, to avoid serious damage to this important natural resource.

Well managed forest lands which are harvested in a timely manner will provide wood products and healthy watersheds from which crystal clear water will be available to future generations.

Science and technology have opened up a new era in agriculture . . . the dawn of a new day of progress. (Photo courtesy of Utah Agricultural Experiment Station)

SECTION IV

Energy and Power Technology

- **Power for Production Agriculture**
- **Electrical Energy in Agricultural Uses**
- **Alternative Energy Sources for Agriculture**

CHAPTER

Power for Production Agriculture

***Power** is force or energy that is controlled to accomplish work. For thousands of years the main source of power for performing agricultural tasks was physical labor from humans and animals. The development of new power technologies has taken the harnesses off the teams of horses, mules, and oxen and put them on new sources of energy to produce power for farming.*

OBJECTIVES

After completing this chapter, you should be able to:

- account for the shift from energy sources produced on the farm to fossil fuel products for production of agricultural goods.
- assess the importance of steam engines in the development of power sources for agricultural production.
- describe the types of petroleum fuels and define their importance to the agricultural industry.
- forecast the role of ethanol fuels in providing energy for agricultural production.
- evaluate the contribution of the internal combustion engine to production agriculture.
- explain the operation of four-cycle and two-cycle engines.
- compare efficiency ratings for four-cycle diesel engines, four-cycle gasoline engines, and two-cycle gasoline engines.
- define and give examples of mechanical power.
- identify similarities and contrast differences between hydraulic and pneumatic power.
- relate the use of force converter devices to power applications.

TERMS FOR UNDERSTANDING

The following vocabulary terms should be studied carefully as you read:

- power
- steam engine
- petroleum
- ethanol
- biodiesel
- torque
- axis
- diesel engine
- diesel
- gasoline engine
- gasoline
- compression ratio
- four-cycle engine
- two-cycle engine
- fulcrum
- hydraulic power
- hydraulic ram
- pneumatic power
- force converter

ENERGY SOURCES

Draft Animals

Prior to the days of modern tractors, thousands of horses, mules, and oxen were required to produce agricultural crops. Animals that were used to pull implements were called draft animals, Figure 11–1. Their energy source was hay and grain. Much of the farm land was devoted to production of crops to be fed to the teams of draft animals.

Steam Engine

One of the problems associated with the development of the first tractor was designing a machine that harnessed an efficient form of energy to produce power. The **steam engine** was one of the earliest power sources adapted to agricultural uses, Figure 11–2. The tractors were large and difficult to maneuver in the fields, and large amounts of wood or coal were required to produce steam,

FIGURE 11–1 Draft animals have been used to provide power for farming for many centuries. Since the beginning of the 20th century there has been an accelerating change to mechanical power instead of horse and mule power.

FIGURE 11–2 Steam power was one of the first energy sources used to replace the physical labor of humans and animals. (*Photo courtesy of Utah Agricultural Experiment Station*)

Figures 11–3 and 11–4. Animal teams were more mobile than steam tractors, and they did not require a constant source of on-board fuel to maintain the steam supply. Steam powered tractors never did gain much popularity.

The first mechanical tractor was built in 1769 by Nicolas-Joseph Cugnot. It was powered by a steam engine, and it traveled at about two miles per hour. A heavy load of wood and water was required to produce the steam that powered this huge machine. When the tractor was fully loaded with fuel, it was very heavy. Power steering had not been invented yet, so it was also very difficult to steer. A large area was required to turn the tractor around, because it did not have a sharp turning radius. The steering problem finally resulted in the destruction of the machine when it crashed into a wall.

FIGURE 11–3 Steam tractors were developed as power sources for agriculture as early as 1769.

FIGURE 11–4 Although steam tractors were manufactured and sold for many years, they never were widely used.

Fuel

Petroleum products such as gasoline and diesel fuels have become the main sources of fuel for farm tractors. Other petroleum and natural gas products such as distillates and propane have also been used successfully.

Crude oil is refined in large amounts by heating it. This causes the different petroleum products such as gasoline, diesel, and oils to evaporate from the crude oil. Each product evaporates at a different temperature, and the products are recovered in separate storage areas.

A new fuel source that is being used successfully is **ethanol**, Figure 11–5. It is an alcohol that is produced by fermenting grain and other farm products rich in carbohydrates. The liquid is distilled to recover the alcohol produced by this process, Figure 11–6. One advantage for ethanol fuel is that it is a renewable fuel that can be produced from grain and other products that are raised on farms.

Several new oil crops are being studied to determine whether they can be developed as alternate sources of fuel for internal combustion engines. One of these crops is canola. Oil from the seeds of this plant is now used as cooking oil, but agricultural engineers have developed a product from this oil that is called **biodiesel**. It is very similar to the diesel fuel that is obtained by distilling crude oil. Biodiesel has been produced from several different vegetable oils, and it is a renewable fuel.

FIGURE 11–5 Ethanol fuel is a clean burning fuel that is obtained from plant materials such as corn, other grains, or other materials high in carbohydrates.

FIGURE 11–6 Ethanol is produced by fermentation from carbohydrates in plant materials. This ethanol processing facility produces ethanol from potato waste that is obtained from a nearby potato processor.

INTERNAL COMBUSTION ENGINES

Heat as a Power Source

Heat can be used as a source of power because of the way it affects most substances. Gases, liquids, and solids usually expand or become larger when they are heated. They contract or decrease in size when they are cooled. As a substance is heated, its molecules move farther apart, causing it to expand in size or volume. Expansion and contraction is much greater in liquids and gases than in solids.

When expansion occurs in a closed system, pressure is created. This pressure can be converted to mechanical power by the crankshaft of an engine, or to electricity by a generator. It is the expansion of gases in the cylinder of an engine that drives the cylinder downward when combustion gases are heated by the explosion of the vaporized fuel.

An internal combustion engine burns fuel internally and uses the energy obtained from the expansion of the combustion gases to provide power, Figures 11–7 and 11–8. The first internal combustion engine was developed in 1876 by Nikolaus August Otto, a German inventor. The engine generated power by burning gasoline in a closed chamber. As the heated gases expanded, they exerted pressure on the piston, causing it to move downward. The vertical movement of the piston was converted to rotary motion by a crankshaft,

FIGURE 11–7 The four-cycle gasoline engine is used today to perform a large number of tasks.

FIGURE 11–8 The four-cycle engine is used extensively for lawn mowers and garden tillers. It is a very reliable source of power. *(Used with permission of the Toro Company. "Toro" is a registered trademark of the Toro Company, Minneapolis, Minnesota.)*

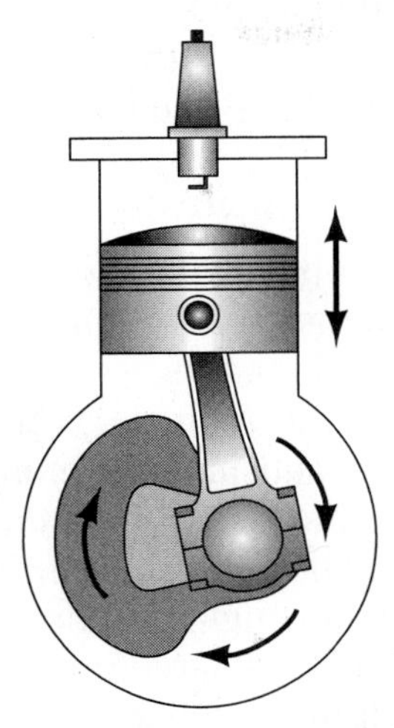

FIGURE 11–9 The vertical up-and-down motion of the engine piston is converted to rotary motion by a crankshaft making it possible to turn wheels using mechanical power.

Figure 11–9. Gasoline engines in use today still have basic components that are similar to those used in early engines.

Torque

Torque is a measurement of the turning power that is exerted on or by an object. The amount of torque that is exerted is determined by the force that is applied to an object to cause it to rotate. Rotation occurs around an imaginary line known as the **axis**. Torque is also affected by the distance from the axis that the force is applied. Torque is increased by lengthening the distance from the axis at which force is applied or by increasing the amount of force. A long-handled wrench applies more torque to a bolt than a short-handled wrench, and a large gear applies more torque to a shaft than a small gear.

The laws of physics apply to all of the mechanical devices that are used on agricultural machines. For example, one of the laws of circular motion states that the speed at which an object rotates is determined by the amount of force that is applied to the object and the distance from the axis that the force is applied.

A different type of internal combustion engine known as the **diesel engine** was developed by Rudolf Diesel in 1893. Both the engine and the fuel oil called **diesel** that it burned were named after Mr. Diesel. The design of the engine was similar to the gasoline engine except that it had no spark plugs and the fuel was ignited by injecting it into the hot compressed air inside the combustion chamber of the engine.

Gasoline and diesel powered tractors were smaller in size than steam tractors, and the fuel supply did not take up very much space in comparison with the wood and coal that provided fuel for steam engines. These tractors were much easier to control than the large steam tractors had been. They gradually replaced horses and mules during the years before World War II.

Gasoline Engines

The development of the **gasoline engine** was one of the most important events in the history of agriculture. It used a petroleum product called **gasoline** for fuel, and its acceptance led to major changes in farming practices. It replaced draft animals as the major source of power for farm machinery. As the horse and mule populations were reduced, fewer acres of land were needed to produce feed for draft animals. Much of this land was planted to crops that could be sold to generate farm income.

Gasoline engines of several types and sizes are used on modern farms to perform many different tasks. They provide power for tractors, harvesters, and other self-propelled machines, Figures 11–10 and 11–11. They are also used in farm trucks, airplanes, and cars. They provide power for irrigation pumps, hay elevators, electrical generators, and grain augers. In fact, gasoline engines can be found performing nearly any task that requires a power source, Figures 11–12 and 11–13.

Diesel Engines

Diesel engines are usually used for jobs that require a lot of power or constant use. Most large farm tractors, trains, and many large trucks are powered by diesel engines, Figure 11–14. This is because diesel engines are more efficient

FIGURE 11–10 Gasoline engines have gained wide acceptance as power sources for tractors.

FIGURE 11–11 Gasoline engines also power many other agricultural machines.

FIGURE 11–12 Gasoline engines are used to perform a variety of tasks such as harvesting turf grass.

FIGURE 11–13 Many small gasoline engines are used on farms to provide the power required to move irrigation pipes across the field.

FIGURE 11–14 Large tractors such as this one are usually powered by diesel engines because they operate at lower engine speeds and last longer than gasoline engines.

than gasoline engines, and they operate at lower speeds. Diesel engines also offer the advantage of burning fuel oils, such as diesel, that are usually less expensive than gasoline.

Diesel engines are constructed of heavier materials than gasoline engines because they operate with higher **compression ratios**. This means that the volume of gases inside the cylinder is compressed more, and the pressure inside the combustion chamber when the fuel mixture is ignited is higher than it is in a gasoline engine.

Diesel engines usually last longer than gasoline engines. This is due in part to lower operating speeds and more precise engine construction methods. Engines in semi-trucks frequently provide service for a million highway miles or more.

Four-cycle Engines

Engines are classified according to the way they operate. Regardless of size or the type of fuel that is used, an engine described as a **four-cycle engine** will always operate through the same sequence of events. Each piston will operate through four strokes, and four different events will occur inside each cylinder during each complete cycle. Engines of this type are also classified as Otto-cycle engines.

A stroke is defined as the movement of the piston from the top to the bottom or from the bottom to the top of the cylinder. An engine cycle begins with the piston at the top of the cylinder, Figure 11–15. As the piston begins to move downward, the intake valve opens. This allows a mixture of air and gasoline to be drawn into the cylinder. When the piston reaches the bottom of the cylinder, the intake valve closes and seals the gases inside the combustion chamber. As the piston moves back to the top of the cylinder, the fuel mixture is compressed into a much smaller space than it originally occupied.

The fuel mixture is ignited when the piston reaches the top of its stroke. The heated gases expand to provide a power stroke as the piston moves again to the bottom of the cylinder. As the piston starts to move upward in the

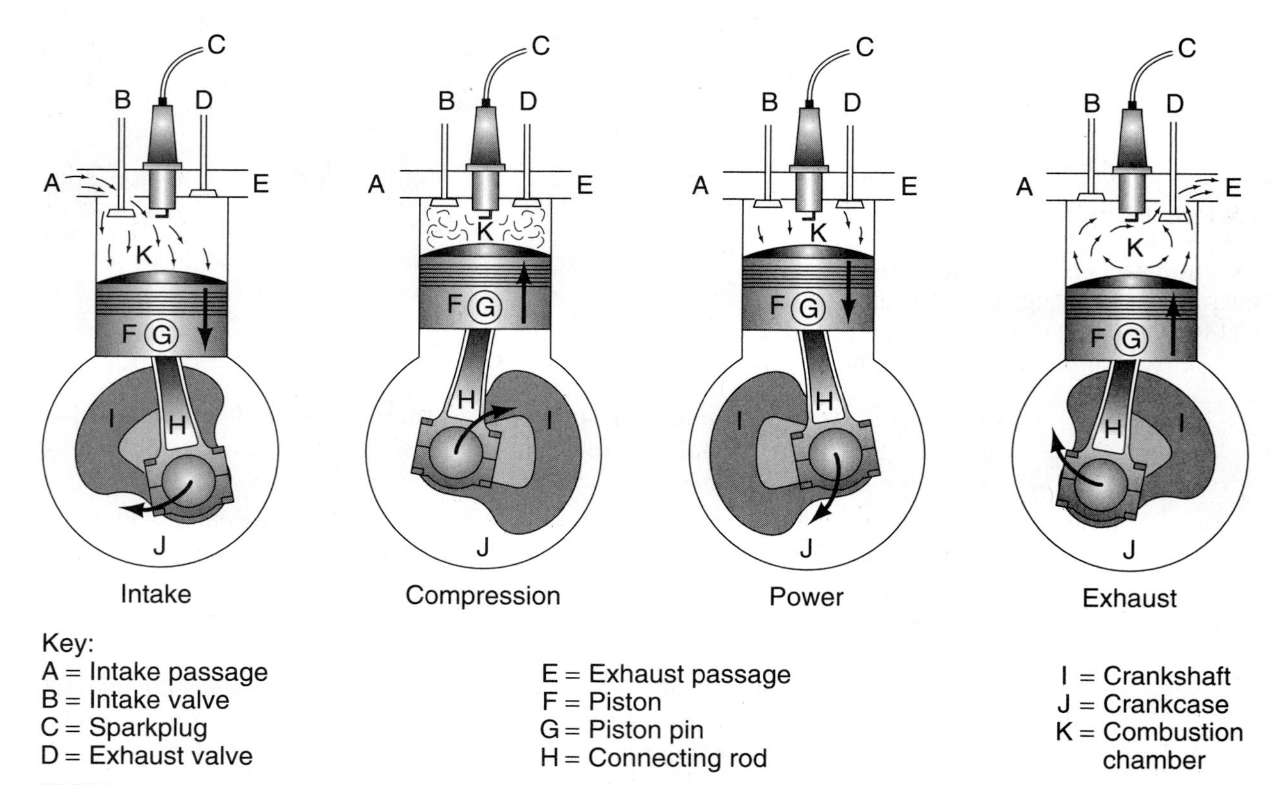

FIGURE 11–15 Operation of a four-stroke-cycle engine *(Adapted from Small Gasoline Engines Student Handbook, The Pennsylvania State University, Department of Agricultural and Extension Education)*

cylinder, the exhaust valve is opened and the piston drives the burned exhaust gases out of the combustion chamber. The cycle has been completed and it will be repeated many times each minute while the engine is operating.

The four strokes in the cycle are called the intake stroke, compression stroke, power stroke, and exhaust stroke. If an engine has more than one cylinder, each cylinder functions at a different phase of the operating cycle.

Diesel engines are usually four-cycle engines. They operate differently from gasoline engines in the way the fuel enters the combustion chamber, and in the way the fuel is ignited. Air is drawn into the chamber during the intake stroke. When the air is compressed at the top of the compression stroke, a small amount of diesel is injected into the chamber where it ignites immediately in the hot compressed air. Diesel engines have no spark plugs like gasoline engines have. The heat required to burn the fuel is generated by compressing the air in the cylinder so much that heat is produced.

Two-cycle Engines

The **two-cycle engine** combines the events of a four-cycle engine into two events. It does not have intake and exhaust valves. The fuel mixture enters the combustion chamber through a port or opening near the bottom of the cylinder. The fuel mixture is under pressure created by the downward movement of the piston. As the piston descends below the intake port, the fuel mixture rushes into the combustion chamber forcing the exhaust gases out through the exhaust port on the opposite side of the cylinder.

As the piston rises on the compression stroke, it blocks off the intake and exhaust ports and seals the combustion chamber. The fuel mixture is ignited at the top of the compression stroke, and the piston moves downward on the power stroke. As the piston reaches the bottom of the stroke, the ports are uncovered and the cycle starts over, Figure 11–16.

Two-cycle engines are less efficient than other engines, but they do have the advantage of a power stroke with every complete cycle of operation. Engines of this type are frequently used on motorcycles, lawn and garden equipment, and chainsaws, Figure 11–17.

Two-cycle engines do not contain oil in the crankcase. Oil must be mixed with gasoline to lubricate the engine. Failure to mix the correct proportions of oil and gasoline can result in damage to the engine.

MECHANICAL POWER

Mechanical power is force that is generated by using a tool or device to transfer or apply energy to a task or to convey a force from its source to the place where it is used.

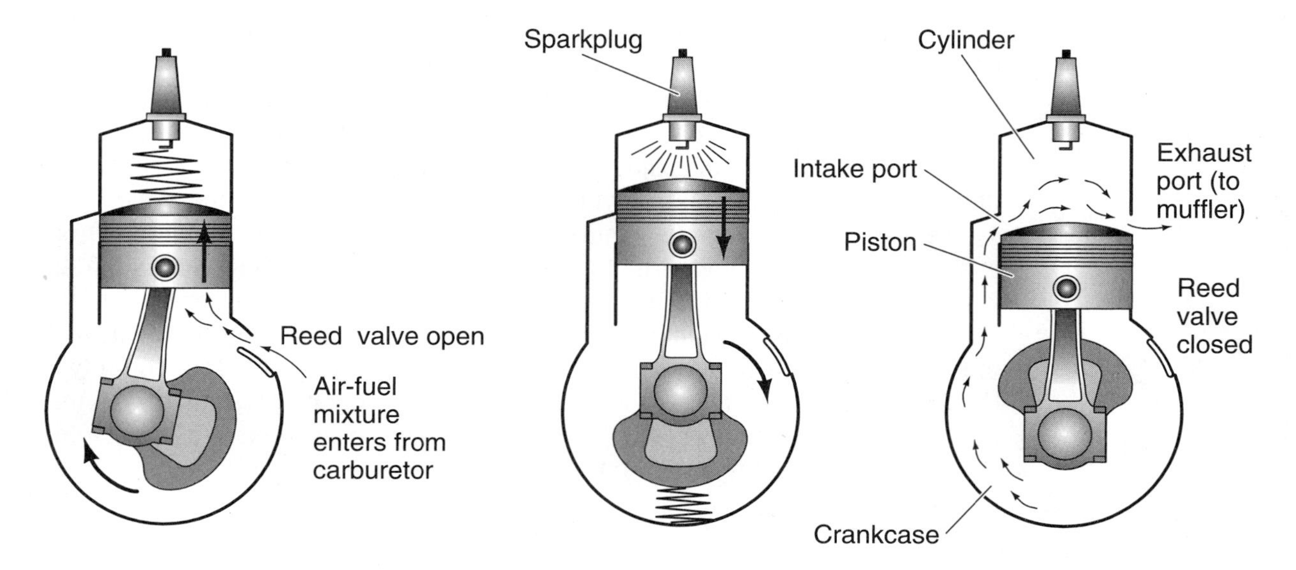

(A) Upward stroke (Compression)
- Compression in cylinder
- Crankcase fills with air-fuel mixture
- Reed valve open

(B) Downward stroke (power)
- Ignition at or near TDC
- Expanding gases drive piston downward
- Reed valve closed
- Piston compresses air-fuel mixture in crankcase

(C) Action at bottom dead center (exhaust and intake)

FIGURE 11–16 Operation of a two-stroke-cycle engine *(Adapted from Small Gasoline Engines Student Handbook, The Pennsylvania State University, Department of Agricultural and Extension Education)*

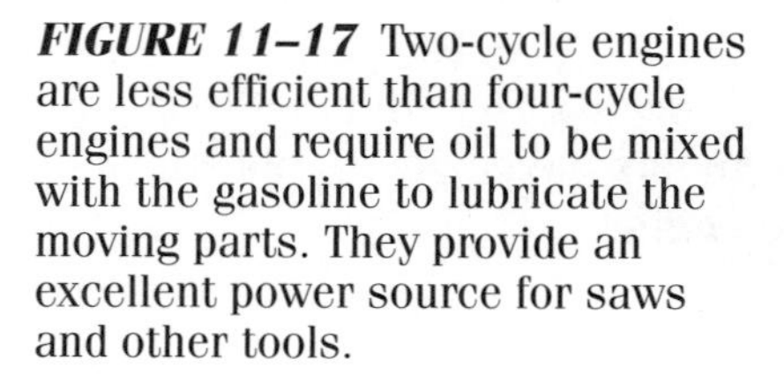
FIGURE 11–17 Two-cycle engines are less efficient than four-cycle engines and require oil to be mixed with the gasoline to lubricate the moving parts. They provide an excellent power source for saws and other tools.

The Principle of the Fulcrum

The use of a pry bar to create a fulcrum is an example of mechanical power. It is an extension of the physical laws that were described earlier in this chapter during the discussion about torque.

CAREER OPTION

Farm Equipment Dealer

A career as a farm equipment dealer involves daily interactions with mechanics, technicians, salespeople, customers, office staff, and the entire farm community, Figure 11–18. During years when farming is profitable, the business is likely to do well, but during the low profit years, a farm equipment dealership will often take financial losses. The success of the business is tied directly to economic conditions on the farms and ranches.

Many farm equipment dealers have entered the business as sales people or mechanics, until they are able to generate enough credit to buy into the ownership of the business. Large amounts of capital are needed to maintain the inventories of farm equipment and parts.

Success in a farm equipment dealership requires a good understanding of the machines that are used in producing crops. A farming background is useful in understanding the kinds of conditions in which farm machines are operated. Formal education and strong backgrounds in business practices are important in managing farm equipment dealerships.

Good public relations skills and a real interest in providing services to people are important for success in this agricultural business.

FIGURE 11–18 A farm equipment dealer has a large investment in the business. In addition to the new and used farm equipment kept in the inventory, replacement parts must be purchased and stocked for resale. *(Photo courtesy of USDA #036)*

A **fulcrum** is the support point at which a lever is placed to pry or lift an object, Figure 11–19. This principle is based on balance. If two children of equal weight are seated on a teeter-totter, the fulcrum is located an equal distance between them. If one of the children is heavier than the other, the fulcrum must be located closer to the heavier child than it is to the smaller child. The teeter-totter is really a lever. The small child is able to exert the same amount of force at the fulcrum by sitting on the long side of the "lever" as the larger child exerts due to his/her weight advantage. The small child uses distance from the fulcrum to balance weight.

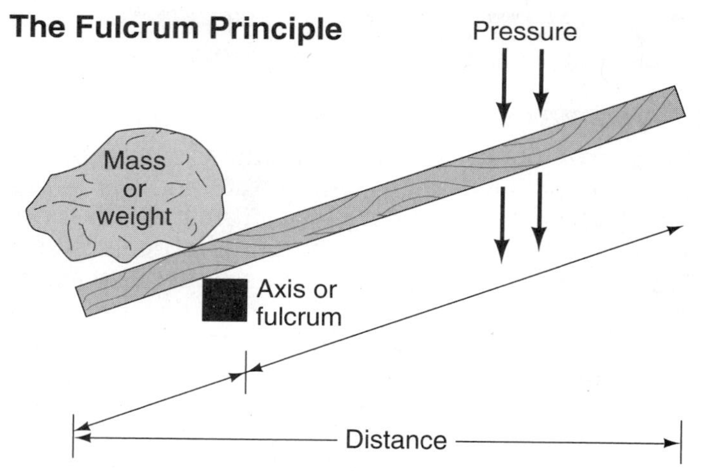

FIGURE 11–19 The principle of the fulcrum is to balance weight with leverage.

The principle of the fulcrum is used in many different applications on farm machinery. For example, a hydraulic ram is often attached to the long side of a lever, Figure 11–20. This has the effect of multiplying the amount of power that can be exerted. This application of mechanical power is used to perform a variety of tasks such as raising and lowering the wheels on tillage equipment by extending or retracting the hydraulic ram, Figure 11–21.

Hydraulics

The use of a fluid material to transfer force from one location to another location is **hydraulic power**, Figure 11–22. When force is applied to liquid in a tube, the pressure is exerted at the opposite end of the tube. This occurs because liquids are not easily compressed. They transfer the force to all of the surfaces that they are in contact with. The brake systems for many tractors and cars are

FIGURE 11–20 Hydraulic systems take advantage of the fulcrum principle to transfer power from hydraulic pumps to large implements.

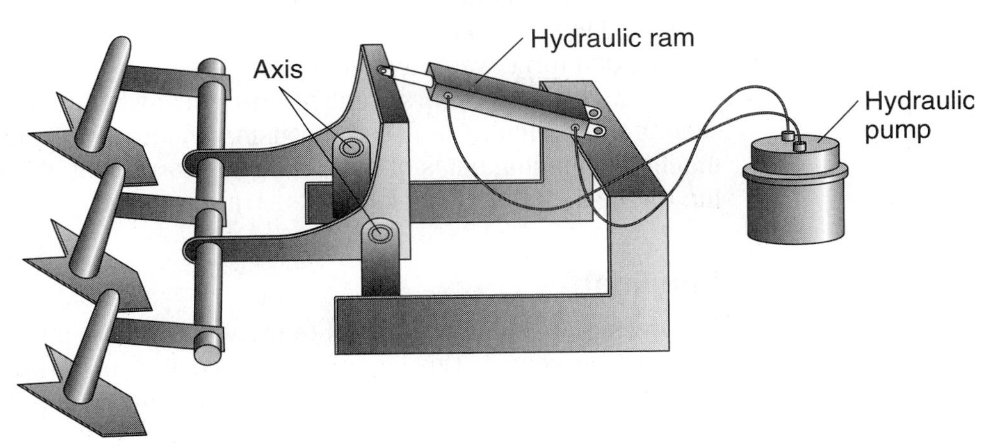

FIGURE 11–21 Hydraulic power allows the operator of machinery to raise or lower the implement without getting down from the tractor seat.

hydraulic systems. When the operator pushes on the brake pedal, hydraulic fluid is forced through a tube to the brake mechanisms located on the wheels. The pressure of the hydraulic fluid forces a small hydraulic ram to push the brake pads against the disk or drum on the wheel to stop the vehicle.

A **hydraulic ram** is a device that converts fluid force to mechanical force, Figure 11–23. This is done by telescoping a solid metal shaft into a heavy pipe that is connected to the hydraulic line or tube. When mechanical pressure is

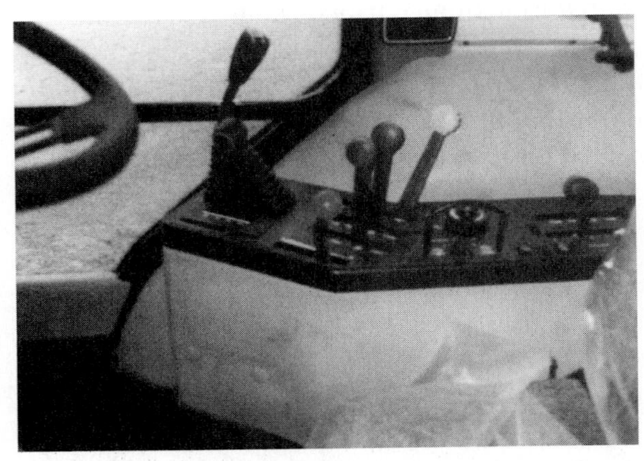

FIGURE 11–22 Hydraulic power is generated when fluid is pumped through a closed tube, hose, or pipe. This occurs when the controls on the pump are activated. As pressure builds up near the pump, it is exerted on moving parts throughout the system.

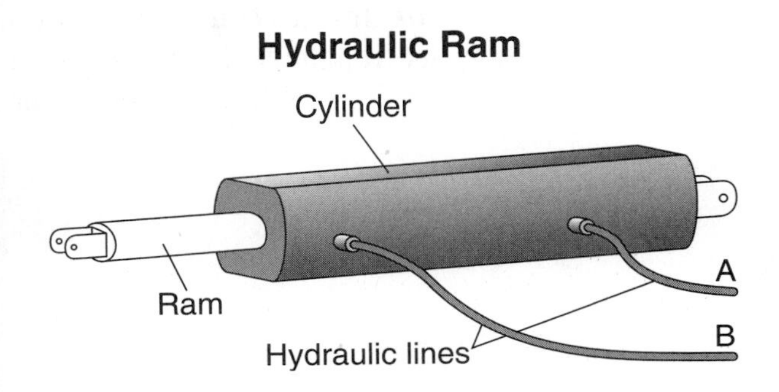

FIGURE 11–23 A hydraulic ram is a force converter that converts fluid pressure to lateral movement. When oil is pressurized in line A, the ram is extended. Pressure in line B retracts the ram back into the cylinder.

applied to fluid at one end of the hydraulic line, the telescoping shaft is forced to move out of its position by the hydraulic fluid. This movement can be converted to mechanical force at a location some distance from the place where the original force was exerted.

Farmers, processors, and others use hydraulics for many purposes. Among the agricultural uses are raising and lowering plows and other heavy machines, loading bales of hay, lifting tractor-mounted front end loaders, lifting the beds of dump trucks, and operating gates.

Pneumatics

Air pressure is sometimes used to deliver power to a location that is remote to the power source. This type of power is called **pneumatic power**. Unlike liquids, air can be compressed. In its compressed state, it can provide a source of "stored" power. A pneumatic tool or machine uses compressed air as an energy or power source. Many shops are equipped with air lines that supply compressed air to different areas of the shop. Power tools such as pneumatic chisels, grinders, sanders, and impact wrenches are available for routine construction and repair jobs.

Pneumatic power is used in a variety of ways. It is used in the brake systems for many farm trucks. Air brakes operate with compressed air, and they are very effective with heavy loads. A vehicle that has air brakes must be equipped with an air compressor and a high pressure air storage tank.

Pneumatic power is used for a variety of tasks in processing plants. It is used on processing lines to trip gates on conveyor systems, separate materials, and to agitate fluids. It is used in manufacturing systems as a propellant and power source for painting and for cleaning surfaces using a sandblaster.

Air Jet Sorting Device. A unique technology is used to separate high quality french fried potatoes from those that are damaged or dark in color, Figure 11–24. The french fries are propelled through the air at a rapid rate where they

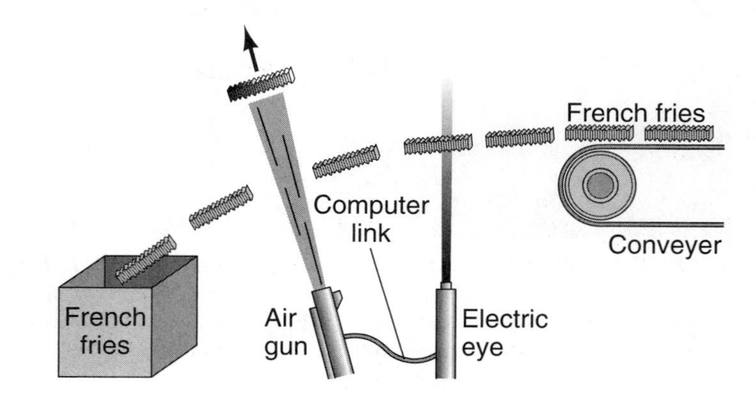

FIGURE 11–24 Compressed air can be used to perform many tasks. It can blast a seed or food product out of the production line or even provide power for pneumatic tools.

are inspected "in flight" by a laser equipped scanner. When the electronic eye detects a discolored french fry, it is sorted out from the high quality product by a short burst of compressed air.

Force Converters

Any device that is used to convert an energy source to motion or that converts one form of motion to another can be classified as a **force converter**. Many forms of energy are available that provide power for agriculture, but the greatest difficulty encountered in harnessing some of these potential sources of energy is to convert the power to a form that can be used to perform a task.

Steam power is a good example of this kind of problem and its solution. One force converter that was developed was a turbinelike device that converted steam pressure to rotary motion. A different device converted steam pressure to horizontal or vertical motion by driving a pistonlike device back and forth. Once energy has been converted to motion, it can easily be converted to other forms of motion using different kinds of force converters.

Horizontal or vertical motion is used for such tasks as moving a knife blade back and forth to cut the stems of grain or hay crops. Rotary motion is used to drive wheels of vehicles, turn the shafts in machines, and provide power to machines through the power takeoff attachment of a tractor.

The explosive force in the combustion chamber of an engine is converted to rotary motion by a crankshaft attached to a piston. Other types of force converters include belt drives, chain drives, pulleys, gears, hydraulic drives, pneumatics, electric motors, generators, solar panels, and hydroelectric turbines.

CHAPTER SUMMARY

Power is controlled energy that is used to perform tasks. Many different power sources have been developed to perform agricultural tasks. New technologies to adapt power sources to agricultural uses continue to be developed. As power sources for agricultural production have shifted from physical labor by humans and animals to internal combustion engines, the source of energy for production purposes has shifted from feeds grown on the farm to fossil fuel products such as petroleum. Future energy sources may include alcohol derived from farm crops rich in sugars and carbohydrates.

Internal combustion engines may be classified by the type of fuel they burn (gasoline, diesel), and by the way they operate (four-cycle, two-cycle).

Mechanical power is generated by using a device to apply energy to a task or to deliver power from a source to its point of use. It includes the use of levers, gears, hydraulics, pneumatics, shafts, pulleys, drive chains, turbines, and other mechanisms.

CHAPTER REVIEW

Discussion and Essay Questions

1. Describe how the shift from using draft animals to using tractors for farm power affected the amount of land that was available to produce marketable crops.
2. In what ways were steam engines important sources of power for agricultural production?
3. List the main types of petroleum fuels that are used by farmers, and describe why they have gained wide use in the production of agricultural products.
4. What advantages do ethanol fuels have as energy sources for production in agriculture?
5. How important are internal combustion engines as power sources on farms and ranches?
6. Chart the similarities and differences between four-cycle and two-cycle engines.
7. Compare efficiency ratings for four-cycle diesel engines, four-cycle gasoline engines, and two-cycle gasoline engines.
8. Define mechanical power and list three examples of how it is used.
9. Identify ways that hydraulic power and pneumatic power are similar, and describe some ways that they are different.
10. Explain how force converter devices such as gears or drive chains are used to adapt specific power sources to tasks that require other forms of power. (Example: conversion of vertical motion to rotary motion.)

Multiple Choice Questions

1. Force or energy that is controlled to accomplish work is called:
   A. latent energy.
   B. power.
   C. hydraulics.
   D. hydrology.

2. The primary source of agricultural power prior to the 1930s was:
   A. steam engines.
   B. gasoline engines.
   C. draft animals.
   D. diesel engines.

3. Which of the following fuels is not a petroleum product?
   A. diesel
   B. ethanol
   C. gasoline
   D. propane

4. Biodiesel is a renewable farm fuel that is produced from:

   A. vegetable oils.
   B. crude oil.
   C. coal.
   D. natural gas.

5. Heat can be used as an energy source because when it is applied to most substances they tend to:

   A. contract.
   B. intrude.
   C. expand.
   D. dehydrate.

6. A German inventor named Nikolaus Otto invented the first:

   A. car.
   B. internal combustion engine.
   C. diesel engine.
   D. spontaneous combustion engine.

7. Internal combustion engines convert the vertical movements of the pistons to rotary motion using a force converter called a:

   A. crankshaft.
   B. torque converter.
   C. gear.
   D. belt.

8. The measurement that describes the strength of the rotation of an object is called:

   A. compression.
   B. expansion.
   C. circumference.
   D. torque.

9. An engine that ignites fuel *without* the use of spark plugs is the:

   A. diesel engine.
   B. two-cycle engine.
   C. gasoline engine.
   D. Otto engine.

10. An engine that has a power stroke each time the piston moves downward in the cylinder is the:

    A. diesel engine.
    B. two-cycle engine.
    C. four-cycle engine.
    D. rotary engine.

11. A system in which oil is used to transfer force or power from one location to another is called:

    A. pneumatics.
    B. hydraulics.
    C. petromatics.
    D. fluidomics.

LEARNING ACTIVITIES

1. Locate a retired farmer in the community who began his career using draft animals to pull his implements. Ask him to share his memories and experiences with the class including old photographs, humorous incidents, team-related accidents, and routine daily work schedules. Ask him to make comparisons between his early farming experiences using teams and his farm experiences after he got his first tractor. (You might want to record this activity using a camcorder.)

2. Challenge your students to develop an Agricultural Science Fair project that demonstrates at least two different uses of mechanical power in performing agricultural tasks. Display and demonstrate the completed projects in a public location.

CHAPTER

Electrical Energy in Agricultural Uses

Electricity is essential to the agricultural industry. It is used to perform many tasks associated with life on the farm. Electric motors, lights, and heating devices are important tools to the entire industry of agriculture. They are used on the farms and ranches, they are used by agricultural processors, and they are used by the industries that support agricultural production.

OBJECTIVES

After completing this chapter, you should be able to:

- explain in general terms what electricity is and how it works.
- define basic terms associated with electricity.
- distinguish between incandescent and fluorescent lamps.
- discuss the principles by which an electric heater operates.
- suggest several criteria that should be considered in the selection of an electric motor.
- identify several types of electrical controls that are used to automate electrical systems.

TERMS FOR UNDERSTANDING

The following vocabulary terms should be studied carefully as you read:

electricity
conduction electrons
conductor
generator
voltage
potential difference
amperage
resistance
ohms
short circuit
Ohm's Law
direct current (DC)
alternating current (AC)
cycle

- sine wave
- incandescent lamp
- filament
- fluorescent lamp
- refrigeration
- heat elements
- arc welding
- magnetism
- lines of force
- magnetic field
- electromagnet
- armature
- alternator
- induction
- nameplate
- duty rating
- service factor
- temperature rise
- solenoid
- switch
- variable switch
- relay
- sensor
- thermostat
- humidistat
- photoelectric cell

GENERAL PRINCIPLES OF ELECTRICITY

Electricity is a form of energy that is associated with movement of electrons between particles. Each electron carries a negative charge and is capable of flowing through the network of atoms that makes up the wires and cables that are used in electric power lines. Electrons that flow in this manner are called **conduction electrons**. Materials through which electrons are capable of flowing are called **conductors**. Electricity behaves very much like water that is under pressure. The pressure in a water system is maintained by using a pump, and the pressure in an electrical system is maintained by a generator.

An electrical **generator** converts energy from combustible fuel, flowing water, nuclear fuel, geothermal heat, or other sources into electrons or electricity. A generator maintains electrical pressure in the power line. The measurement of electrical pressure is called **voltage**. When the voltage at one end of a conductor is different than the voltage at the other end, electrons flow through the conductor from the area of high voltage (high electric potential) to the area of low voltage (low electric potential). This difference in voltage is called **potential difference**. Electrons flow in a conductor every time that potential differences exist, Figure 12–1. When the same electric potential exists at both ends of the conductor, electricity no longer flows.

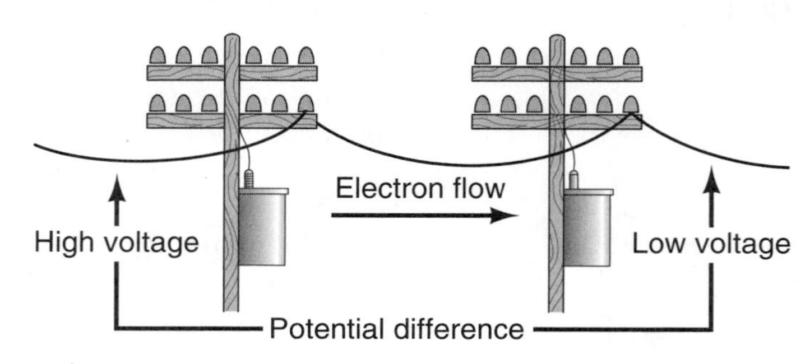

FIGURE 12–1 Electricity flows from high to low voltage each time a potential difference exists between two locations that are joined by a conductor.

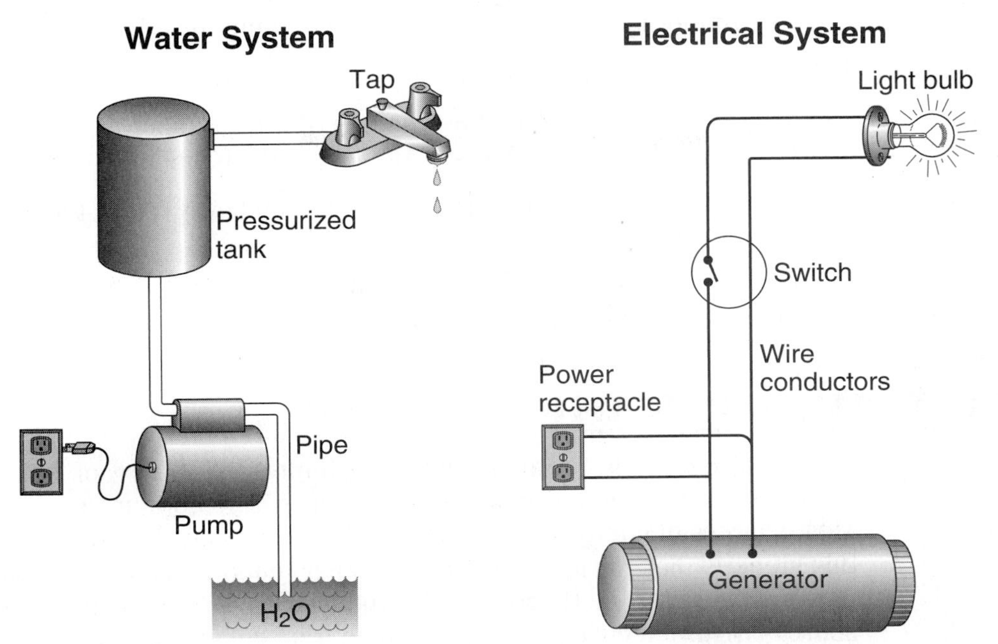

FIGURE 12–2 Electrical systems and water systems have several similar functions. Both systems require conductors with water pressure driving water through the pipe, and voltage pressure moving electricity through wires or cables. Pumps and generators maintain pressure in the systems as water and electricity is consumed or released.

When a water tap is opened, water flows out of the system and water pressure is lost. The water pressure is recovered when the water pump puts water back into the system. When an electrical switch is opened, electricity flows out of the system and must be replaced to maintain voltage. Electrical pressure is replaced by a generator, Figure 12–2.

The rate at which water flows through a pipe is measured in gallons per minute. The rate at which electricity flows through a circuit is measured as **amperage**. In both water and electrical systems, friction slows the rate of flow. The rate of flow depends on the size, length, and composition of the pipe or conductor. A reduction in the rate of flow due to friction is called **resistance**. The resistance within a circuit to the flow of electrical energy is measured in **ohms**.

A leak in a water pipe results in low pressure in the system. A **short circuit** or leak in an electrical line allows electricity to flow out of the desired path. This can cause damage to the electrical system and poses a potentially dangerous situation to people and animals.

Ohm's Law

George Simon Ohm discovered an important principle of physics that describes the relationships of amperage, voltage, and resistance. It is called **Ohm's Law**. He described it in this formula:

Amperage = voltage/resistance OR amperes = volts/ohms

Letter symbols are often used to write the formula for Ohm's Law:

E = Volts I = Amperes R = Resistance
E = IR (Volts = Amperes x Resistance)

When any two of these values are known, we can always solve the equation for the unknown value:

I = E/R (Amperes = Volts/Resistance) and R = E/I (Resistance = Volts/Amperes)

AGRICULTURAL APPLICATIONS

Electrical energy is probably the most important source of power in the world today. It has many applications in agriculture both on and off the farm. Most sources of energy can be converted to electricity, Figure 12–3, and it is this property that has made electricity such an important source of agricultural power, Figure 12–4. Electricity is produced from several different energy sources. Hydroelectric power is obtained from generators that are turned by the force of moving water driving the fanlike turbines located at the base of a dam. Nuclear power is electricity that is generated using the heat obtained from a controlled nuclear reaction. Coal and gas are other sources of heat that are used to generate electricity. Some electricity is produced by generators powered by wind turbines. Geothermal wells sometimes produce water that is so hot it turns to steam when it is released from its high pressure chamber underground. Steam-driven turbines turn the generators to produce geothermal power.

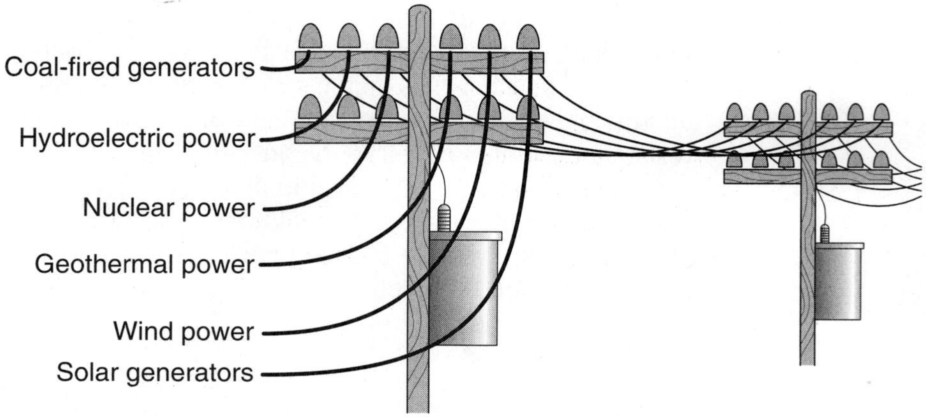

FIGURE 12–3 Electrical energy is important to agriculture because it is used to perform many tasks. Most other energy sources can be converted to electricity.

FIGURE 12–4 Electrical substations are made up of transformers that step-up the voltage for more efficient transmission of electrical energy over long distances, and that step-down the voltage near the site where the electricity will be used.

Electricity is generally classed as either direct current (DC) or alternating current (AC). **Direct current** is electricity that flows in one direction. It is stored in cells and batteries or generated by a DC generator. **Alternating current** is electricity that changes its direction of flow in the conductor many times each second. Most sources of alternating current in North America produce alternating current that is 60 cycle current. This electricity reverses directions 60 times per second. A separate **cycle** occurs each time the armature of the generator completes a full rotation. Each cycle produces a wave of electrical current called a **sine wave**, Figure 12–5. Each sine wave consists of a positive pulse and a negative pulse of electricity that is generated as the generator armature rotates past the positive and negative poles of a magnet.

Alternating current can be converted to direct current by using a transformer. Examples of these applications are evident when a battery charger converts AC to DC electricity to charge a battery.

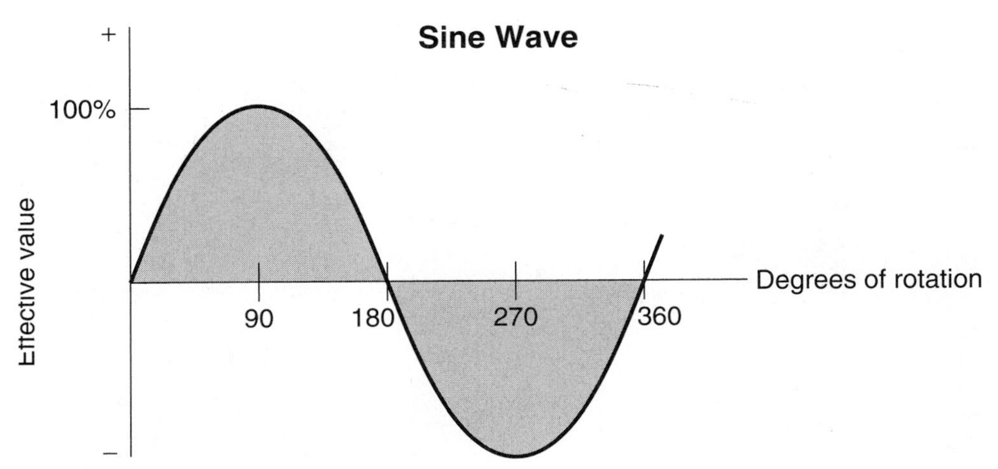

FIGURE 12–5 The alternate positive and negative pulses of electrical current produced during one 360 degree rotation of a generator armature constitute a sine wave.

Lighting

The most obvious agricultural use of electricity is for lighting homes, barns, shops, and yards. Thomas Edison developed the first electric lamp for commercial use in 1879. Since that time, lighting technology has improved, and more reliable lightbulbs have been developed.

Conversion of Energy to Light

Light is closely associated with heat. When electricity passes through a conductor that has high resistance to the flow of electricity, heat is created. Such a conductor may become so hot that it begins to glow, and materials get so hot that they glow brightly. In the presence of oxygen, the filament of a lightbulb would burn out, but when oxygen is not present, the energy that is responsible for heating the filament is given off as light.

Most electric lights in use today are **incandescent lamps**, Figure 12–6. This kind of light uses a small tungsten wire to form a **filament** or conductor sealed inside a glass bulb that is filled with argon and nitrogen gas. Resistance to the flow of electricity through the filament converts electrical energy to heat, causing the filament to glow.

Another type of lightbulb is the **fluorescent lamp**, Figure 12–7. It is constructed using a glass tube that has been coated on the inside with a material called phosphor. The bulb is filled with mercury vapor that acts as a conductor. Radiation from the electric arc causes the phosphor to glow. This type of lamp provides light that is very efficient and nearly equal in quality to natural lighting.

The installation of lights on tractors, trucks, and other farm machinery has modified the work habits of many farmers. It is now possible to operate machinery and perform field work at night. In dry climates, some crops such as alfalfa hay or dry beans are harvested at night when the dew prevents

FIGURE 12–6 The incandescent lamp is a commonly used source of lighting for homes and buildings.

FIGURE 12–7 The fluorescent lamp is a highly efficient source of light that is used in many agricultural lighting situations.

FIGURE 12–8 Modern farm equipment is fitted with lights that make it possible to work effectively at night.

the leaves and seed pods from shattering. Farmers are able to perform many field operations nearly as effectively at night as they do during the daylight hours, Figure 12–8.

Lighting as a Management Tool

Agricultural uses for lighting go far beyond providing light to make it easier to see at night. They also include some specialty uses of light as a management tool for plants and animals.

The natural breeding seasons for sheep and some other animals are sometimes modified by controlling the amount and quality of light to which breeding animals are exposed. By restricting the amount of light exposure, conditions can be created that simulate those of the natural breeding season. The advantage of this technology is that it is possible to reduce the time interval between births and to increase the number of offspring born during the life of the female.

The female organs of some sheep and goat breeds are inactive during some seasons of the year. This period of reproductive inactivity is called anestrus. The breeding cycles of sheep and goats are controlled by the amount of time each day that the females are exposed to light. As daylight hours begin to decrease in late June, the reduced amount of light exposure acts as a signal to the nerves and glands causing them to begin producing hormones that initiate the breeding cycle.

Experimental control of the photoperiod, or period of light exposure, has been used to induce breeding during seasons when female goats (does) and female sheep (ewes) are experiencing the anestrus phases of their breeding cycles. Light control is achieved by confining does and ewes in dark buildings,

FIGURE 12–9 Artificial lighting is used to simulate optimal conditions resulting in increased production for some plants and animals. *(Courtesy of USDA/ARS)*

underground caves, or other dark structures. Artificial lights may be used to control the length of the photoperiod, or the animals may be turned out into the sunlight for the correct amount of light exposure. The general effect of managing the photoperiod of these two species is to induce reproduction and to shorten the intervals between parturition dates. The cost of implementing this practice has limited its commercial use.

Light has been used as a management tool for laying hens and flowering plants for many years. The lights are turned on early in the morning, and left on until late evening to simulate long summer days. Hens respond to this treatment by laying more eggs during their productive lives than they would under natural lighting conditions.

Plants also respond to modified lighting environments. Some flowers are produced under greenhouse conditions during seasons of the year when it would be impossible for them to bloom naturally. By modifying the lighting environment with special fluorescent lamps, some plants respond as though they were flowering naturally, Figure 12–9.

The agricultural processing industry depends heavily on electric lighting systems. Many food processing factories operate day and night during the harvest season. This allows perishable crops to be processed when they are at the peak of quality. For many fruit and vegetable crops, quality begins to decline very rapidly after they have matured.

The availability of lighting from electrical energy has played a major role in agricultural production and processing. It has added flexibility to the agricultural industry, making it easier to produce, harvest, and process the abundant crops that supply food and fiber.

Refrigeration

One of the most important agricultural uses of electricity is **refrigeration** or cooling of perishable foods and other products. Harvest and storage losses of food products continues to be a big problem in developing countries of the world. Sometimes more than half of the perishable food supply is lost because of poor harvest practices and spoilage during storage.

Bacteria thrive when environmental conditions are moist and temperatures are warm. They can spoil fresh meat or milk in a few hours, and many fresh fruits, vegetables, and processed foods spoil in a few days without refrigeration. When these foods are cooled to a few degrees above freezing, they can be stored for much longer periods without significant spoilage. This is because low temperatures restrict reproduction rates in bacteria.

Refrigeration can be as simple as the household refrigerator or as elaborate as a drive-in freezer, Figure 12–10. Food products such as milk and eggs require refrigerated farm storage with enough capacity to store the fresh products for short periods, Figure 12–11. Fresh milk is seldom stored on the farm for longer than 48 hours, and eggs are shipped to markets within a week of the time they are produced.

The storage life of all fresh food products depends to a large extent on how quickly the products are cooled after they are produced or harvested. Warm milk is cooled by refrigerated coolers immediately after it passes through the pipeline from the cow. Most large egg farms clean and package the eggs almost as fast as they are produced by the hens. They go immediately into refrigerated storage or into refrigerated trucks for delivery to markets.

Fresh meat requires several hours to remove body heat from the carcasses. During that time, bacteria multiply rapidly unless steps are taken to speed the cooling process. Poultry carcasses are usually immersed in ice water to quickly remove heat. Large animal carcasses are often placed in coolers where super-cooled air is circulated around the meat for several hours. All meat products require constant refrigeration until they are cooked and consumed, or until they are processed and canned, Figure 12–12.

FIGURE 12–10 Refrigeration is an important agricultural use of electrical energy. Modern food processing facilities often include large walk-in freezers in which processed products are stored.

FIGURE 12–11 Stainless steel bulk tank used for refrigerating liquids such as milk or juice *(Photo by Michael Dzaman)*

FIGURE 12–12 Perishable food products require special arrangements for rapid cooling to immediately reduce temperatures during the early stages of processing.

Fresh vegetables are often packed in ice immediately after they are harvested. Rapid cooling at this time can add several days to the shelf life of products like lettuce and sweet corn. Fruits and vegetables of all kinds can benefit from cool storage conditions. Some products, such as potatoes and apples, can be stored until the next harvest season in modern refrigerated storage structures.

Air conditioning is used in some large livestock confinement facilities to help maintain a constant temperature inside the buildings. Air conditioning makes use of refrigeration principles to remove heat from the air in the confined space. Heat is conducted from the air into liquid coolants through thin metal conductors through which the hot air is circulated. The liquid coolant circulates through pipes to the environment outside the building. There the heat is transferred from the coolant to the atmosphere.

Heaters

An important use of electricity is the production of heat. **Heat elements** are special conductors that are used to convert electrical energy to heat. Electricity is passed through the conductor, which becomes hot and gives off energy in the form of heat, Figure 12–13. Many types of electric heaters operate using this principle.

Electric heat is used in many agricultural applications. Among its uses are heating for homes, incubators, buildings, and livestock water supplies. Heat is often provided for the comfort of animals during cold weather. Heat lamps and special heat pads are available for use by baby pigs, poultry, and newborn animals. The pads work on the same principle that is used to make electric blankets.

Agricultural processors use electric heat for many purposes. They include cooking, pasteurizing, drying, and dehydrating of food products. Water is heated for use in cleaning and sanitizing equipment and food products.

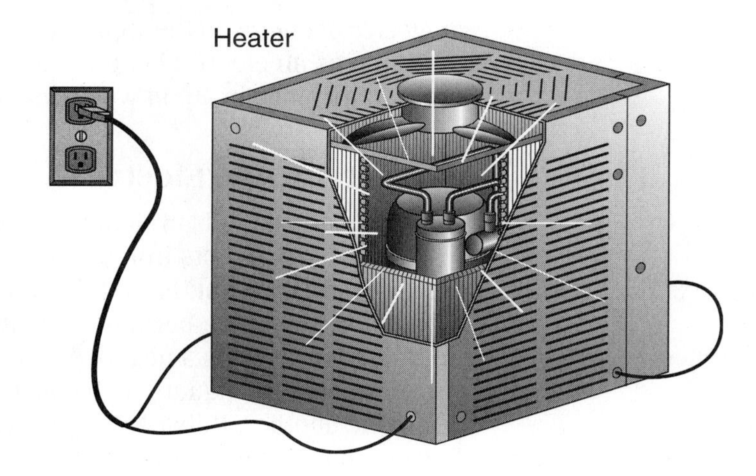

FIGURE 12–13 Electric heaters are used for many different agricultural purposes. Some of these include cooking, preheating cold engines to facilitate starting, and providing heat for plants, animals, and people.

Packages are sealed, and work areas are heated using heat derived from electrical energy.

Arc Welding

Arc welding is a process that applies heat to metals, causing them to fuse or melt together, Figure 12–14. It is used to construct and repair many types of metal farm machinery.

The heat that is required to arc weld metal is obtained from an electric arc or flame. The intensity of the arc is controlled by the amperage or rate of flow

FIGURE 12–14 Arc welding is a process in which electricity is used to heat metals above their melting points and to fuse metal parts together. It is a basic construction process.

of electrical energy to the weld area. An arc welder operates on much the same principle as an electric heater, except that the electrical circuit is completed through an electric arc in welding and through a heating element in a heater.

Magnetism and Electricity

A close relationship is evident between magnetism and electricity. **Magnetism** is a force that attracts iron or steel. It occurs when electricity flows through a conductor. Material that retains its magnetism after the electrical current is turned off is called a permanent magnet.

Each magnet has a negative and a positive pole called the north and south poles. Opposite magnetic poles attract one another, but poles that are alike repel one another. Between the two poles are invisible **lines of force** that follow curving pathways from one pole to the other pole. The area that is occupied by these lines of force is also known as a **magnetic field**.

An electric current can be produced by moving a conductor such as a copper wire up and down repeatedly across the lines of force that arc between the positive and negative magnetic poles, Figure 12–15. In a similar manner, passing an electric current through a conductor creates a magnetic field around the conductor. This magnetism can be intensified by coiling a conductor such as copper wire around a mild steel core. Such a magnet is an **electromagnet**, Figure 12–16. When such a magnet rotates inside a motor or generator, it is called an **armature**.

The use of mechanical power to cause the poles of an armature to rotate across the lines of force in the magnetic field of a larger magnet causes an electric current to be produced. Electricity flows out each time the magnetic

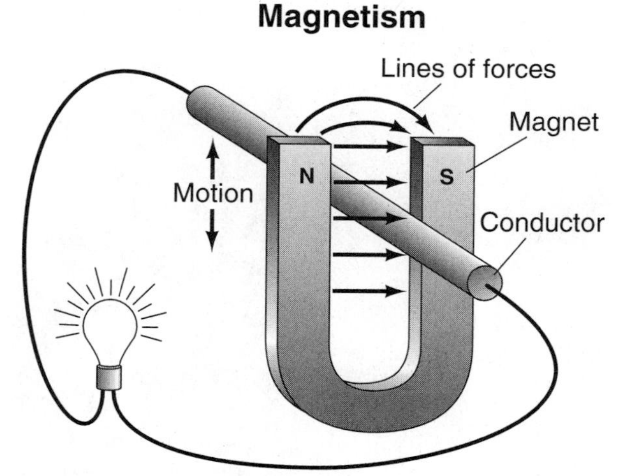

FIGURE 12–15 Electricity is produced when a conductor is moved across magnetic lines of force.

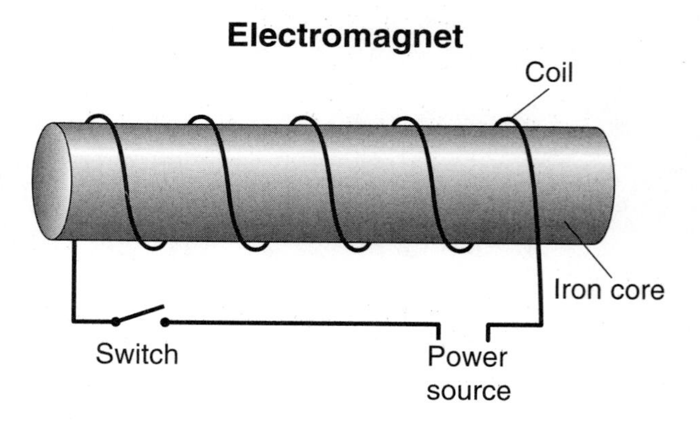

FIGURE 12–16 An electromagnet consists of a conductor coiled around an iron core. When the electrical circuit is activated, lines of force are created and a magnetic field exists.

poles pass one another. A device of this kind is called a generator or **alternator**. A source of power such as steam, flowing water, wind, etc. is required to cause the armature to rotate. This same kind of device will function as an electric motor when a constant supply of electric current is passed through the armature. This arrangement causes the armature to turn with sufficient torque to provide power to machines of many kinds. It is these opposing forces that alternately attract and repel the poles of an armature that causes the shaft of a motor to rotate.

Motors

The process by which electrical or magnetic lines of force cause electricity to be generated or magnetism to occur in a conductor is called **induction**. This force is closely related to the forces of attraction between positive and negative magnetic poles and the tendency of positive poles to repel positive poles and negative poles to repel negative poles. Induction makes it possible to convert electrical energy to rotary or lateral motion using inductive devices such as motors and solenoids.

An electric motor is a specialized machine that converts electrical energy to motion, Figure 12–17. It uses magnetic forces to create rotation of the motor shaft. This rotary force is called torque. Some types of motors develop more starting torque than others.

The amount of torque or turning power required for a task should be considered when a motor is selected. Another factor that should be considered in selecting a motor is where the motor will be located. Special types of motor housings or covers are available for wet or dusty conditions.

The motor **nameplate** is a metal plate that is fastened to the motor, Figure 12–18. It contains information that is useful in selecting the proper motor. The **duty rating** of a motor rates the ability of a motor to operate under a load. A continuous-duty rating means that the motor can operate under an appropriate load without interruption and not burn out. A limited-duty rating

FIGURE 12–17 An electric motor is used to convert electrical energy to motion. The motion is used to provide power for many agricultural tasks.

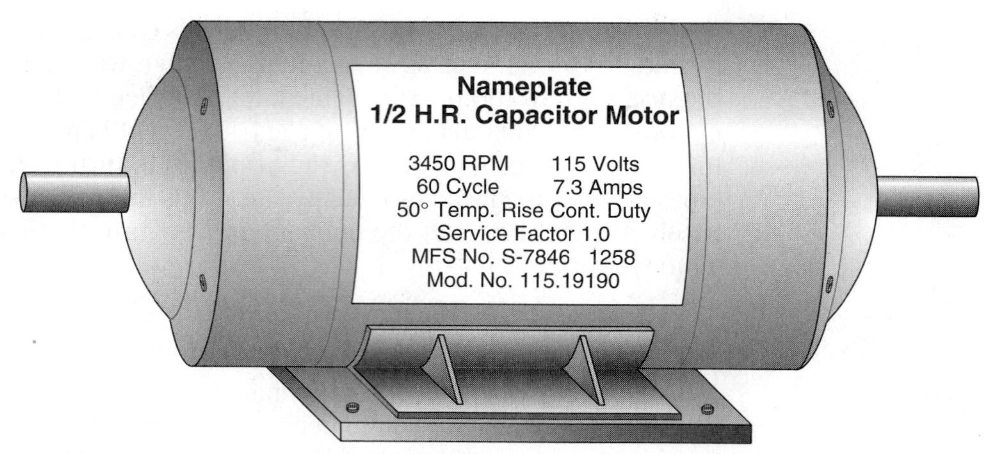

FIGURE 12–18 A motor nameplate provides important information about the motor and the kind of work it can be expected to perform.

is a warning that the motor may burn out if it is operated under load for a longer period of time than it is rated for.

The **service factor** of a motor is a rating of the capacity of a motor to handle an overload. A service factor of 1.0 means that the motor does not have the capacity to handle a continuous load greater than its rated horsepower.

Temperature rise is another motor rating that is found on the nameplate. It rates the motor according to how hot it is capable of running above the temperature of the surrounding environment. When proper selection procedures are followed, motors are capable of providing long service.

Many different types of motors have been developed for specialized tasks on farms and in agricultural processing. They are available for both AC and DC power supplies. Some motors are capable of starting under very heavy loads, while others are designed for fast operation under light loads. Agricultural uses for electric motors include providing power for water pumps, fuel pumps, sludge pumps, air compressors, and milking machines, Figures 12–19 and 12–20. Motors are used to operate electric tools to construct, maintain, and repair agricultural equipment. They are used to operate fans for cooling, ventilation, and humidity control. Motors are the sources of power for refrigeration, air conditioning, and forced air heating systems. They provide power to feed mills, conveyer systems, elevators, and augers, along with many other electrical applications.

Solenoids

A **solenoid** is similar to a motor in that it uses magnetic forces to convert electricity to a form of motion that can be used to do work. It is unlike an electric motor in that it produces forward and backward motion instead of rotary motion.

FIGURE 12–19 Electric motors perform many tasks in agricultural settings. Some motors are large and are capable of pumping large volumes of water from deep wells.

FIGURE 12–20 Electric motors have taken over some of the menial tasks of agriculture. Modern milking machines are usually powered by electric motors, making it possible for a farmer to manage large numbers of dairy cows. Hand milking is no longer practiced on most farms.

A solenoid is made by winding a coil of wire around a tube. An iron bar is aligned with the tube in such a manner that magnetic forces can attract the iron bar or plunger causing it to move into the tube, Figure 12–21. A spring connected to the outer end of the iron bar pulls the bar back out of the tube when the power is turned off.

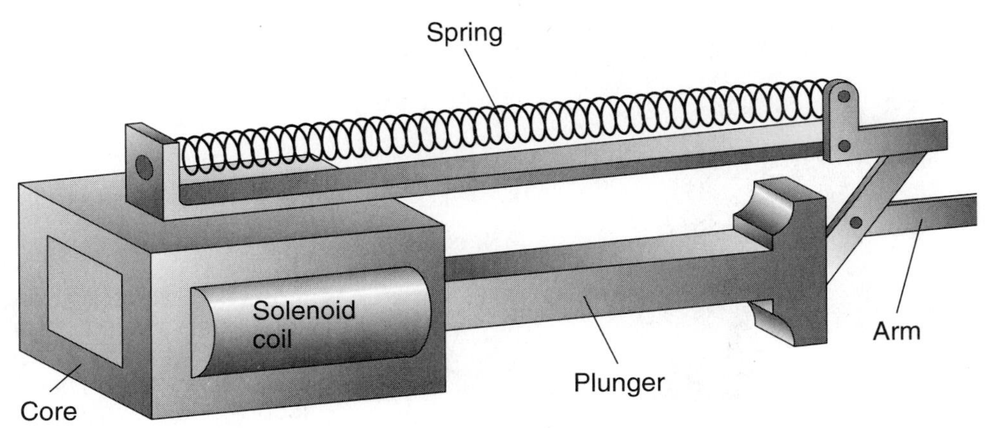

FIGURE 12–21 A solenoid is an inductive device that converts electrical energy to lateral motion. The arm of the solenoid is used to perform work.

A solenoid is used on the starter motor of a large engine to push a small gear into position to engage the large gear on the engine flywheel. When the ignition switch is turned to the start position, the solenoid is activated. It is disengaged when the engine starts, and the ignition switch directs the electricity to the electrical system of the engine.

A relay is another inductive device that is used to switch power on or off from distant locations.

CONTROL DEVICES

Electrical control devices were discussed in general terms earlier in this text (Chapter 9). Electrical controls can be used to direct power to any machine or device that is powered by electricity.

CAREER OPTION

Electrical Control Technician

A person who engages in a career as an electrical control technician works in manufacturing or processing industries in which equipment is controlled automatically, Figure 12–22. A technician is responsible for the installation, troubleshooting, maintenance, and repair of all types of electrical control devices.

A career as an electrical control technician requires specialized electrical training, which is usually obtained through an educational program at a vocational technical school or college. Employment opportunities are good due to the trend toward highly sophisticated electronic control systems in the processing industry.

FIGURE 12–22 Technicians who work with and understand electrical controls are employed in many agricultural processing and manufacturing plants and factories. *(Courtesy of DeVry Inc.)*

Switches

An electrical **switch** is a device that opens or closes a circuit through which electricity flows to deliver power. The most simple switches are either on or off. When such a switch is in the on position, full power is delivered. When it is in the off position, the flow of electricity stops completely. The most common switch of this type is the toggle switch.

Variable switches control the amount of current that is allowed to flow, and can be easily adjusted up or down as desired. A full range of power is available for use. When such a switch is installed in a light circuit, the light can be operated at any level of intensity.

A **relay** is an inductive switch that is used to turn power on or off from locations that are remote to the controls. An example is the relay switch located near a water tank that controls an electric pump located at the bottom of a well.

Sensors

Electrical **sensors** are devices that are used to detect changes in environments or conditions. Sensors can be devised to react to nearly any situation in which changes occur. The most common types of sensors for agricultural purposes are used to detect changes in temperature, humidity, light, pressure, time, and weight. Sensing devices are installed to automatically activate a switch when a change occurs in the condition being measured, Figures 12–23 and 12–24.

FIGURE 12–23 Many farm chores such as feeding can now be performed by machines. Electrical devices are able to activate auger systems that move feed from storage bins to the feeding areas. *(Photo courtesy of Utah Agricultural Experiment Station)*

FIGURE 12–24 Automated feeding systems make it possible to care for large numbers of birds on this modern turkey farm. *(Photo courtesy of Utah Agricultural Experiment Station)*

FIGURE 12–25 A thermostat is an electronic control device that detects changes in the temperature of an environment and activates electric switches on heating and cooling equipment. *(Photo courtesy of Honeywell)*

A sensing device used to detect changes in temperature is a **thermostat**, Figure 12–25. It can be used to activate heating or cooling equipment when the temperature fluctuates from the desired range. A common use for a thermostat is to control the heating element that provides heat for baby chickens and other newborn animals.

Humidistats are sensing devices that are activated when the moisture in the air gets above or below desired levels. They are used to activate ventilation systems in confinement housing for livestock, or to add humidity to storage facilities for fresh fruits or vegetables. One agricultural use for a humidistat is to control the precise level of humidity inside an egg incubator. If the humidity inside the incubator is not maintained at precise levels, the developing embryos die inside the eggs.

Changes in light intensity can be detected using a **photoelectric cell**. This device is used to activate an electrical circuit either in the presence or the absence of light. It is also capable of comparing the color of an object to a standard color and triggering a switch in a sorting process that rejects objects not in conformity to the standard. It is used to turn outdoor yard lights on in the evening and off in the morning.

Pressure control switches are frequently used in agricultural settings to maintain water pressure in the water system. As water is drained from the line, water pressure drops until the switch is activated and the water pump is turned on. When the desired water pressure is restored to the system, the switch turns the pump off. Most pressure control switches operate within a predetermined range.

Special Sensor Application: Automatic Milker Removers. Milking cows is easier today than it has ever been. New technologies in milking equipment are capable of performing every part of the job. A modern improvement in milking equipment is the equipment for removing milker units when the milking process is complete, Figure 12–26.

A special sensor in the milker unit is triggered when milk flow from the cow is no longer detected by the sensor. The milker unit is automatically detached from the cow by a retractable cord.

Timing Devices

Many electrical devices are controlled by time clocks. They can be set to activate a switch at any time of the day or night. Many applications of this device are found in agriculture. The alarm clock that wakes the farmer up each morning is such a device. Time clocks are used to operate automatic feeding systems, and to turn the lights on in livestock facilities. Nearly any activity that is repeated on a regular schedule can be controlled by a time clock, Figure 12–27.

Several different devices are available that are used to control the amounts of materials desired in a particular location. Some of these devices weigh materials to determine amounts. Other materials are metered in precise amounts. Still other devices respond to pressure exerted against pressure plates.

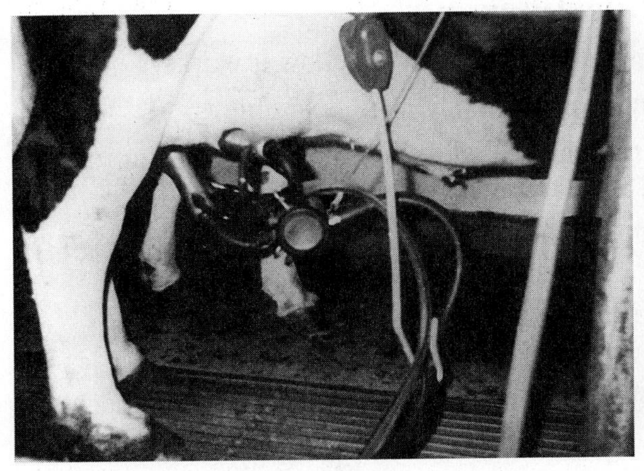

FIGURE 12–26 A special electronic sensor has been designed to remove automatic milkers from dairy cows. When the flow of milk from the cow stops, the sensor causes a small cable to retract and pull the milking unit off the cow.

FIGURE 12–27 An electronic timing device is used to activate equipment that needs to be turned on or off at specific times.

A common electrical control device that is used in agricultural processing is the laser. It can be used to activate a switch when the light beam is interrupted. This type of device is often used in agricultural processing to maintain a constant supply of products on conveyor lines.

Control devices play important roles in promoting high levels of efficiency in agricultural production and processing. They make it possible for activities to occur and tasks to be performed automatically.

CHAPTER SUMMARY

Electric power is very important to the industry of agriculture because it is used to perform many tasks that used to require human labor. Electricity flows through a conductor much like water flows through a pipe. It flows under pressure, and its energy can be expressed as heat, light, or motion.

Many types of electric lights, heaters, inductive devices and control devices are available, and each has unique applications. Care should be taken in the selection of electrical motors, appliances, and devices to be sure that the selected equipment is appropriate for the use to which it is put.

Electrical control devices installed in a circuit are capable of measuring changes in the surrounding environment and triggering electrical switches when changes occur. This allows equipment to be operated automatically. Many electrical systems operate independently of human control.

CHAPTER REVIEW

Discussion and Essay Questions

1. Explain what electricity is, and briefly describe how it works.
2. Define the following terms: voltage, amperage, and ohms.
3. Describe the basic differences between incandescent and fluorescent lightbulbs.
4. What uses do electric lights fulfill on farms and in agricultural processing industries?
5. Why is refrigeration important in the industry of agriculture?
6. Discuss the principles by which an electric heater operates. Give several examples of heating applications on farms and in agricultural processing industries.
7. How is induction used to convert electrical energy to motion?
8. Identify several factors that should be considered in selecting an electric motor.
9. List some agricultural uses for electrical sensors.

10. Classify and describe several types of electrical control devices according to the functions they perform.

Multiple Choice Questions

1. A material through which electrons are capable of flowing is called a:

   A. conduit.
   B. conductor.
   C. turbine.
   D. capacitor.

2. A measurement of the rate at which electricity flows through a circuit is:

   A. amperage.
   B. resistance.
   C. voltage.
   D. wattage.

3. A law of physics that describes the relationships of amperage, voltage, and resistance is:

   A. Bernoulli's Principle.
   B. Quantum Theory.
   C. Principle of Equivalence.
   D. Ohm's Law.

4. Electricity that changes its direction of flow in the conductor is:

   A. direct current.
   B. reverse polarity.
   C. alternating current.
   D. straight polarity.

5. A type of lightbulb that uses mercury vapor as a conductor is the:

   A. fluorescent lamp.
   B. halogen lamp.
   C. incandescent lamp.
   D. Edison bulb.

6. Which of the following heating devices *does not* derive heat from electricity?

   A. heat elements
   B. heat lamps
   C. heat pad
   D. heat pump

7. Lines of force between the north and south poles of a magnet can also be called a:

   A. magnetic field.
   B. conductor.
   C. solenoid.
   D. torque converter.

8. Which information *is not* available on the nameplate of a motor?

   A. duty rating
   B. gear ratio
   C. temperature rise
   D. service factor

9. An inductive device that converts electricity to rotary motion is a:

   A. solenoid.
   B. relay.
   C. motor.
   D. generator.

10. A sensing device that is activated when changes occur in the amount of moisture in the air is called a:

    A. pressure switch.
    B. photoelectric cell.
    C. thermostat.
    D. humidistat.

LEARNING ACTIVITIES

1. Take a field trip to an agricultural processing facility to observe and record the tasks that are performed using electric power. Also observe and record the types of electrical controls that are used to operate the machinery. Follow up in the classroom by reporting and discussing what was observed during the tour.

2. Teach students to construct a simple electrical circuit complete with a switch, an outlet, and a light socket.

CHAPTER 13

Alternative Energy Sources for Agriculture

*The agricultural industry has become very dependent on **fossil fuels** such as coal, petroleum, and natural gas as sources of fuel. These fuels are called **nonrenewable resources** because they are formed slowly and cannot be replaced once they are gone. Petroleum products are used almost exclusively as the fuel for internal combustion engines. Fossil fuels are also used to generate electrical power for agricultural uses.*

OBJECTIVES

After completing this chapter, you should be able to:

- describe the importance of fossil fuels to the agricultural industry.
- distinguish between renewable and nonrenewable resources.
- identify technologies that are used to take advantage of solar energy.
- appraise the importance of alcohol fuels now and in the future.
- assess the role of hydroelectric power plants in providing electricity for present and future agricultural uses.
- weigh the advantages and disadvantages of producing electrical power using nuclear reactors.
- describe research using whole grain as a source of fuel.
- evaluate the potential of wind as a source of power for agriculture.
- describe the source of geothermal energy and explain how it is used for agricultural purposes.
- identify energy sources that are by-products of animal production, and explore methods used to collect and utilize them.

TERMS FOR UNDERSTANDING

The following vocabulary terms should be studied carefully as you read:

- fossil fuel
- nonrenewable resource
- renewable resource
- solar energy
- greenhouse
- solar cell
- semiconductor
- solar panel
- crude oil
- distillation
- gasohol
- anaerobic respiration
- glycolysis
- pyruvic acid
- hydroelectric power
- turbine
- nuclear energy
- nuclear waste
- radiation
- windmill
- geothermal energy
- aquaculture
- methane gas

It is important that renewable energy sources are identified for agriculture if efficient production is to continue, Figure 13–1. As supplies of petroleum and other nonrenewable energy sources are reduced, the costs of these fuels will increase. Research efforts have intensified to identify **renewable resources** for fuel from plants and other sources. These energy sources are needed to supplement or replace petroleum as the fuel source for agriculture. Energy sources must be identified that can be replaced by natural cycles as they are used.

SOLAR ENERGY SYSTEMS

Solar energy is obtained directly from the sun. It is used in a variety of applications in agriculture. Early livestock farmers learned that livestock shelters

Energy Sources

Nonrenewable	Renewable
Coal	Solar (sun)
Petroleum	Wind
Natural gas	Geothermal
	Ethanol (grains)
	Hydro-Power (water)
	Nuclear
	Methane gas
	Wood fuel

FIGURE 13–1 Energy comes from both renewable and nonrenewable sources.

FIGURE 13–2 A greenhouse is a structure that provides a protected environment for plants. It traps heat from the sun and makes it possible to raise plants in artificial environments when outside conditions are not favorable to plant growth.

were warmer in the winter if they had a southern exposure. Heat from the sun was radiated into the building.

Modern energy efficient buildings derive heat from the sun in the same way today. They are constructed to allow the sun to shine into the structure. Masonry products or water are often used to absorb the heat from the sun. During the hours of darkness, the heat is radiated back into the room from the storage materials.

A **greenhouse** is a structure designed to create an indoor growing environment for plants, Figure 13–2. It is designed to capture energy from the sun in the form of heat to allow plant growth during unfavorable seasons of the year. A greenhouse also allows plants to capture light energy using the process of photosynthesis. Although we do not often think of photosynthesis and solar energy being closely related, all food energy is captured from the sun and stored in plant tissues.

A **solar cell** uses silicon to generate electricity directly from light. Silicon is a nonmetallic element that occurs in nature as sand and other compounds. It is a **semiconductor**, meaning that it is a poor conductor until it is acted upon by heat, light, or electricity. Silicon and some other materials give up electrons when light strikes them. It is this characteristic that makes a solar cell function. When radiant heat in the form of sunlight strikes the surface of a solar cell, an electrical current is produced.

Solar cells will require a lot of testing and development before they are capable of becoming major sources of electricity. The efficiency of solar cells is improving, however, and it is reasonable to expect that a time will come when solar cells will become major sources of electrical energy. Their use is limited to low voltage applications at the present time.

Solar panels are heat exchange devices that have been developed to trap heat from the sun, Figure 13–3. They are often used to heat water or other liquids from which the heat can be extracted for other uses. Some solar panels

FIGURE 13–3 Solar panels are used to convert the energy of the sun to electricity or to capture heat from the sun.

FIGURE 13–4 Electricity is generated by huge solar generating plants that focus the energy of the sun to concentrate light and heat. *(Photo courtesy of Southern California Edison)*

are used to generate electricity, that can be stored in batteries for later use, Figure 13–4. Electricity from solar sources is sometimes used in remote locations to pump water for livestock.

ALCOHOL FUEL

Alcohol fuel has the potential to solve at least two major problems associated with petroleum fuels. Petroleum fuels will be used up someday, and they are a source of pollution to the atmosphere when they are burned. Evidence of air pollution is most easily seen in populated areas, but pollution also occurs in rural areas.

Gasoline and diesel fuels come from **crude oil** that is pumped from huge underground reservoirs. Crude oil is a nonrenewable resource, which means that once it is used up there will not be any more to replace it. Although vast supplies of oil are still available for our use, we must begin to develop new sources of fuel that are renewable. The time will come when petroleum products will be available in smaller amounts and at much higher costs than we pay for the products today.

Alcohol fuels come from renewable resources, Figure 13–6. They are obtained from the sugars and carbohydrates that are produced by plants. Every new generation of plants produces another supply of the raw materials needed to produce alcohol fuels, and many plant by-products can be used for this purpose.

CAREER OPTION

Agricultural Engineer

A person who chooses a career as an agricultural engineer will work with agricultural problems to find practical solutions, Figure 13–5. Agricultural engineers design agricultural machines, livestock facilities, storage facilities, field improvements, and other structures and systems that are associated with agricultural problems.

A career as an agricultural engineer requires a university degree. The course of study is rigorous and will require a strong background in math and science. Much of the development of new energy sources for agriculture is the work of agricultural engineers.

FIGURE 13–5 An agricultural engineer solves agricultural problems by designing machines, structures, and processes that provide solutions to the problems. *(Photo courtesy USDA/ARS #K–3396–6)*

FIGURE 13–6 Alcohol fuel is produced by fermenting plant materials containing carbohydrates and sugars. This ethanol plant uses potato waste from a nearby potato processing plant as a source of raw materials. The alcohol is mixed with gasoline to make gasohol.

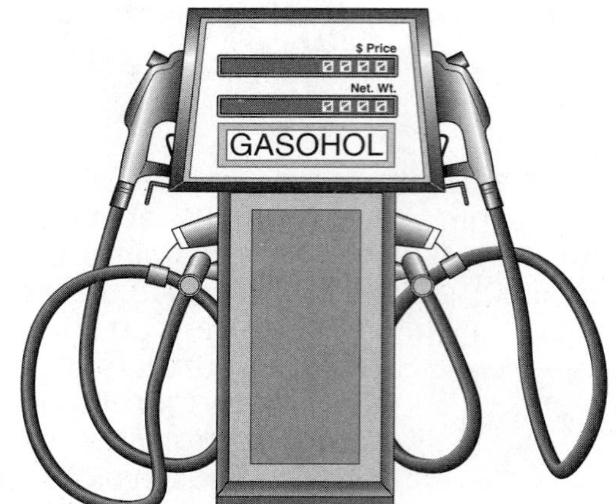

Advantages of Alcohol Fuels:

- Renewable source of energy
- Burns clean
- Provides new markets for crops
- Improves octane rating of gasoline

Disadvantages of Alcohol Fuels:

- Expensive to produce
- Negative image

FIGURE 13–7 Alcohol fuels are likely to gain favor for agricultural and commercial uses as the cost of petroleum products increases and more emphasis is placed on reducing pollution to the environment.

Air pollution is a serious problem in our world. It is true that much of the pollution comes from cars and factories, but some of it comes from agriculture's use of fossil fuels. These fuels are used to operate the many large engines used to perform agricultural tasks. They are also used to generate electricity for use on farms and in agricultural processing plants. When alcohol fuels are used, air pollution is minimal because alcohol is clean burning, Figure 13–7.

Ethanol

Ethanol is an alcohol that is produced by fermenting products that contain large amounts of carbohydrates and sugars, Figure 13–8. The most common farm product used in this process is grain. Once the fermentation process is complete, the alcohol is separated from other liquids in the solution by a process called **distillation**. The solution is heated just above the temperature at which ethanol boils while remaining below the temperature at which water boils. The ethanol vapor is condensed back to a liquid in a separate container.

The most common use of ethanol as a fuel is to mix it with gasoline to form a product called **gasohol**. This mixture contains about 10 percent ethanol and 90 percent gasoline. It is a clean burning, efficient fuel that is available from many commercial sources.

Ethanol Production

Ethanol production for fuel is a relatively new concept that has become commonplace since the 1970s, Figure 13–9. The process by which ethanol is produced is known as **anaerobic respiration** or fermentation. Oxygen is not present

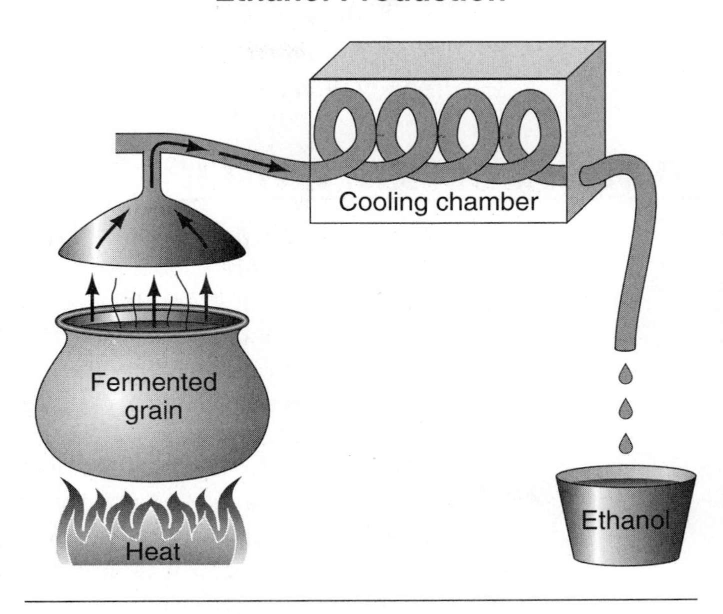

FIGURE 13–8 Ethanol is produced by fermenting grains and other carbohydrates to form alcohol. The alcohol is recovered by boiling it out of the solution and cooling the alcohol vapor to condense it into a liquid.

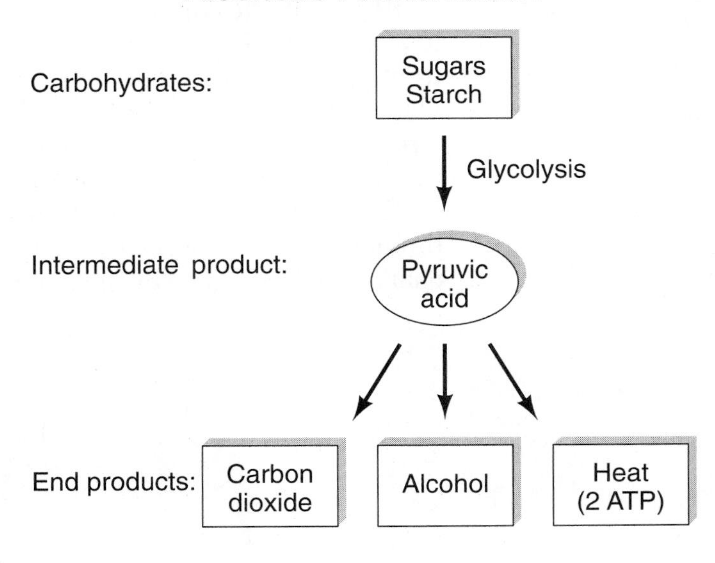

FIGURE 13–9 Alcohol is a product of fermentation or anaerobic respiration. Most of the energy obtained from the raw materials is concentrated in the alcohol fuel that is produced.

during any phase of the anaerobic respiration process. The first stage of the fermentation process is called **glycolysis**. During glycolysis, sugar molecules break down to form organic acid molecules called **pyruvic acid**.

In plant cells and yeasts, electrons are added to the pyruvic acid molecules, and ethyl alcohol, also known as ethanol, is produced. This fuel is clean burning, and ethanol does not add pollution to the atmosphere as it is used. Water molecules are formed as by-products during burning. One disadvantage of pure ethanol as a fuel is that it tends to be unstable, and it explodes more readily than gasohol.

HYDROELECTRIC POWER

Hydroelectric power is electricity that has been generated from the energy of moving water. It is a major source of electrical power in areas where rivers and streams are plentiful and where dams and hydroelectric generators have been constructed, Figure 13–10.

Most hydroelectric facilities are designed so that falling water strikes the curved vanes of a **turbine**, causing it to turn an electrical generator to produce electricity.

Hydroelectric power is relatively clean and inexpensive to produce. It is usually a reliable source of energy, but in years when drought conditions occur, electrical power output is severely limited by reduced stream flows. The use of hydroelectric power is limited in areas that are unsuited for the construction of dams. Limitations also exist at sites that are inappropriate for large reservoirs and lakes.

Hydroelectric power has been blamed in recent years for many of the problems associated with the decline in populations of migrating fish. Steelhead trout and several species of salmon live in rivers and streams for several months after they are hatched. In their second year, they move down the streams to the ocean where they live until they are mature adults. Eventually they migrate back up the rivers to the streams where their lives began.

The construction of dams has disrupted the flow of the major rivers through which these fish migrate. Fish ladders have been constructed at dams to accommodate the inland migration of mature fish, but the juvenile fish

FIGURE 13–10 Many large dams have been constructed for production of hydroelectric power. The large lakes that are created are useful for recreation, and they provide habitats for fish and wildlife.

populations are being subjected to heavy losses on the outward migration. Many of these fish are killed when the water current carries them through the blades of the turbines that provide power to electrical generators. Others lose their way to the ocean in the quiet backwaters of the dams.

River currents keep the young fish moving in the right direction, as they make their outward migration to the ocean. Fast river currents also move them quickly past such predators as squawfish and seals. When the river currents are slowed by dam construction, the migrating fish lose the advantage of speed in their trip to the ocean. Some salmon populations have become so small that the fish have been placed on the list of endangered and threatened species.

Efforts have been made to maintain populations of these migratory fish by capturing the adult fish, collecting their eggs, and raising their offspring in fish hatcheries. This is done when they return to their native streams to spawn, but these efforts have not been very successful. Too many of the young fish die as they migrate down the rivers. Large amounts of stored water have been released in recent years in efforts to flush the young fish through the backwaters of lakes and reservoirs, but success has been minimal. This practice uses large amounts of water needed to irrigate farms, generate electricity, and maintain commercial shipping lanes on the inland rivers.

NUCLEAR ENERGY

The use of nuclear reactors for any purpose is a controversial issue in many parts of the world. Some people believe that the technology is unsafe. They are concerned that a nuclear accident would endanger the lives and property of many people.

People who favor the use of nuclear power defend the industry by describing the safeguards that are in place to regulate the industry. They believe nuclear power is a safe, clean, reliable source of energy.

Nuclear energy is heat that is released from an atom during a nuclear reaction. It is frequently used to generate electricity. The agricultural industry is a major consumer of electrical energy. It is used to perform many tasks on farms and ranches, and to provide power for processing agricultural products. Large amounts of electricity are used to pump water for irrigation in arid regions.

Nuclear energy will probably continue to provide large amounts of electricity for use in the cities, towns, and rural areas of the world. It is a source of power that will still be available long after the fossil fuels are gone.

The most difficult issue that has been raised in the debate over nuclear safety is what should be done with the spent fuel. This material is called **nuclear waste** and it still emits radiation. **Radiation** consists of high energy particles from the nuclei of atoms of certain elements such as plutonium, uranium, radium, and a few others. These particles are known to be very dangerous to living organisms. Exposure to low levels of radiation is known to cause cancer and birth defects in humans. High levels of exposure cause severe burns and sometimes death.

Unlike most other waste materials, nuclear waste is slow to degenerate into harmless debris. The nuclear waste that is generated in our lifetimes will still be unsafe to our grandchildren. Much of this waste is stored in metal containers at nuclear facilities such as the Idaho National Engineering Laboratory. Nobody knows how long the metal storage containers will last before radiation leaks are detected.

If we are to continue to use nuclear power, permanent storage methods and sites must be developed for nuclear wastes. Sites have been constructed in New Mexico and Nevada as permanent storage facilities for radioactive materials, but neither site has yet accepted any nuclear waste shipments.

GRAIN AS A FUEL SOURCE

Grain contains a high concentration of energy that has the potential to produce heat. Experimental work is being conducted using grain as a coal substitute. The researchers are evaluating the efficiency of mixing corn with coal and burning it to provide heat and to generate electricity for a large midwestern university. If the technology proves to be efficient, whole grains, which are renewable resources, could be used to replace some of the nonrenewable resources such as coal and other fossil fuels.

Besides providing a new source of heat, the use of grain as a fuel source could increase the demand for corn and other small grains. Higher demand for grain and grain products would tend to stimulate higher market prices for grains.

WIND POWER

The industry of agriculture has used the wind as a source of power for many years, Figure 13–11. **Windmills** represent an old technology that has been revived for modern uses. They are constructed using many different kinds of blades or vanes that rotate in the wind, turning the shaft to which they are attached.

Windmills have been used in some areas to provide power for grinding grain, but the main use of windmills has been to pump water to the surface from underground sources for use by humans and their animals. Wind power is a valuable resource in areas where electrical power is unavailable. Windmills are used today in many of the desert livestock ranges where surface water is scarce.

In recent years, windmills and wind turbines have been used to turn generators in the production of electricity, Figure 13–12. Some government agencies require power companies to purchase surplus electricity from individual citizens who generate more power than they need using windmills, water power, and other generating technologies.

FIGURE 13–11 Humans have used natural energy sources such as wind for many years. In addition to generating electrical power, wind is still used to pump water in remote desert areas. *(Photo courtesy of Utah Agricultural Experiment Station)*

FIGURE 13–12 The Darrieus windmill will turn regardless of the direction the wind blows. It is used to turn a generator to produce electricity. *(Photo courtesy of the United States Department of Energy)*

Power companies are researching ways to generate electricity using the wind as a source of energy. In some areas, large numbers of windmills are located in a single windy zone. Substantial amounts of electrical power can be generated at these sites and is available for commercial, agricultural, and domestic uses.

Wind powered generators are not always reliable power sources because the wind cannot be depended upon to blow all of the time. Some solutions to this problem include using generators powered by coal or petroleum products at times when the wind is not blowing. Another way to solve this problem may be to use electricity stored in large batteries when wind conditions are too weak to generate adequate supplies of electrical power.

GEOTHERMAL ENERGY

Geothermal energy is obtained from the interior of the earth in the form of heat, Figure 13–13. When the molten material that forms the core of the earth is near the surface, underground water is heated to high temperatures. Geysers and hot springs are the result of this condition. Sometimes the water

FIGURE 13–13 Geothermal energy occurs as hot water created when water is heated by the hot core of the earth's molten center. Heat and energy can be recovered from geothermal sources, using a variety of technologies to capture the heat from the water.

FIGURE 13–14 Agriculture includes the production of fish. This modern aquaculture facility uses solar and geothermal resources to create a warm water environment for producing tilapia fish.

is hot enough to produce large amounts of steam that can be used to generate electricity.

Hot water is a good source of heat. Sometimes it is used to heat concrete floors in livestock buildings or to heat homes, greenhouses, and other farm structures during cold weather.

Most hot springs produce water that is a constant temperature during all seasons of the year. When water from these sources does not contain excessive amounts of dissolved minerals, and when the temperature of the water is moderate, it may be used for **aquaculture** or fish farming, Figure 13–14.

Some species of fish such as tilapia thrive in warm water. Consumption of fish in the United States is expected to rise during the next few years, and the aquaculture industry is expected to expand. The availability of fresh water heated with geothermal energy is an important natural resource to this growing industry. Fish production using tilapia and other warm water species of fish will become an important industry in northern climates as favorable water sources are identified.

METHANE GAS

Methane gas is a component of natural gas. It is a product that is formed when vegetable matter decomposes. Methane is produced in natural settings such as marshes and swampy areas where plant materials build up and begin to decay. The damp, warm conditions that are found in these areas during the summer season are favorable to the production of methane gas. Large amounts of

FIGURE 13–15 Methane gas is produced commercially by decomposing organic materials. It is a natural product that is generated wherever decomposition occurs.

this gas are released into the atmosphere from this source. It is used most efficiently as a fuel for heating.

Methane gas is produced by ruminant animals during the digestion process. This gas is usually lost into the atmosphere when cattle, sheep, and other ruminants burp to relieve the pressure of the gases that form as forage is digested. Attempts to recover methane from this source have met with limited success. Methane is lighter in weight than many of the other gases in the atmosphere. It has a tendency to rise to the highest point in an enclosed space. Efforts to recover methane have used special building designs that allow methane to rise to collection areas near the peak of the roof.

One successful method that can be used to obtain methane gas from agricultural sources is to collect it from decomposing animal manures. This is done by storing manure in large closed or sealed storage areas, Figure 13–15. This practice has been used on a limited scale in some countries as a source of heat for houses. Animal waste and other vegetable matter is fermented in large, sealed plastic bags or other enclosures from which methane gas is collected. Hoses and tubes connected to the bags transport the gas to stoves and heaters.

CAUTION:
People should avoid entering closed manure storage areas because gases which are present in such facilities can be fatal. People lose their lives every year when they enter manure storage areas.

FUEL FROM ANIMAL WASTES

Dried animal manure has been used as a source of fuel for centuries in many parts of the world. People still use it to heat their homes and to cook their meals in areas of the world that have limited supplies of coal, woody plants, and other fuels.

Researchers have developed a method for removing oil from animal manure. The product appears to have properties that are similar to those of petroleum products. This new oil product may prove to be useful as a source of fuel or lubricants at some future time.

CHAPTER SUMMARY

The agricultural industry has become dependent on petroleum and other nonrenewable fossil fuels as energy sources. Alternative fuels will be needed as supplies of petroleum, coal, and natural gas are used up. Among the alternative sources of energy that are available for agricultural uses are solar energy, alcohol fuel, and electricity from generators powered by water, wind, and nuclear energy.

Energy for some agricultural purposes is recovered in the form of heat by burning grain mixed with coal, and from water heated by geothermal energy. Oil and methane gas have been recovered as by-products of animal metabolism.

CHAPTER REVIEW

Discussion and Essay Questions

1. Name the fossil fuels that are important to agriculture, describe how they are used today, and predict their importance in the future.
2. What is the difference between renewable and nonrenewable resources?
3. List and describe technologies associated with agricultural uses of solar energy.
4. How are alcohol fuels used in agriculture today, and how are they likely to be used in the future?
5. What are the advantages and limitations of hydroelectric power?
6. Discuss the arguments for and against using nuclear energy as a major source of electrical power.
7. How is whole grain used as a source of power?
8. What is the potential of wind power as a source of energy for agriculture?
9. What is the source of geothermal energy, and how is it used for agricultural purposes?
10. What are the sources of methane gas, and how can it be recovered in livestock production?

Multiple Choice Questions

1. Which of the following energy sources is nonrenewable?

   A. natural gas
   B. methane gas
   C. ethanol
   D. solar power

2. A resource or source of energy for which a new supply is constantly becoming available even as it is consumed is (a):

   A. coal.
   B. petroleum.
   C. renewable resource.
   D. nonrenewable resource.

3. Solar energy is obtained from:

   A. sunlight.
   B. hot water.
   C. petroleum.
   D. greenhouses.

4. A material that produces and conducts an electrical charge when it is exposed to heat, light, or electricity is known as which of the following:

   A. insulator
   B. conductor
   C. magnet
   D. semiconductor

5. An alcohol fuel that is produced by a process known as anaerobic respiration is called:

   A. ethanol.
   B. pyruvic acid.
   C. glycol.
   D. buterol.

6. Electricity that is produced by the action of falling water on a turbine that turns the armature of a generator is known as:

   A. geothermal power.
   B. nuclear energy.
   C. solar power.
   D. hydroelectric power.

7. Nuclear waste is a potentially dangerous by-product of nuclear power generation because it emits energized particles through a process known as:

   A. amatorization.
   B. radiation.
   C. fermentation.
   D. distillation.

8. The most limiting factor associated with obtaining energy from the wind is:

   A. lack of technology to "harness or capture" the energy from wind.
   B. undependable winds that sometimes stop blowing.
   C. inadequate methods of storing energy obtained from the wind.
   D. the high cost of maintaining equipment.

9. Heated water obtained near the molten mass of material that makes up the center of the earth provides energy that is often referred to as:

   A. nuclear energy.
   B. solar energy.
   C. geothermal energy.
   D. radiant energy.

10. Which of the following *is not* a source of methane gas for fuel?

    A. swamps and marshes
    B. stomach gases from ruminant animals
    C. decaying animal manure
    D. fermented plant materials

LEARNING ACTIVITIES

1. Assign each class member to research a different kind of energy source for agriculture. Assign three students to play the role of commission members who will choose a new source of energy for agricultural use in your area. Hold a mock public hearing in the class with teams of students who were assigned the same energy source acting as corporate attorneys making presentations to promote the advantages of the energy source they researched. The three students who are acting as commission members should make a decision based on the evidence and report their findings to the class.

2. Invite a representative from a local electrical power supplier to make a presentation to the class describing how electricity in the area is used by agriculture. Ask the presentor to provide information on the following:

   a. source of local power
   b. the cost of producing power using several alternatives
   c. agricultural uses of electricity in the local area
   d. plans for meeting future power demands

SECTION V

Computer Aided Management

- **Computer Aided Management in Marketing and Business**
- **Computer Aided Production Management**

CHAPTER 14 Computer Aided Management in Marketing and Business

The ***computer*** *is a machine that is capable of performing mathematical calculations in a programmed sequence, Figure 14–1. It is able to interpret large amounts of information and apply it to marketing and management situations. The computer has become a very important tool for agricultural marketing and business management. The information provided by a computer is as good or as weak as the information placed in its memory.*

OBJECTIVES

After completing this chapter, you should be able to:

- describe how computers are used to access and provide management information.
- explain what a computer network is and how it can be used in the agricultural industry.
- list examples of information sources available through agricultural computer networks.

FIGURE 14–1 The computer has become an important business tool. It is used to keep records, manage inventories, and track products from processing through the marketing process. *(Photo courtesy of Utah Agricultural Experiment Station)*

- appraise the value of electronic mail services available through computer networks.
- assess the value of computerized records to agricultural businesses.
- define the importance of computers for enterprise analysis.
- examine the roles of computers in agricultural marketing activities.
- discuss the use of computers in managing agricultural business inventories.
- explore the role of computers in tax management.

TERMS FOR UNDERSTANDING

The following vocabulary terms should be studied carefully as you read:

computer
hardware
laptop
notebook
software
hard drive
CD-ROM
documentation
electronic mail
home page
database
computerized records
enterprise analysis
asset
inventory management
depreciation
tax management

COMPUTER HARDWARE

Computer **hardware** consists of the mechanical and electronic structures and devices that are used to make a computer. This includes the monitor, keyboard, disk drives, modem, printer, and any other similar components found in a computer system. There are many kinds of computer hardware, Figure 14–2. A good

FIGURE 14–2 Computers are available in many styles and sizes adapted to a variety of uses and conditions. *(Courtesy of Michael Dzaman)*

computer consultant can be very helpful in assisting to match computer hardware to the needs of the computer buyer.

Computer hardware should be matched to the conditions in which it is expected to operate. For example, a computer keypad that is located in a dairy barn must be designed to be waterproof, and it must be easily cleaned. Computer equipment that is located in an office can be designed quite differently.

Computers that are used to perform marketing and business management functions require sufficient memory capacity to run complicated analysis programs. Large amounts of computer memory are needed for this purpose. The computer programs that are used will determine the kind and capacity of computer hardware that is needed.

Some of the most versatile computers are called **laptop** or **notebook** computers. These are small computers that can easily be carried in a briefcase or large purse. They are equipped with battery packs that make it possible to use them when a power source is not available. Many business managers carry one of these computers with them as they travel so that they can use their travel time efficiently. In many instances, a computer of this type that is equipped with a modem is used to access office and business computer files from distant locations.

Printers are available with the capability of printing in black and white or color. Professional quality printing can be easily attained when a high quality printer is used. Much of the work that is done in marketing and business management, including the design of high quality marketing materials, can be done using a home or business computer system and printer.

COMPUTER SOFTWARE

Computer **software** includes programs, data, and information that is stored on magnetic tape, disks, or CD-ROM systems, Figure 14–3. A huge industry has

FIGURE 14–3 Most business computers contain internal storage disks, but information is also stored on CD-ROM, floppy disks, compact disks or on tape.

grown up to support the use of computers. Computer programmers design and write programs that direct the computer to perform specific functions upon command. These programs are saved on computer disks that are sold through retail stores and mail order companies in locations throughout the world.

Computer program disks that are properly packaged and handled can be shipped almost anywhere, Figure 14–4. The ability to save computer programs on disks has made it possible for the very best in computer technology to be marketed and distributed widely.

Several different technologies have been developed to store computer information. Magnetic tapes were used with early computers, but the tapes became obsolete for most purposes when computer disks were developed. More recently, internal computer **hard drive** information storage has become a standard feature in computers. It allows vast amounts of information to be stored permanently inside the computer. The use of disks has become a convenient way to transfer information from one computer to another, or to provide a backup information storage system, but they are seldom used as primary information storage sites.

A new technology that is used to make high quality music recordings has now been adapted to storage of computer information. It is called **CD-ROM**, which stands for Compact Disk-Read Only Memory. The technology involves storage of information on plastic disks coated with vaporized aluminum or silver, Figure 14–5. Information is inscribed on the disk by a laser beam, and the intensity of reflected laser light is measured by an optical sensor to retrieve the information.

Much more information can be stored on a CD-ROM disk than on a standard computer floppy disk. A single CD-ROM disk is capable of storing vast

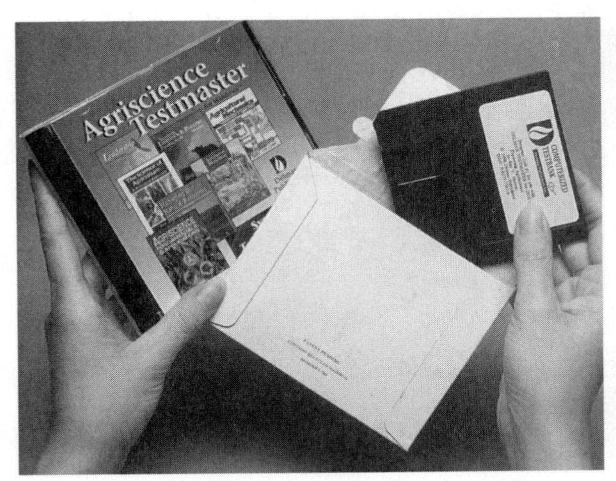

FIGURE 14–4 Special packages make it possible to ship information and computer programs stored on disks to distant parts of the world.

FIGURE 14–5 A CD-ROM disk can be used to store written text, video, photographs, slides, and sound.

amounts of printed materials, but it can also be used to store video, photographs, slides, and sound. Agricultural extension educators in some states store all of their extension bulletins and booklets on CD-ROM, and they print copies only as they are needed. This technology is capable of changing the way we file and retrieve all types of information.

An experienced user of computers is often able to use computer software with minimal instruction. **Documentation** that consists of written instructions on how to use the computer program should be available with quality software at the time it is purchased.

AGRICULTURAL NETWORKS

One of the most important functions of a computer is to access and organize information. It is important to remember that information can be delivered and interpreted, but it cannot be created by the computer. All information with which the computer works has been selected and entered into the computer by people.

A computer network is a delivery system that links two or more computers together and allows them to have access to the same information, Figure 14–6. A network can be developed within an agricultural business or processing operation using electrical cables between computers.

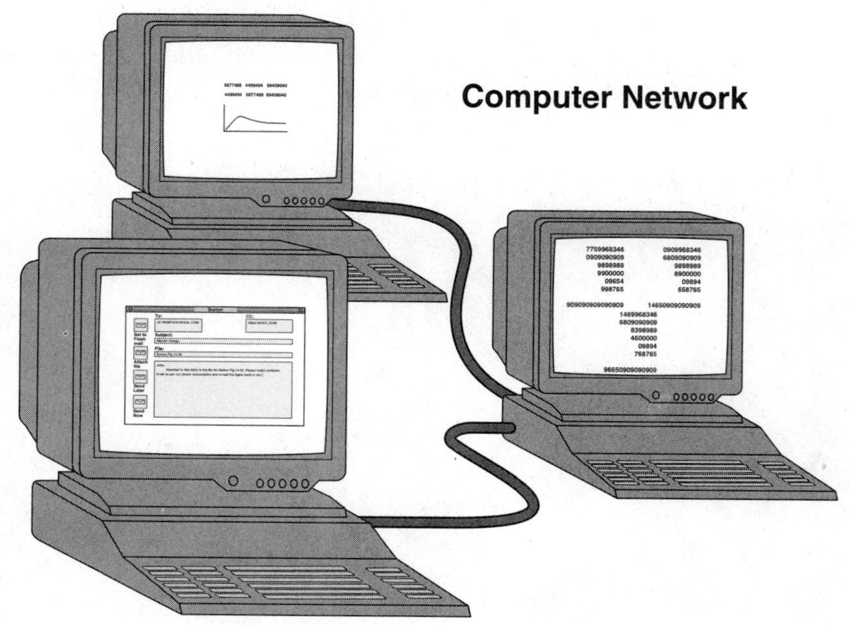

FIGURE 14–6 A computer network consists of two or more computers linked together to allow them to have access to the same information. They may be connected by wires, cables, telephone lines, or a variety of other methods.

FIGURE 14–7 A modem is used to link a computer to a telephone line. It converts electronic signals produced by the computer to signals that are used by telephone companies to transmit information. Some modems are now internal.

Computer networks can also be set up using a device called a modem to connect the computer to a telephone line, Figure 14–7. A modem is a device that allows computer data to be transmitted over telephone lines. The data is changed by the modem to a format that is compatible with the telephone data system. A modem at the other end of the phone connection transforms the data back to its original form. The network is completed when computers using modems at distant sites retrieve information from the telephone line. This is done by calling a designated phone number that allows the caller to access the computer on which the desired information is stored. This type of computer network is used for many agricultural applications.

Many types of information and services are available using computer networks. They include market prices, educational materials, electronic mail, information searches, commodity market information, and other information and services. Computer networks provide price updates for many different agricultural commodities. Information available includes prices for futures contracts, options contracts, cash sales at major markets, and other vital market information.

Internet

The internet system has emerged as the most widely used information network in the world. Many smaller networks can be accessed using the internet, and much of the world's important information is available to anyone who cares to view it. The libraries and databanks of the world are at our fingertips through a simple connection with the internet system.

An internet connection is available by subscribing to a connecting service through one of the business organizations that is established for this purpose. Many people use the internet only to obtain information or to communicate with other network users by **electronic mail**, Figure 14–8. Electronic mail allows a person to receive messages by obtaining an electronic mail address.

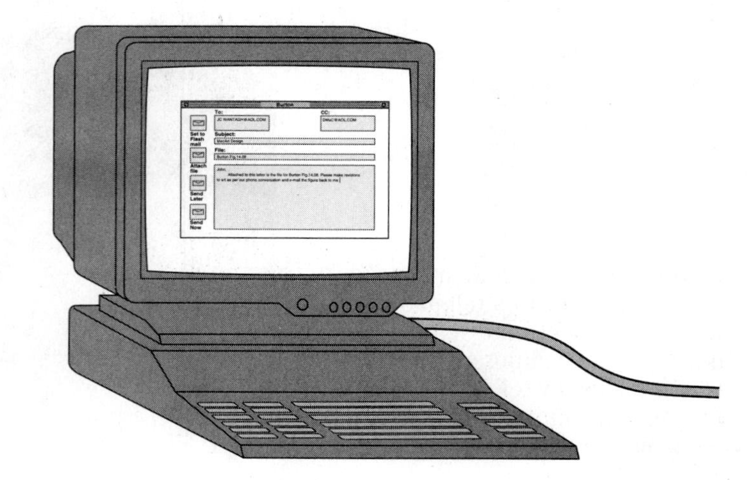

FIGURE 14–8 Electronic mail is used to send and receive messages to and from persons who are served by the same computer network.

Electronic mail can be sent to anyone on the internet system whose electronic address is known.

Individuals or institutions desiring to provide information to internet users are able to do so by establishing a **home page** on the system. A home page provides a menu of the information that is available, and it allows subscribers to obtain the information by accessing it from computer files. These files and the information that they contain are called computer **databases**.

ACCOUNTING SYSTEMS

A major task of every farm and agricultural business is to keep financial records. **Computerized records** are important tools that are useful for financial management. They provide a method of accounting for the financial success of the business. Dependable financial accounting procedures are required for taxing purposes. Financial institutions that provide funds to operate and expand agricultural businesses require good accounting procedures. Good financial records provide information about the strengths and weaknesses of a business.

The most important use of a good accounting system is to provide accurate information upon which business decisions can be based. Poor quality records often lead to poor business decisions. Owners of farms and other agricultural businesses can obtain computer accounting programs that are helpful in interpreting and analyzing the financial fitness of their business enterprises. When accurate records are maintained on a high-quality computerized system, information needed to make good decisions is readily available, Figure 14–9.

The most critical function of a business manager is to make wise decisions based on accurate information. Part of this process is to know when to buy and when to sell. For example, a large purchase such as a tractor should be carefully studied before the purchase is made. Computerized records can be used to answer important questions such as: How large should the tractor be in order

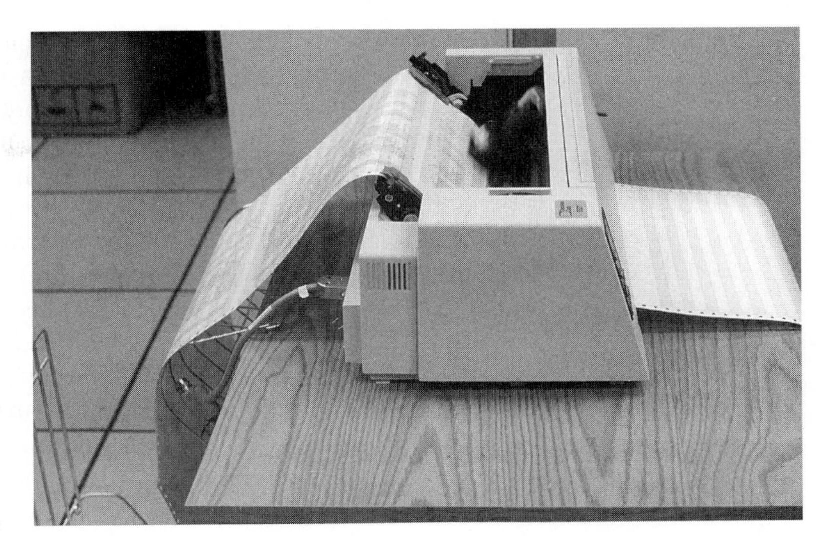

FIGURE 14–9 Computers are capable of storing large amounts of information. They are also used to generate reports that aid managers in making sound business decisions.

CAREER OPTION

Agribusiness Manager

The manager of an agricultural business provides expert advice, supplies, or services to the agricultural industry. He/she needs a good educational background in business and accounting procedures, Figure 14–10. A college degree in an agricultural discipline is often required.

Most agribusiness managers work their way up in the management system of the company or cooperative with which they are employed. As their management skills improve, they are given added responsibilities.

Successful agribusiness managers are paid a good wage for their services. They usually work in rural communities near the farms and ranches with which they do business.

FIGURE 14–10 Many agricultural businesses offer products and services to farmers, ranchers, and other agricultural businesses. Managers and employees of these organizations must understand the agricultural industry and be skilled in the use of effective business practices.

to do the work that is required? When will income be available to make the purchase? How will this debt affect the ability of the business to obtain operating loans? How much additional income must be generated to make the payments? Should a new tractor be purchased, or would a good used model do the job? These and many other questions can be answered when a good set of records is available.

Computer software packages are available that provide excellent financial accounting systems for agricultural farm and business uses. Computer programs can be used to prepare balance sheets, net worth statements, and cash flow projections that are required by banks and other lending agencies. They are useful in determining how much financing will be needed and when it will be required. These kinds of computer programs are valuable to both the owner of the agricultural business and the lender, because they make it easier to manage the cash flow and to minimize risk.

ENTERPRISE ANALYSIS

It is important to know which parts of a business are profitable and which parts are resulting in losses. **Enterprise analysis** is an accounting procedure that is used to examine the profitability of a segment of the business. For example, many farms produce both crops and livestock. In many cases more than one crop or livestock enterprise is produced.

It is best to analyze each component of the business separately from all other components of the business. This is because unprofitable enterprises can remain hidden from view when profit or loss is calculated for the entire business. Enterprise analysis exposes weak business enterprises and identifies strong ones, Figure 14–11.

Sometimes farmers spend their lives believing that they are making a profit on livestock when the crop enterprise was the part of the business that was profitable, while the livestock enterprise may have produced little or no profit. In other cases, the farmer may do better to concentrate on raising livestock and buy the feed instead of raising it.

A good computer program capable of performing an enterprise analysis is an important tool for farm business managers.

Enterprise Analysis

	Enterprise			Total Farm
	A	B	C	
Gross Income	42,710	14,374	64,211	121,295
Total Expense	–36,514	–19,710	–41,512	–97,736
Net Income	6,196	–5,336	22,699	23,559

FIGURE 14–11 Enterprise analysis is a method of comparing the different components within a business with each other. Sometimes the total business will show a profit even though one or more enterprises is losing money.

MARKETING DECISIONS

Many factors affect the decision to sell agricultural products. Farmers and ranchers must decide whether it is most profitable to sell their produce, or to store it in the hope that commodity prices will increase faster than storage costs accumulate.

The manager of an agricultural business must compare the costs of product storage versus the present value of the products. It may be to the advantage of the seller to accept a reduced price for a commodity when the cost of storing the product is high.

At each step in the marketing process, the owner of the product faces possible losses. Once animal products, fruits, or vegetables reach ideal market condition, quality begins to decrease. The producer has only a limited time in which to sell most products. Value of products declines sharply once quality begins to diminish.

Computer programs are useful for determining when a product should be sold to achieve maximum net profit, Figure 14–12. When the expense of producing or storing a product becomes as great as the potential income that is generated by storage or additional yields, the time to sell has arrived. When the costs of production and market prices are known, computers can be used to predict the ideal time that a product should be sold to maximize profits.

FIGURE 14–12 Computers are useful tools in determining the appropriate time to buy or sell a product to achieve the greatest profitability. It can quickly compare costs and rates of production, current market prices, and market trends to predict the ideal time to purchase or sell.

INVENTORY MANAGEMENT

Business **assets** consist of anything that is owned and can be exchanged for something of value. **Inventory management** is a system for keeping track of business assets. Inventories are used by insurance companies to compensate for losses. It is important that inventories are up-to-date and accurate. Computers can be used to simplify inventory management, Figure 14–13.

The first step in inventory management is to compile a list of business assets by physically inspecting and recording all of the assets of the business. Once the inventory has been compiled in a computer program or on paper, it will be necessary to update the inventory at regular intervals. New assets should be added to the inventory listing, and assets that have been disposed of should be removed from the list.

Computer software is available that can be used to automatically reduce the value of the inventory assets as their value diminishes or as the tax laws allow it. This reduction in the value of business assets is called **depreciation**. Inventory management includes preparing depreciation schedules.

Agribusinesses that carry large inventories consisting of a large number of different items for resale are able to keep track of the inventory at the check-out stand, Figure 14–14. In some instances, the items are marked with computer bar codes that can be read by passing a laser operated wand across the code at the time of the sale. The computer can be programmed to automatically remove the item from the business inventory. It also prepares lists of materials that are needed to restock the shelves and maintain the product inventory.

Computers are used by agricultural processors to keep track of the raw product inventory, and to determine how efficiently the raw product is converted into

Inventory Management

FIGURE 14–13 An important part of every business management plan is to maintain an accurate inventory record. A computer is an ideal tool to use for this task because it is easily updated by adding or deleting items. A current inventory record can be printed at any time.

FIGURE 14–14 The task of managing the inventory of items that are stocked for resale is made easier by computerized equipment at checkout stands. *(Courtesy of USDA/ARS #K–2656–2)*

FIGURE 14–15 Farm equipment businesses provide valuable services to the agricultural industry by maintaining an inventory of replacement parts and a staff of trained mechanics and technicians who maintain and repair equipment.

FIGURE 14–16 One very important agricultural business is the insurance company. Many different insurance companies provide services for agricultural businesses and farms by assuming some of the financial risks associated with accidents, bad weather, and other adverse conditions that affect agriculture.

manufactured products. A computerized inventory of the processed products is maintained, and products are deleted from the inventory as they are shipped.

There is a broad array of agricultural businesses. They range from farms and ranches to business organizations that support the agricultural industry. They include suppliers and processors of agricultural products, Figures 14–15, 14–16, and 14–17. The vast transportation system includes businesses that transport agricultural products. Marketing companies exist for the purpose of bringing buyers and sellers of agricultural goods together to do business. Inventory management is important to all of them.

FIGURE 14–17 Feed and seed companies exist in nearly every rural community. They provide quality seeds at planting time and feed processing services throughout the year.

TAX MANAGEMENT

The use of computers for keeping agricultural business records has made it possible for managers to use the assets of the business in a much more effective manner. **Tax management** is a procedure for managing business assets. It should not be a tax avoidance procedure. Tax management simply shifts profits from high income years to years when income is lower than normal. Agricultural managers are able to use their records to determine when it is to their advantage to delay receiving income until another tax year. When taxable profits are higher than usual, they can be used to buy supplies that will be used the following year. In this manner, tax advantages can be gained.

Computer accounting programs have been developed to compile specific financial information that is needed to complete tax forms. Tax laws are always changing, and computer software that is used to calculate and understand taxes should be updated as often as necessary to reflect the changes in the tax law. Some tax software companies provide annual updates for customers who have purchased their basic tax software packages.

Some computer record systems can generate the financial data that is required for tax forms directly from the agricultural business accounts. Once these initial inputs have been completed, computers are able to perform the calculations that are required to calculate the amount of taxes owed. Computer records have greatly simplified the computation of taxes, and they have improved the abilities of agricultural managers to manage business assets.

CHAPTER SUMMARY

Computers have become important tools for agricultural managers. They are used to perform many of the routine tasks associated with marketing and management activities of agricultural businesses. Many different kinds of information related to the success of a business are readily available when business records are managed using computers.

Several useful computer programs are available to managers. They are used to keep financial records, update inventories, gather market information, communicate, analyze business options, and gather data for marketing decisions and tax management purposes.

CHAPTER REVIEW

Discussion and Essay Questions

1. Describe how computers are used to gather useful information for agricultural management decisions.
2. Explain what a computer network is and describe how it is used by the agricultural industry.
3. List examples of the kinds of information that can be accessed through computer networks.
4. What information can the internet system provide that is valuable to agricultural managers?
5. What is the value of electronic mail service and how is the service used?
6. In what ways are computerized record systems valuable to agricultural businesses?
7. Why are computers so useful in the process of enterprise analysis?
8. Describe the roles of computers in marketing agricultural products.
9. Explain how computers are used to compile and manage agricultural business inventories.
10. In what ways are computers used in compiling information for tax management purposes?

Multiple Choice Questions

1. Which of the following computer items is considered to be hardware?

   A. monitor
   B. computer program
   C. computer paper
   D. disk

2. Which of the following computer items is considered to be software?

   A. monitor
   B. keyboard
   C. modem
   D. disk

3. A computer information storage device that uses laser technology is called a:
   A. magnetic tape.
   B. 5¼" floppy disk.
   C. CD-ROM.
   D. 3½" computer disk.

4. Written instructions for the operation of a computer system or program is called:
   A. a service manual.
   B. documentation.
   C. hard copy.
   D. a notebook.

5. A network is a system that links two or more computers together using electrical cables, or that links computers together using a telephone line and a device called a:
   A. modem.
   B. circuit breaker.
   C. condenser.
   D. module.

6. The internet computer system provides access to information all over the world through a computer system called a:
   A. home page.
   B. database.
   C. tutorial.
   D. network.

7. Which of the following functions cannot be performed by a computerized record system?
   A. make business decisions
   B. calculate a net worth statement
   C. prepare a balance sheet
   D. project cash flow revenues

8. A record analysis procedure that evaluates the profit earned by each product that is marketed or each service that is offered is called:
   A. cooperative analysis.
   B. analytical depreciation.
   C. market analysis.
   D. enterprise analysis.

9. A business practice that sometimes uses computer software for the purpose of tracking assets that belong to the business is called:
   A. computer sleuthware.
   B. inventory management.
   C. cooperative analysis.
   D. enterprise analysis.

10. The *most important* consideration in the purchase of computer software for tax management purposes is:
    A. cost of the software.
    B. compatibility of the tax software with the accounting software.
    C. the availability of updated software that takes into account the changes in the tax law.
    D. compatibility of the software with the computer hardware.

LEARNING ACTIVITIES

1. Invite an agricultural loan officer to the class to discuss the importance of business records to lenders during the approval process for agricultural loans. Prior to the visit, students should learn to define simple financial terms such as balance sheet, net worth, cash flow, assets, liabilities, collateral, etc. Ask the loan officer to relate the financial forms used by the lending agency to the records the students are expected to keep on their own agricultural enterprises.

2. Take your class to the computer laboratory in your school. Load a simple accounting program into the computers, and have the students make the appropriate entries into the program from a simple record keeping problem provided to them.

CHAPTER 15 Computer Aided Production Management

Technology has entered the decision making process on farms and ranches. It has provided new tools that are useful in farm management. The computer is one of these tools, Figure 15–1. New computer software is available that is used to simulate the effects of changes in costs, yields, nutrient content, and other variables that affect management decisions in production agriculture.

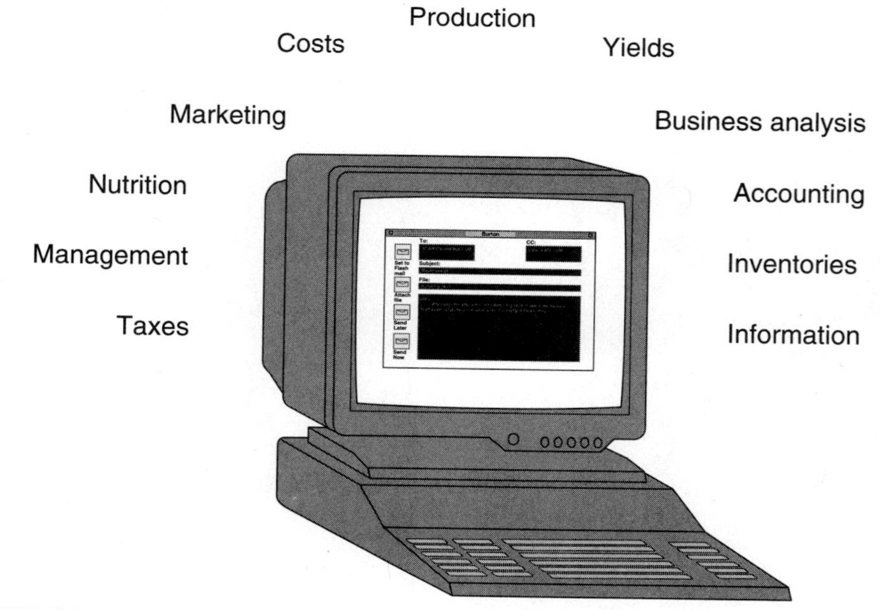

FIGURE 15–1 Computer technology has many uses in agriculture. Large amounts of information can be accessed, stored, and analyzed using a computer.

OBJECTIVES

After completing this chapter, you should be able to:

- explain the role of computers in identifying superior animals.
- describe how computers are used to identify health problems in breeding herds.
- explore the uses of computers by breed associations for livestock management purposes.
- discuss the importance of computers in selecting and culling breeding animals.
- justify the use of computers to analyze and formulate livestock rations.
- assess the uses of computers for managing growing crops.
- evaluate the importance of computers in managing farm machinery.

TERMS FOR UNDERSTANDING

The following vocabulary terms should be studied carefully as you read:

seedstock
pedigree
mature equivalent
fertility
breeding record
parturition record
breed association
progeny
progeny testing
ration
balanced ration
least cost ration
prescription farming

LIVESTOCK MANAGEMENT

Animal agriculture is a big industry. It begins with the purebred livestock breeder who produces **seedstock** for the industry. Seedstock consists of high quality animals that exhibit particular genetic characteristics. These animals are used to produce the foundation of breeding animals used in commercial herds. Animals of seedstock quality are identified using production records. Once a superior family of animals has been identified, it is often possible to select superior animals by studying their **pedigrees**. The pedigree of an animal is the record of its ancestors. The presence of superior animals in a pedigree is a good indication that an animal possesses superior genetic traits, but pedigrees are of little value unless production records are also available for each animal.

Many of the livestock record systems used in the purebred industry are available on computers. When the production records have been entered in the system, a computer is capable of comparing the merits of individual animals.

This kind of information is used to select breeding animals that exhibit superior traits and to identify mediocre animals.

Dairy farmers have been leaders in establishing the value of dependable production records. The Dairy Herd Improvement Association (DHIA) was organized to provide unbiased production testing for dairy cows. Milk samples are gathered each month, and the production of each cow is recorded for a 24 hour period. The technician also gathers other information such as breeding and parturition dates. All of the data is entered into a computer database, and monthly reports are generated for the dairy owners.

DHIA computer printouts rank each cow in comparison with her herdmates. A cow in her first lactation period is compared with mature cows in the herd by calculating her **mature equivalent**, or ME Index. This index takes into account her expected production as a mature animal. The index is based on the known production trends of all of the cows that have been production tested by DHIA. The ME Index allows the estimated genetic value of young animals to be compared against the known production of mature cows. DHIA records are also used for comparisons between individual cows in different herds and to identify outstanding animals in the breed.

Other services that are provided by DHIA include detailed statistics about each cow. These statistics include fertility information such as the number of days between calving and conception of the next calf. They alert the dairy manager to do pregnancy checks on cows that have skipped a heat period, and they identify cows whose production appears to be different than what was expected.

Computerized livestock production records are useful in analyzing the health of animals. When the milk production of a dairy cow, Figure 15–2, or the egg

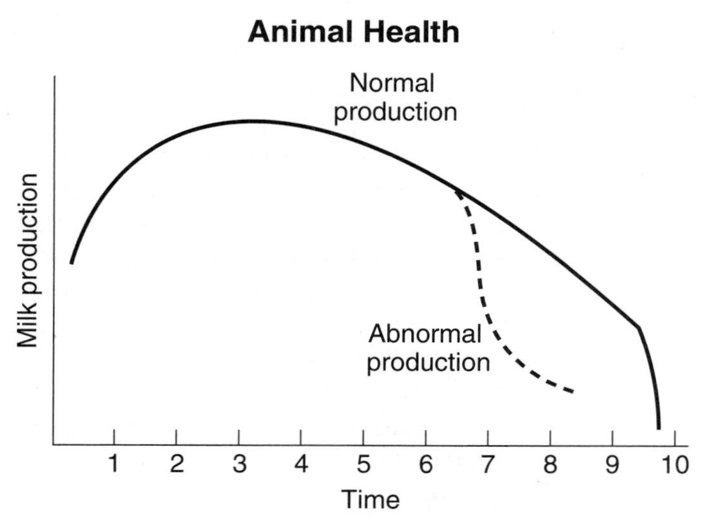

FIGURE 15–2 Poor health can sometimes be detected in a dairy cow when her milk production suddenly drops below expected levels. Computers are useful in creating graphs and other production information that can be used for this purpose.

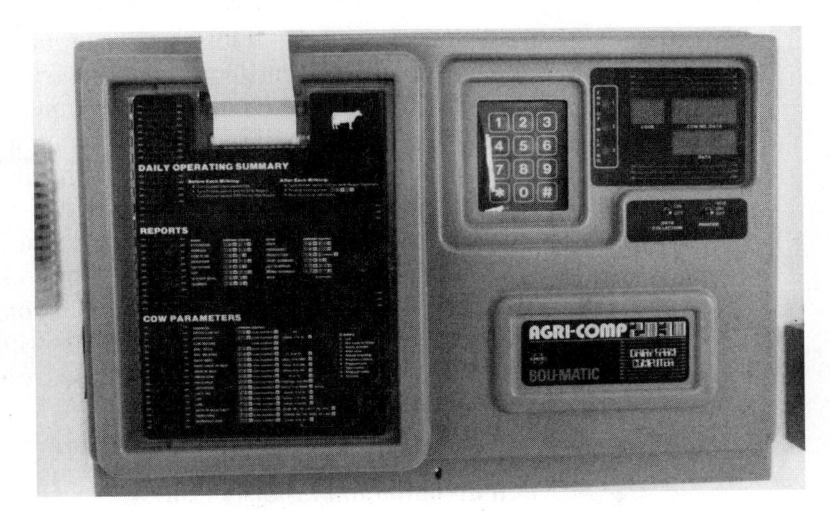

FIGURE 15–3 Computer terminals located in strategic areas make it possible to enter data in the records of individual animals for use in making management and culling decisions. *(Photo courtesy of Utah Agricultural Experiment Station)*

production of a flock of hens drops below the expected levels, a computer printout can be useful in identifying a particular event or date when a suspected health problem began, Figure 15–3. This kind of information is useful in treating health problems before they become advanced.

Breeding and Parturition Records

Fertility is the capacity of an animal to produce offspring. It is one of the most important traits for which breeding females are selected. **Breeding records** include the date of the mating and the identities of the sire and dam. **Parturition records** list the identity of the offspring and the date of birth. Breeding and parturition records are used to measure the fertility of a breeding animal.

An animal that requires several matings to conceive, or is not able to consistently give birth to live, healthy offspring is described as being infertile or low in fertility. Computer analysis of the breeding and birth records of an animal can provide information that is useful in identifying and treating fertility problems.

Animals that do not conceive readily may have an infection in the reproductive organs that can be treated successfully. Birth records that indicate a history of long intervals between births may be an indication that nutrition is inadequate or that other serious management problems exist.

Selection of Replacement Females

Production of animal products has increased at a brisk rate as production records of animals have become widely used. The records have made it possible to identify and preserve superior female breeding stock. Several types of records have proven useful. They include birth weight, rate of gain, weaning weight, frame size, health, carcass characteristics, and records of other traits the owner considers to be important.

Animal **breed associations** are organizations that keep records of animal pedigrees and issue certificates of registry, that verify the pedigrees of animals. They have developed computer programs that are used to compare the performances of individual purebred animals with other animals in the respective breeds. A comparison between an individual animal and other animals within a herd or a breed is assigned a numerical value called an index. An index value of 100 is used to represent an animal that is of average value.

A herd index is useful as a selection tool. Replacement females should be selected whose index values are high in comparison with their herdmates. By replacing low value animals with the offspring of animals having high index ratings, the production of the herd can be expected to improve at a steady rate.

Culling Management

A cull is an animal that is removed from the breeding herd and sold. An animal whose production index rating is low in comparison with other females in the herd should be considered a candidate for culling.

The computer is an important tool in culling management. It is used to calculate the production index of each female in the breeding herd. A computer can also be used to identify animals that are poor producers on the basis of their production index ratings. Once an animal enters the breeding herd, it is time to develop a production index based on her mothering ability, breeding performance, and ability to produce high quality products.

The production index should be based on the fertility of the female and the average weaning weights (adjusted for age) of her **progeny**, or offspring. Fertile animals will produce healthy offspring at appropriate intervals. High weaning weights for the progeny tend to measure the milk production of the mother. Both traits are important in measuring the productive performance of individual females in the breeding herd.

Sire Selection

Computers frequently play a major role in evaluating the performance of prospective sires, Figures 15–4 and 15–5. Large amounts of data are compiled on sires in the livestock industry. Much of this information has been gathered in attempts to determine the values of individual sires before they are widely used by purebred livestock breeders or artificial breeding organizations.

Test stations are established in many locations to test the relative value of prospective sires in the beef and swine industries. Some of these stations are located at land grant universities as part of the research facilities, but a growing number of these stations are owned and managed cooperatively by the livestock breeders who use them. Computers are used to assemble test data and to analyze the results of the test.

A bull or boar testing station usually accepts only a limited number of animals at the facility during the testing period. Animals of similar weights are

FIGURE 15–4 Computerized performance records are frequently used to compare the merits of prospective sires. *(Photo courtesy of David Burton)*

FIGURE 15–5 Production records of valuable sires are frequently compiled using computer technology. *(Photo courtesy of Columbia Sheep Breeders Association of America)*

AGRI-PROFILE

Selecting a Super Bull

Selection of a super bull begins before the bull calf is even born. In fact, the process begins before he is even conceived. Computers are used to compile data on the production performances of the top cows within the breed. These cows must also be sound in conformation and type. The strengths and weaknesses of each cow are analyzed, and a bull is selected for the mating. Each of these steps is conducted using computer technology.

A bull calf that is produced through this process is given the best of care, Figure 15–6. Only healthy bulls can be used in artificial breeding programs. When the young bull reaches puberty, he is bred to females in several herds and locations. Now time must pass while his daughters mature and begin to produce. Each daughter is evaluated to see how well she performs in comparison with the production of her mother and her herdmates. This process is called **progeny testing**.

After many months the proofs are complete. Bulls whose female progeny have performed in an exceptional manner are placed in the stud barn. During their active breeding years they will sire thousands of calves in many parts of the world. Bulls whose progeny fail to perform well are hauled away to become hamburger.

FIGURE 15–6 Computer matings are used to produce superior bulls for commercial use. Stud farms and breeding associations frequently select prospective sires before they are even born.

entered at the station where they are tested for rates of weight gain, feed efficiency, carcass characteristics, and other traits that are believed to be important. Computer printouts of test results are made available to owners as the testing period progresses. Production testing is usually followed by auction sales where the prospective sires are sold as breeders.

Breeding associations are interested in marketing semen obtained from the best sires available. To ensure that only the best sires are used, young sires are selected from among the progeny of the best animals in the respective livestock breeds. If the young sires produce offspring that perform better than their dams did at similar ages, they will be put into active service. Sires that fail this test are usually marketed through the meat industry.

Balancing Livestock Rations

A **ration** is the amount of feed that an animal eats in a day, Figures 15–7 and 15–8. A **balanced ration** provides an animal with nutrients that are of the proper quality and in the proper amounts based on production to satisfy the nutritional needs of the animal, Figure 15–9.

Computers are used to select a balanced supply of nutrients in livestock feeds. Sources of feed ingredients are identified and tested to determine their nutrient contents, Figures 15–10 and 15–11. This information is keyed into a computer. Computer software programs are available that are capable

FIGURE 15–7 Computer terminals located in livestock barns make it possible to dispense measured amounts of feed to individual animals. *(Photo courtesy of Utah Agricultural Experiment Station)*

FIGURE 15–8 Computer chips that identify individual animals are used to provide measured amounts of feed for each cow. These cows carry computer chips on their neck chains.

FIGURE 15–9 This cow carries a computer chip on her neck chain that controls her access to feed. Another chip is located in her left ear. It provides her identity to the computer, which records her milk production. The amount of grain she receives depends on the amount of milk she produces.

FIGURE 15–10 Computers play important roles in calculating feed ingredients. Feed samples are tested in the laboratory, and the data from the tests is fed into computers to be analyzed.

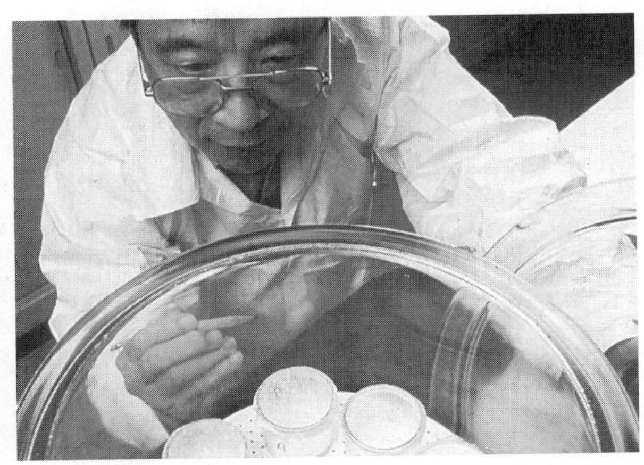

FIGURE 15–11 Agricultural laboratories perform many services such as analysis of feeds, soils, and plant tissues. Most of these facilities use computers in the scientific processes that they perform. *(Photo courtesy of USDA/ARS #K–4472–8)*

of selecting ration ingredients in the proper amounts. The computer is used to formulate a ration matched to the needs of the animals to which it is fed.

When several feed ingredients are available, the cost of the ration should be considered as the computer selects ingredients. **Least cost rations** are composed of feed ingredients that satisfy the needs of the animal at the lowest cost. Many livestock managers have paid for a computer system with the money that is saved through least cost selection of ration ingredients.

CAREER OPTION

Farm/Ranch Manager

A farm or ranch manager is a person who has excellent practical skills in management, business, and marketing. Some of these skills are learned by working with people, but many skills require formal training. Many farm managers are college or university graduates who have acquired skills through their agricultural experiences and technical knowledge through agricultural studies.

A farm manager may be a farm owner, or he/she may be an employee of an individual or corporation that has ownership of a farm. Farm managers often perform the work required to produce farm products, but the primary responsibility of the farm manager is to gather the records and other information that is needed to make sound business decisions.

A farm manager must be a self-motivated person who likes to work hard. The work days are long, and the work conditions are often demanding and dirty. Farm managers often work outside in the heat of summer and the cold of winter, Figure 15–12. They also spend long hours with farm records—gathering, compiling, and analyzing production costs and marketing information.

A farm manager's family experiences a unique style of living. They usually live in the country where they are away from the crowds, the smog, and the traffic of cities. They have opportunities to observe nature. The children of the family are often given responsibilities on the farm where they learn to work at a young age.

FIGURE 15–12 Farm and ranch managers need to be experienced in performing the work required to produce a product. They also need skills in managing people and conducting business. *(Photo courtesy of Angie Richard)*

CROP MANAGEMENT

Computer aided crop management has become a reality in recent years. Agricultural consulting services and farm managers frequently use computers to monitor the needs of fields and crops. They are used to determine when irrigation is needed, based on soil moisture tests, Figure 15–13. They are also used to make fertilizer recommendations during the growing season, based on the

FIGURE 15–13 Soil moisture is measured using a variety of instruments including the soil tensiometer pictured here. Soil moisture information at two different soil depths is available using this arrangement.

FIGURE 15–14 The lands exhibit, located at the Epcot Center in Orlando Florida's Disney complex, is devoted to researching possible ways to raise food in outer space.

levels of plant nutrients that are present in the leaves of crops. When computers are used in combination with other scientific instruments, the needs of plants can be readily identified, Figures 15–14, 15–15, 15–16, and 15–17.

Remote sensing equipment provides crop and soil information to computer centers that enhances the data and interprets the information. Some work has been done to link computers to sensing devices located on growing plants. The sensing devices are designed to measure tiny electrical impulses that are asso-

FIGURE 15–15 Several methods have been devised for raising large numbers of plants in limited space—the implications of this are far-reaching.

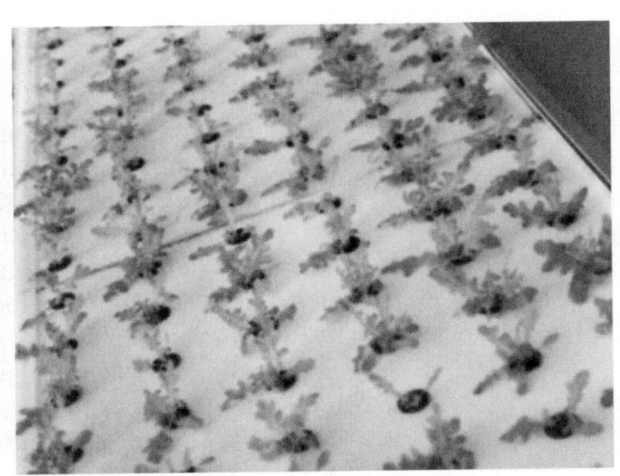

FIGURE 15–16 In space travel the amount of room that can be devoted to food plants is limited. This problem has resulted in an effort to find new ways to raise plants that are used for food production.

FIGURE 15–17 The atmosphere within the lands exhibit is controlled by a variety of electronic devices including computers.

FIGURE 15–18 Field research, such as range analysis and measurement of climatic conditions, uses field-based computers to record and interpret data. *(Photo courtesy of Utah Agricultural Experiment Station)*

ciated with cell metabolism. Stressed plants are thought to give off a different pattern or intensity of electrical impulses than healthy plants. The computer can be used to detect stress in plants and to suggest management options appropriate for the situation, Figure 15–18.

Computers can be used to monitor costs and inputs for a crop during the growing season. This is a record keeping activity, but such information is valuable in determining crop budgets and cash flow plans in the years that follow.

MACHINERY MANAGEMENT

Computer management of machinery is a growing concept. Modern tractors and other types of farm machinery often have computer equipment, Figures 15–19 and 15–20. Digital readings display such information as total hours of use and the length of time since the machinery was last serviced. They can also be used to determine the amount of time required to perform different field operations. This information is important in determining the machinery costs involved in raising various crops.

The computer is also useful in selecting machinery. Programs are available that analyze the costs of owning and operating machines of various types and sizes. This kind of information is helpful in making decisions such as whether to buy a machine or what size of machine should be purchased. Sometimes a computer analysis will indicate that it is more cost effective to hire a neighbor to do the work than it is to own the machine.

The use of computers to control irrigation systems is another important crop management technology. Highly sophisticated irrigation systems such as

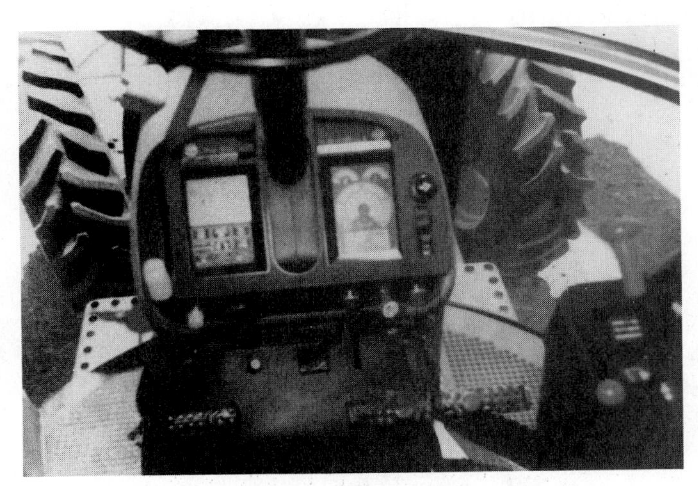

FIGURE 15–19 Modern farm tractors and combines are equipped with a variety of computerized devices.

FIGURE 15–20 Computers are standard on new farm equipment. They are used for keeping maintenance records, calculating acreages, and many other uses.

center pivot and lateral move systems depend on computer technology to keep the system properly aligned as it moves across a field. Even a slight deviation in alignment can damage the system. A computer controls the drive system at each tower to ensure proper alignment of the towers and the water delivery system. Modern engines have become so high tech that we now depend upon computerized equipment to properly maintain and service them, Figure 15–21.

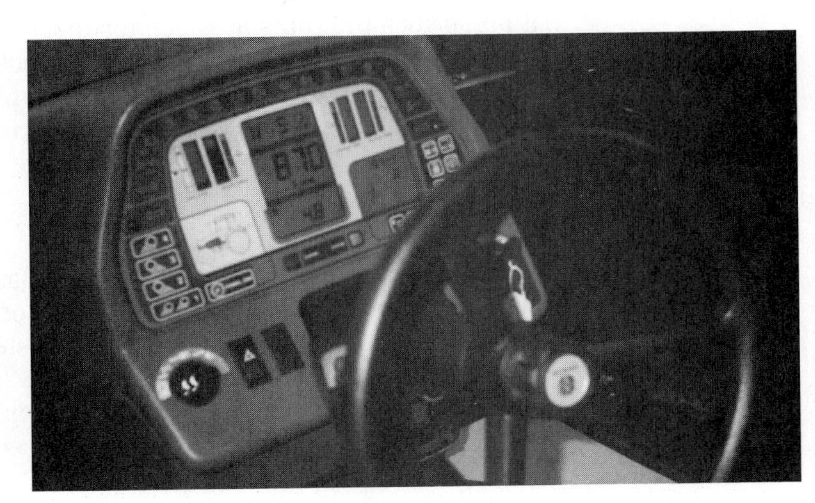

FIGURE 15–21 In addition to traditional tools, mechanics frequently use on-board computers to maintain service records and analyze the performances of modern agricultural machines.

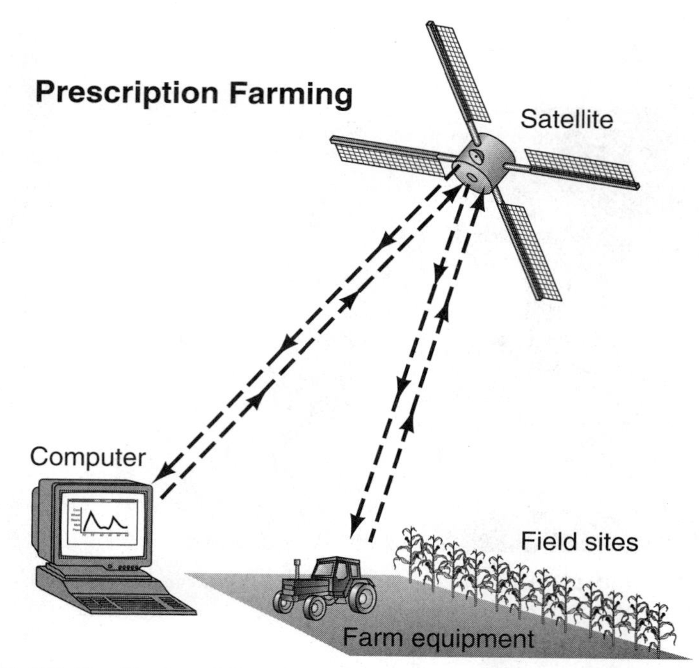

FIGURE 15–22 The National Site-Specific Technologies project is developing space-age engineering technologies for agricultural uses.

The National Site-Specific Technologies for Agriculture project is adapting engineering technologies for agricultural uses that were developed as part of the NASA space program. The integration of global positioning systems (GPS) and geographic information systems (GIS) was discussed in Chapter 8. This project depends on computers to manage the databases that contain the information each field has generated.

The computer system receives information via satellite from field sites and it prescribes the rates at which fertilizers, seeds, and chemicals are applied to each field, Figure 15–22. All of these inputs are influenced by past production at each specific field site. This is an example of the highest level of computer aided management. It is called **prescription farming**.

CHAPTER SUMMARY

A computer is a management tool that has gained wide acceptance in production agriculture. Its greatest value is in organizing, tabulating, and analyzing farm records. The data that is obtained using a computer is useful in making farm management decisions.

Among the agricultural production uses of the computer are identification of animal health problems, selection of breeding animals, formulating and analyzing rations, determining needs of growing crops, and scheduling the maintenance of farm machinery.

CHAPTER REVIEW

Discussion and Essay Questions

1. Explain the importance of computers in identifying seedstock quality breeding animals.
2. Describe how computer records of breeding and production can be used to identify health problems in animals.
3. Identify the types of records that are compiled by breed associations, and describe how they are used by farm and ranch managers.
4. Discuss the value of computers in selecting and culling breeding animals.
5. Describe how computers are used to analyze and formulate livestock rations.
6. List some ways that computers are useful in managing and caring for growing crops.
7. Explain how computers may be used to recognize stress symptoms in growing plants.
8. Identify ways that computers are used on modern farm machines.
9. Defend the practice of using computers to make decisions related to farm machinery purchases.
10. Propose additional uses of computers in the production of livestock and crop products.

Multiple Choice Questions

1. Breeding animals that have been selected from families with superior production records are known as:

   A. progeny.
   B. purebred livestock.
   C. registered stock.
   D. seedstock.

2. A mature equivalent index for a dairy cow refers to:

   A. the age of an animal.
   B. the predicted productive ability of an animal corrected for age differences.
   C. the degree to which cartilage in the skeleton has been replaced.
   D. the number of calves to which the cow has given birth.

3. Sudden unexplained reductions in animal productivity are strong indicators that:
   A. the animal may have a health problem.
   B. the moon has entered a phase that interferes with production.
   C. an error has been made in the collection of production data.
   D. the animal is too old to maintain production.

4. Breeding and parturition records are used as a measurement for:
   A. production.
   B. fertility.
   C. maturity.
   D. progeny testing.

5. Livestock organizations that keep records of the pedigrees of purebred animals are called:
   A. agricultural cooperatives.
   B. Dairy Herd Improvement Associations.
   C. production credit associations.
   D. breed associations.

6. A herd index number of 102 indicates that the relative value of the animal in comparison with herdmates is:
   A. below average.
   B. average.
   C. above average.
   D. far superior.

7. The milk production of commercial breeding females of the meat animal breeds is usually measured by comparing the progeny's:
   A. weaning weights.
   B. birth weights.
   C. mature weights.
   D. feed conversion rates.

8. A sire testing procedure that takes into account the production of the offspring of the sire is:
   A. production testing.
   B. fertility testing.
   C. progeny testing.
   D. efficiency rating.

9. A daily feed supply for an animal that takes into account the proper nutrient quality and amounts in the feed is called a:
   A. balanced ration.
   B. ration.
   C. least cost ration.
   D. rational diet.

10. The function of the computer in a prescription farming program is to:
    A. create new information based on old facts.
    B. store critical crop and soil information and determine the amounts and the timing of seed, fertilizer, and chemical applications.
    C. predict production levels of the field sites.
    D. measure the efficiency levels of the different farm implements that are used.

LEARNING ACTIVITIES

1. Obtain the use of a computer lab and invite an agricultural extension educator to demonstrate the use of farm management computer programs dealing with livestock selection, ration formulation, and machinery costs.

2. Invite a progressive farm manager to speak with the class about ways computers may be used to improve management decisions.

Environmental Technology

- **Controlled Living Environments**
- **Protection of Natural Resources**

CHAPTER 16

Controlled Living Environments

A controlled living environment is an area in which the factors that support life are partially or completely controlled. Such an area must be totally enclosed, and the temperature, humidity, light, and atmospheric gases present in the area are maintained at specific levels.

OBJECTIVES

After completing this chapter, you should be able to:

- identify the essential components of a controlled living environment.
- describe how controlled living environments are used to solve agricultural problems.
- explain the importance and functions of electronic control systems in controlled living environments.
- evaluate the importance of technology in managing particulate and chemical contaminants in controlled living environments.
- suggest ways that controlled living environments can be adapted for crop production and storage purposes.
- propose ways to use controlled living environments for confinement livestock housing.
- relate controlled living environments to the environments in which aquaculture is practiced.

TERMS FOR UNDERSTANDING

The following vocabulary terms should be studied carefully as you read:

environmental management
static environment
Biosphere II
ambient temperature
comfort zone
humidity

humidifier
dehumidifier
visible spectrum
absorption spectrum
particulate matter
chemical contaminants
elemental cycle
denitrification
nitrogen cycle
carbon cycle
dormant
confinement housing
ventilation

ENVIRONMENTAL MANAGEMENT

Environmental management is a system used to control some or all of the factors that maintain a stable life support system. It includes the structures, equipment, procedures, and materials required to monitor and maintain a **static environment** in which conditions do not change.

One example of a controlled environment is the astronaut living area in each of the space shuttles. Air, water, light, and heat are all provided in the environment in the right amounts to maintain a comfortable and healthful living area. These same factors must be controlled in the living environments that are used for agricultural purposes.

Some important activities in the environmental management process are determining the appropriate amounts of humidity, heat, light, and gases that should be supplied in the environment, and designing ways to control them.

The **Biosphere II** project was the most advanced effort ever attempted to create a controlled living environment, Figure 16–1. It was an experimental prototype of an environment for people, animals, and plants located in the Arizona desert. It was engineered to produce food and oxygen for people and animals, and carbon dioxide for plants. Water was recycled by plants and through evaporation. The original plan was to seal it to prevent movement of air or water into or out of the facility for the two-year duration of the experiment.

FIGURE 16–1 Biosphere II is a complex in which all human needs were to be supplied using only the resources that were found inside the sealed structure. Eventually the oxygen level had to be increased and it is widely believed that other supplies were supplemented as well. *(Photo courtesy of Space Biospheres Ventures, Arizona)*

FIGURE 16–2 Confinement housing for animals must provide adequate ventilation and light. Artificial lighting helps to increase egg production of laying hens. *(Photo courtesy of Utah Agricultural Experiment Station)*

Soon after the structure was sealed, however, one person was removed due to an injury. It has been alleged that additional supplies were taken into the system when the person was returned to the facility. After a few more months had passed, the oxygen levels in the closed environment became too low for humans to live comfortably. New air was eventually added to the environment.

During the final months of the experiment, the people raised food for themselves and their animals using only the resources that were available inside Biosphere II. In this first attempt to create an artificial environment, scientists learned just how difficult it is to maintain a balance of life-sustaining resources.

Many different kinds of modified or controlled environments are used in agriculture. They include confinement animal housing such as controlled environment livestock buildings and poultry houses, Figure 16–2. They also include storage areas and shipping containers for fruits, vegetables, grains, milk, meat, and other agricultural products. They include fish ponds, greenhouses, family dwellings, and farm shops. They even include artificial methods of preventing freezing temperatures in orchards and groves to prevent losses in fruit crops.

CONTROL SYSTEMS

Control systems for environmental management include many of the devices that were discussed in Chapter 12. They include electronic sensors that are used to detect high levels of humidity, heat, and gases such as ammonia, carbon dioxide, and oxygen, Figure 16–3.

FIGURE 16–3 Growth chambers are used by plant scientists to simulate different climates. The computerized control panel is used to program the system.

Temperature Control

Ambient temperature is the amount of heat in the environment that surrounds an organism or object. It is a critical factor in a controlled setting. For best results, a constant temperature in a storage or living area is needed. It is accomplished by providing supplemental heating or cooling. These systems are turned on and off as needed by a thermostat.

Livestock and poultry housing should be designed to maintain temperatures that are stable, Figure 16–4. The range of temperatures in which an animal is most comfortable is called its **comfort zone**. When ambient temperatures are maintained within the comfort zone, animals are capable of greater production than when they are too cold or too hot.

Many agricultural uses of controlled environments require that temperatures be reduced to slow plant metabolism. Potato and apple storage facilities are good

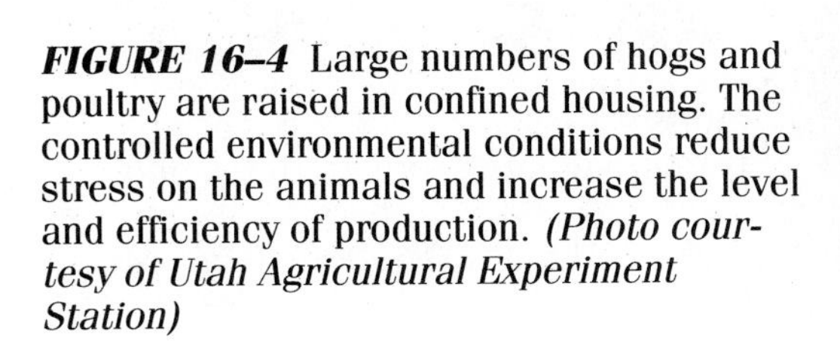

FIGURE 16–4 Large numbers of hogs and poultry are raised in confined housing. The controlled environmental conditions reduce stress on the animals and increase the level and efficiency of production. *(Photo courtesy of Utah Agricultural Experiment Station)*

AGRI-PROFILE

Egg Incubator

An incubator for hatching the eggs of poultry is a good example of a controlled environment, Figure 16–5. Fresh moist air must be constantly brought into the incubator to provide oxygen for the living embryos developing inside the eggs. High humidity is required to prevent the moisture from being depleted inside the egg.

The temperature is maintained near 100 degrees during the first week, and it is reduced slightly as the eggs approach the hatching date. After the fertile eggs have been incubated for 21 days, hatching takes place. If the temperature, humidity, or oxygen levels inside the incubator are not properly maintained, the embryo will die inside the egg, and the egg will rot.

Incubators are also used as a place to keep a baby that has been born prematurely. It provides a place with plenty of moisture to protect the lungs of the baby, and it provides a constant temperature for the comfort of the baby.

Another use for an incubator is to test the germination of seeds, Figure 16–6, or to raise cultures of disease organisms in veterinary labs from which animal vaccines are developed. Incubators have many uses, and they are prime examples of controlled living environments.

FIGURE 16–5 Incubators are used for many purposes such as creating acceptable environments for plants, animals, and premature human babies. This incubator is used to hatch turkey eggs. *(Photo courtesy of Utah Agricultural Experiment Station)*

FIGURE 16–6 Seed testing laboratories evaluate seeds to determine their germination rates. The rate of germination equals the percentage of seeds that sprout and grow.

FIGURE 16–7 Some vegetables and fruits must be consumed soon after they are harvested. Others can be stored in controlled environments that aid in maintaining high quality for extended periods of time.

FIGURE 16–8 A space-age growth chamber for calves protects them from harsh environmental conditions, provides a supply of fresh air, and protects calves from exposure to diseases by isolating them from each other. *(Photo courtesy of Utah Agricultural Experiment Station)*

examples of this technology, Figure 16–7. A potato or an apple is a living part of a plant. Respiration and cell functions continue while it is in storage.

The temperature inside a storage facility must be carefully controlled. If the temperature is not strictly controlled, the starches in fruits or vegetables may be changed to sugars, or sugars may be converted to starches. When this happens, potato products such as french fries or chips may become dark in color when they are cooked. Apples tend to become mushy and lose their quality rapidly when temperatures in storage facilities are improperly maintained.

Humidity Control

Humidity is the amount of moisture that is present in the atmospheric air. In controlled environments it poses some difficult problems. When humidity is too high in a confinement livestock system, animals are likely to have respiratory ailments such as pneumonia, Figure 16–8. When humidity is too low in storage facilities for fruits and vegetables, they tend to dry out and lose quality.

It is important to know what the proper humidity level is for a particular use and to maintain it at that level. Humidity is measured and controlled using a humidistat. Moisture can be added to the atmosphere using a **humidifier** or it can be removed using a **dehumidifier**.

Light Control

The need to control the amount of light to which animals or plants are exposed is an important aspect of the controlled environment concept.

FIGURE 16–9 High intensity light is often provided in greenhouses to raise specialty floral crops in seasons of the year when they do not normally produce flowers.

Different agricultural management situations require different amounts of light, and in some situations it is required at specific times.

The manager of a greenhouse business may need to alter the amount of light that a floral crop, such as chrysanthemums or Easter lilies, receives to simulate a different season of the year. Additional light can be provided by exposing plants to artificial light from special electric bulbs, Figure 16–9. These "grow lights" produce light of specific wavelengths or colors that are most useful to plants.

Plants respond best to light of specific wavelengths. This is because some of the light is absorbed and some of it is reflected, Figure 16–10. Different wavelengths of light make up the different colors that are seen in a rainbow. The colors of a light source may be separated in the laboratory by shining the light through a prism. It refracts the different colors to separate locations just as raindrops refract sunlight to create a rainbow. These colors, or wavelengths, make up the **visible spectrum** of light. The chlorophyll found in plant leaves and stems absorbs light in the red, blue, and violet wavelengths, which it converts to chemical energy. It reflects light that is in the yellow and green wavelengths causing chlorophyll to appear green in color. This reflected light is called the **absorption spectrum**.

Some animal management techniques involve modified light exposures as discussed in Chapter 4. The reproductive cycles of some breeds of sheep are stimulated by a gradual decrease in the amount of daylight to which the animals are exposed. During seasons when daylight is increasing, many breeds of sheep are not sexually active. The breeding cycle can be initiated during seasons of increasing daylight by controlling the length of light exposure to simulate shorter day lengths. This is usually done by keeping the animals in dark buildings during part of the daylight hours.

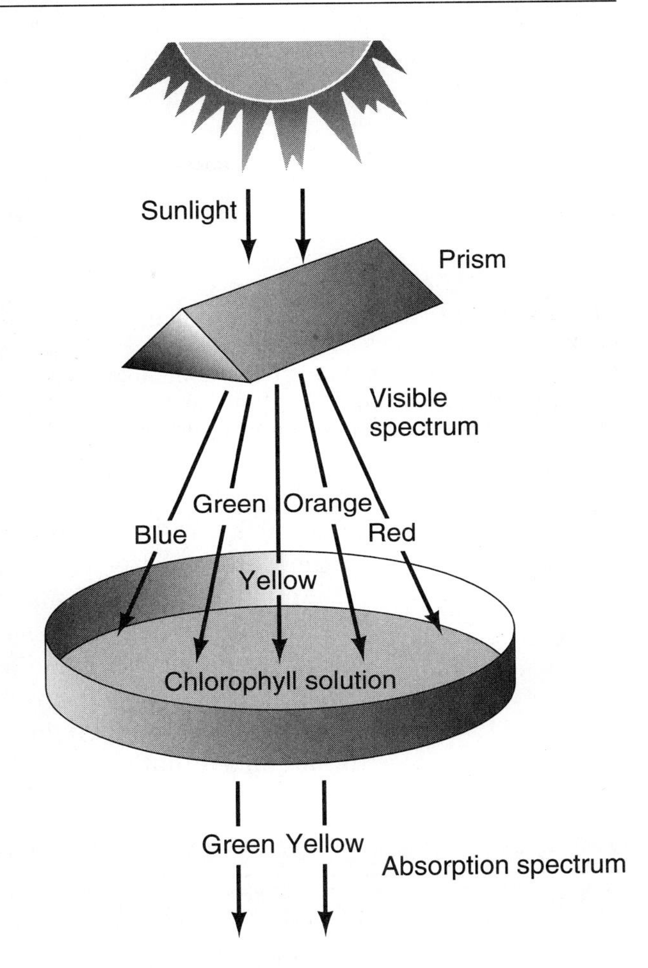

FIGURE 16–10 The chlorophyll in plant leaves absorbs light that appears blue, orange, and red in the visible spectrum. It reflects light that is green and yellow. This causes plant leaves to appear green or yellow in color.

Atmospheric Control

In many instances, atmospheric control is accomplished by using a good ventilation system, Figure 16–11. The outside air that is introduced into the controlled environment results in a fairly stable atmosphere. In some instances, however, other gases are desirable in the atmosphere.

Nitrogen gas is added in some crop storage systems because it slows the metabolism process of living plant tissues such as those found in fruits and vegetables. When fresh air replaces the nitrogen gas at the end of the storage period, metabolism resumes in the stored plant materials.

In the case of fish culture in tanks, the atmosphere is somewhat restricted to the water in the contained area. Air that is bubbled into the water inside fish runs and tanks keeps the level of dissolved oxygen at high levels. Failure to do this can result in the deaths of the fish. Ammonia is produced as a waste prod-

FIGURE 16–11 Ventilation in controlled environments is necessary to maintain healthy plants and animals. This greenhouse is equipped with air inlets to allow fresh air to enter the building as used air is removed by large fans.

uct from fish culture. This must be removed before it builds up to toxic levels, or the water must be replaced with fresh water.

Animals or poultry that are housed in large numbers in confined areas require large volumes of fresh air. It must be moved through the facility to eliminate accumulated carbon dioxide gas, ammonia, and other airborne pollutants and to restore oxygen levels. A shortage of oxygen or a buildup of other gases in the atmosphere of a controlled living environment can become a life threatening situation for animals and the people who care for them.

MANAGEMENT OF PARTICULATE AND CHEMICAL CONTAMINANTS

A major problem associated with confinement housing for livestock is air pollution within the facility. Two forms of air pollution are of concern. Dust can be a big problem when large numbers of animals or poultry are penned in close proximity to one another. These tiny particles that are suspended in the air are known as **particulate matter**. When levels of particulate matter are high, serious respiratory illnesses can occur in the animals and in the humans who care for them. This is caused by a buildup of particulate matter in the lungs.

A different form of air pollution occurs when fumes are produced from animal wastes and other sources. **Chemical contaminants** occur when contamination of the air in an environment is of a chemical nature, such as ammonia or methane gas. Both of these pollutants are products of natural elemental cycles.

The most common solution to both particulate matter and chemical contaminants is to design a good ventilation system, and to keep the facility clean, Figure 16–12. In some cases it may be to the advantage of managers to trap and remove contaminants from the air supply using specially designed equipment. Some contaminants can be trapped by high tech filter systems. Other

FIGURE 16–12 Cleanliness is an important part of reducing air pollution in confinement livestock housing. It reduces the level of particulate and chemical contaminants. *(Photo courtesy of David Burton)*

materials may be removed only by electronic filter systems that apply electrical charges to particles and collecting them on surfaces that carry the opposite electrical charge.

NATURAL ELEMENTAL CYCLES

The largest of all living environments is the atmosphere surrounding planet earth. Natural elemental cycles are known to both pollute and cleanse the atmosphere. Some of the atmospheric contaminants that are of a chemical nature can be attributed to elemental cycles that occur in nature. The most common of these chemical pollutants include ammonia and methane gas. Each of these compounds is an intermediate product in an **elemental cycle**. These elemental cycles occur as elements pass from living organisms to nonliving materials and back, again and again.

Nitrogen is the most abundant element in the atmosphere. It makes up about 80 percent of the air supply. In its elemental form it is a colorless, odorless gas. In this form it cannot be used by plants and animals. It must be combined with oxygen or other elements before it can be used as a nutrient for living organisms. Plants and animals use nitrogen compounds to form protein.

Nitrogen fixation is the process by which nitrogen gas is converted to nitrates. This can occur in several different ways. Nitrogen-fixing bacteria, also known as *Rhizobium*, are able to convert nitrogen gas to nitrates. Some forms of these bacteria live in the soil. Other forms of these bacteria live in nodules on the roots of legumes and some other plants. These types of plants are able to make their own nitrogen fertilizer.

Several industrial processes convert nitrogen gas to nitrates. One of these processes converts nitrogen gas to ammonia as a by-product of steel production. The ammonia is then converted to a form of nitrates that can be used for fertilizer.

Nitrogen fixation occurs naturally when lightning strikes occur. The electrical current that passes through the atmospheric nitrogen converts some of the nitrogen gas to nitrogen compounds that can be used by plants. Nitrates are also produced from ammonia that occurs during decomposition of animal wastes, and from plants and animals that die and decay.

At the same time that nitrates are being produced from nitrogen gas, other nitrates are breaking down to release nitrogen gas back to the atmosphere. This process is called **denitrification**. It occurs when some forms of bacteria come into contact with nitrates. It also occurs when nitrates are carried by runoff water into surface water. Surface water constantly exchanges nitrogen with the atmosphere. The circular flow of nitrogen from free nitrogen gas in the

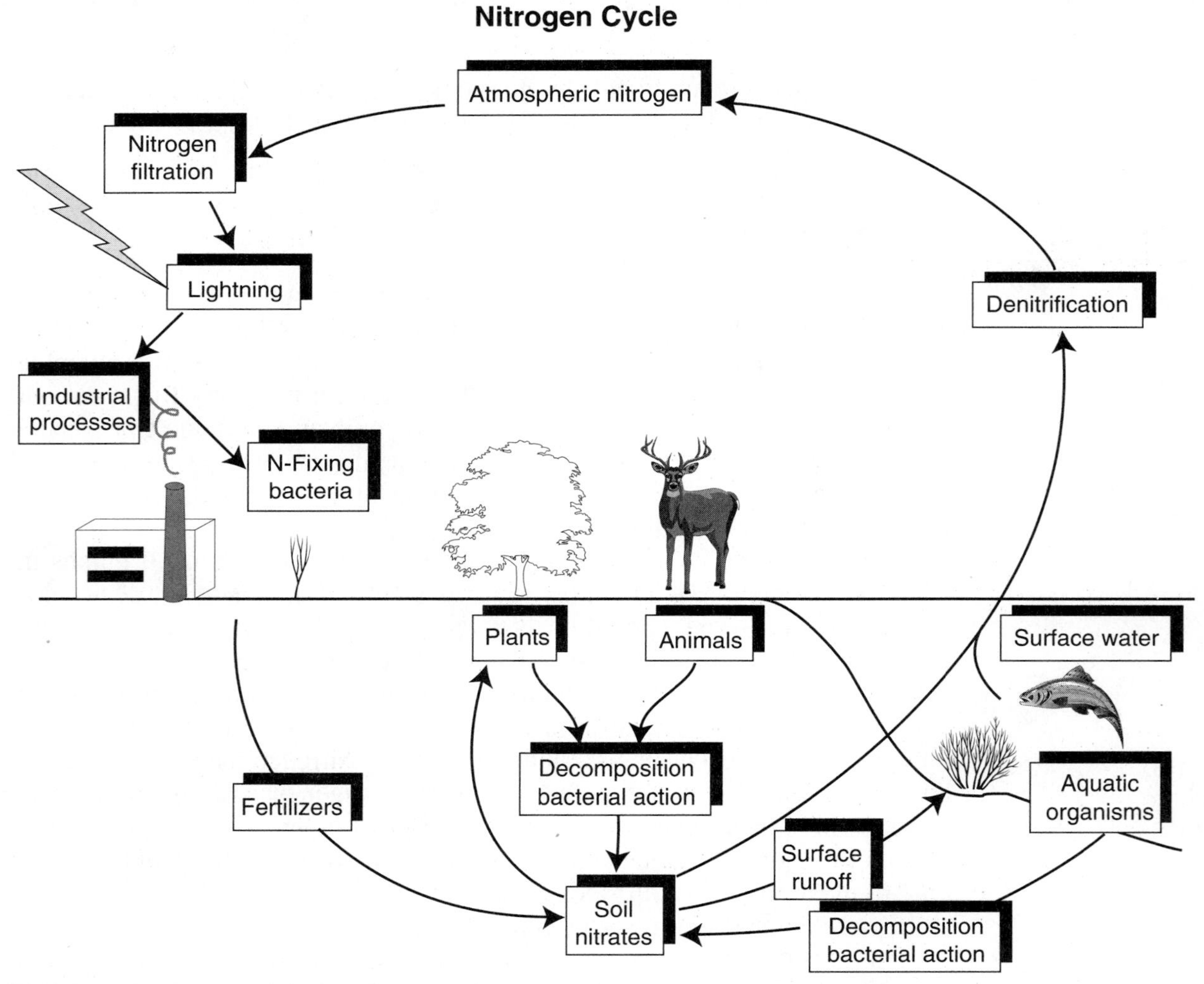

FIGURE 16–13 The nitrogen cycle moves nitrogen from living tissues to nonliving materials and back again, in a never ending flow.

atmosphere to nitrates in the soil and back to atmospheric nitrogen is known as the **nitrogen cycle**, Figure 16–13.

Carbon is the most abundant element found in living organisms. It makes up the framework of the molecules that are found in living tissue. Nearly half of the dry matter found in the bodies of animals or humans consists of carbon. Carbon moves readily between living organisms, the atmosphere, and the soil, Figure 16–14. The respiration process of both plants and animals releases carbon dioxide (CO_2) into the atmosphere. Plants take carbon dioxide from the atmosphere and build it into plant tissue. When plant tissue decays, carbon dioxide is released back to the atmosphere as methane or natural gas, or is converted to fossil fuels such as crude oil or coal. Sometimes plant materials are eaten by animals. When animals die, their bodies decompose and release carbon dioxide to the atmosphere as a gas. The carbon from the bodies of dead animals may also be converted to natural gases or fossil fuels.

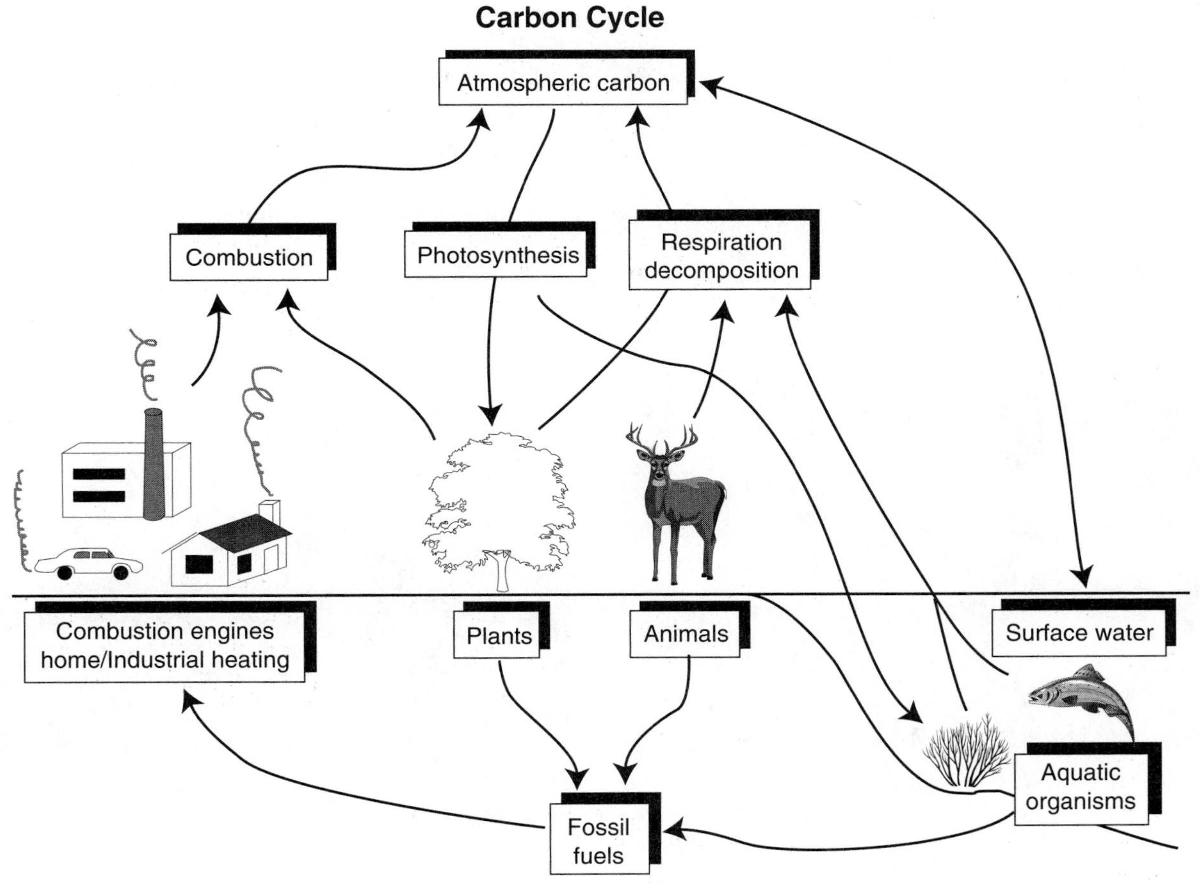

FIGURE 16–14 The carbon cycle moves carbon from living tissues to nonliving materials and back again in a continuous flow.

People mine or extract fossil fuels from the surface of the earth for use as fuels and for other purposes. When these materials are burned, the combustion process releases carbon dioxide to the atmosphere.

The oceans absorb large amounts of carbon dioxide from the atmosphere when the carbon content of the atmosphere is high, and they release carbon dioxide to the atmosphere when atmospheric carbon dioxide decreases. The movement of carbon from living tissue to nonliving materials is known as the **carbon cycle**.

ADAPTATIONS FOR PLANT PRODUCTION

A greenhouse is a structure that provides protection to plants during seasons of the year when adverse growing conditions exist, Figure 16–15. The use of greenhouses is the most common form of controlled environment for managing plant production. They are designed to protect plants from cold temperatures without restricting their access to sunlight, Figure 16–16.

The use of greenhouses is usually restricted to high value crops such as "hot house tomatoes," flowers, and other horticultural plant varieties, Figure 16–17. This is due to the high cost of the building, labor, and supplemental heat. The advantage of growing plants in a greenhouse is that they are protected from killing frosts during early spring months. The effects of a limited growing season can be reduced, and the length of the growing season can be extended using a greenhouse.

Another example of a controlled environment for crop production is a mushroom farm. In this instance, the crop is grown in the dark in an artificial envi-

FIGURE 16–15 A greenhouse provides an environment in which temperature, humidity, and light can be controlled to meet the growing requirements of plants. *(Photo courtesy of Utah Agricultural Experiment Station)*

FIGURE 16–16 Large commercial farms for specialty crops sometimes raise farm products inside large balloonlike structures that are inflated by fans. *(Photo courtesy of Utah Agricultural Experiment Station)*

FIGURE 16–17 Many ornamental and vegetable crops are raised in commercial greenhouses for later use in homes and gardens. *(Photo courtesy of Utah Agricultural Experiment Station)*

AGRI-PROFILE

The Lands

Before humans can successfully travel to distant regions in space, a food production strategy must be developed. A special exhibit, called "The Lands," has been established at Walt Disney World's Epcot Center. It features an experimental food production area in a futuristic greenhouse where experiments are conducted on production of food in space, Figure 16–18. Many problems must be overcome before plants can be raised successfully in zero gravity conditions. One of the problems is keeping water and nutrients on the roots of the plants instead of floating about in the greenhouse.

Among the plant culture methods that are being tested is the practice of planting through small holes in large plastic pipes. The pipes contain the plant roots, and nutrient solutions flow over the plant roots at regular intervals. No soil is used. This practice is called hydroponics.

Another experimental cultural practice is to leave the plant roots exposed to the air inside a closed container. A nutrient "mist" is sprayed on the roots at regular intervals. This practice is called aeroponics.

FIGURE 16–18 A large futuristic greenhouse is used to house the plant science experiments at Walt Disney's Epcot Center.

FIGURE 16–19 Some environmental control over low temperatures is achieved in fruit orchards by mixing warm air from upper air layers with freezing air near the ground using fans. Sometimes this is adequate to prevent fruit blossoms from freezing. *(Photo courtesy of Utah Agricultural Experiment Station)*

FIGURE 16–20 Special plant chambers are used by scientific researchers to protect crops from birds and insects. On other occasions, the same chambers may be used to confine insects to an experimental area.

FIGURE 16–21 Many vegetable crops are planted through plastic. This helps protect the plants against weeds. It also results in warmer soil temperatures, and it aids in directing the flow of irrigation water.

ronment. Optimum growing conditions for mushroom culture can be simulated in underground caves or specially designed structures.

Many plant production technologies have been developed that do not control the environment, but simply modify it, Figures 16–19, 16–20 and 16–21.

ADAPTATIONS FOR CROP STORAGE

Storage is an important part of the effort to provide high quality food products throughout the year. When storage facilities are properly designed and managed, many fruit and vegetable crops can be stored with minimal spoilage until the new crop is available the next year.

A properly designed storage facility allows for total control of the environment surrounding the crop. Light is restricted, cool temperatures are maintained, humidity is controlled, and the mixture of gases in the air is some-

FIGURE 16–22 Some crops must be stored in an environment that supports life. Modern fruit storage facilities are designed to keep the fruit alive and to slow the life processes. *(Photo courtesy of Utah Agricultural Experiment Station)*

FIGURE 16–23 Modern grain storage facilities blow air through the storage areas to reduce moisture. This creates an environment that reduces storage losses. *(Photo courtesy of Utah Agricultural Experiment Station)*

times altered. Each of these modifications to the environment is designed to reduce spoilage of the crop, Figures 16–22 and 16–23.

Some crops such as potatoes, fresh fruits, and grains consist of living plant parts. The life processes continue during the storage period. One of the purposes of storage is to create artificial conditions that cause the plant to become **dormant** by slowing down the life processes. Dormancy in plants is very similar to hibernation in some animal species. It is a time of inactivity. Life processes continue, but at a much slower pace.

ADAPTATIONS FOR CONFINEMENT LIVESTOCK HOUSING

Confinement housing for livestock is a management system in which the living environment is completely controlled, Figure 16–24. This is done to create

FIGURE 16–24 A properly designed confinement hog facility maintains a comfortable and healthy environment for the animals. *(Photo courtesy of David Burton)*

an ideal environment for the animals in which stress is minimal, and living conditions are stable. A good livestock confinement facility eliminates many of the stress factors associated with weather and climate.

The benefits that are gained from confinement livestock housing in comparison with traditional livestock facilities include improved efficiency in the use of feed and labor. Many of the livestock chores can be automated, and more animals can be managed without increasing labor requirements.

Ventilation is the movement of fresh air into a closed environment and the removal of fouled air from the environment. A good ventilation system is an important component of a confined livestock facility, Figure 16–25. This is due in part to the high humidity levels that occur when large numbers of animals are assembled in a closed environment. The air in such a situation becomes loaded with moisture and gases. High humidity occurs when moisture is absorbed from the lungs of the animals as they breathe. Humidity and fouled air are also derived from animal wastes. An emergency generator should be available to provide electrical power for ventilation during times of power loss. This is needed to prevent suffocation and death of animals in confined housing facilities that depend on forced air ventilation.

Care must be taken to be sure that the facility is not drafty. This is especially important in the winter when the temperature of the outside air is often

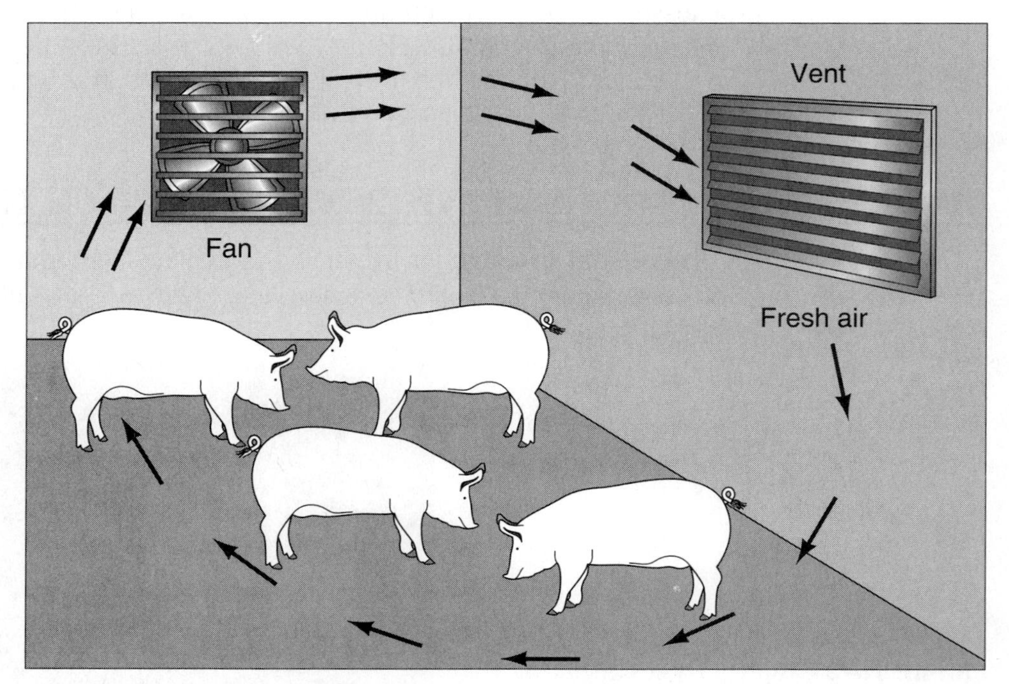

FIGURE 16–25 Total confinement of animals results in a buildup of body heat and fouled air unless adequate ventilation is provided. Large volumes of air must be moved continuously through the facility.

FIGURE 16–26 Fish farming is an important part of aquaculture. The water environment must be carefully controlled to eliminate wastes and to provide an adequate oxygen supply.

much lower than that of the air inside the building. When large volumes of air are brought into a building, it should enter in such a manner that it does not blow directly on the animals. When drafty areas are not corrected, the health of the animals is at risk.

Aquaculture is the practice of raising and managing fish, aquatic animals, or aquatic plants in a water environment, Figure 16–26. Aquaculture is generally

CAREER OPTION

Agricultural/Structures Engineer

An agricultural engineering career specializing in structures will involve designing farm buildings and facilities that are compatible with modern farming methods, Figure 16–27. It includes confinement livestock structures, crop storage facilities, greenhouse structures, facilities for aquaculture, and many different kinds of specialty structures.

This career will require a strong background in biology and science in addition to engineering skills. It will require a practical understanding of agricultural problems and a creative approach to solving structural problems associated with plants and animals.

Preparation for this career should include a good background in math and science in addition to practical agricultural experiences.

FIGURE 16–27 Farm buildings must be designed to be as useful as possible. For this reason, many agricultural buildings are constructed for specific uses. An agricultural structures engineer designs facilities to solve specific problems.

associated with fish farming. In many ways it is similar to raising animals in confinement housing. The main difference is that water takes the place of air. The pond or stream defines the boundaries of the enclosed living space.

The flow of clean fresh water through the fish ponds is very much like a large ventilation system. It removes waste materials from the environment as the water flows out of the enclosure. The fresh water also helps to maintain a constant temperature. Most of the management practices used in confinement livestock operations can also be applied to aquaculture.

CHAPTER SUMMARY

A controlled living environment is one in which environmental factors supporting the lives of plants and/or animals are controlled. These factors include temperature, humidity, atmosphere, and light. Special facilities are required to control these factors. Environmental management is a systematic approach to controlling the structures, equipment, and procedures that are used to maintain a stable life support system.

Controlled living environments that are used in agriculture include livestock and poultry housing, storage areas, greenhouses, fish enclosures, and family work and living areas. Special equipment is needed to monitor and maintain stable environments and to control particulate and chemical contaminants.

CHAPTER REVIEW

Discussion and Essay Questions

1. What are the essential components of a controlled living environment?
2. List agricultural applications that use controlled living environments, and describe how each example has helped to solve an agricultural problem.
3. What are the major control systems that are used to maintain stable conditions in controlled living environments?
4. What are the differences between particulate and chemical contaminants and how does technology contribute to controlling them in confinement housing?
5. In what ways have controlled living environments been adapted to crop production?
6. How are controlled environments used to successfully store perishable crops for extended periods of time?
7. Why are some farm animals raised in controlled living environments?
8. How is a controlled living environment for farm animals similar to the environment in which aquaculture is practiced?

Multiple Choice Questions

1. An environment in which conditions do *not* change is a(n):

   A. temperate environment.
   B. ambient environment.
   C. aquatic environment.
   D. static environment.

2. The ambient temperature in an environment refers to:

   A. the amount of heat in the environment.
   B. the temperature of the water that is found in an environment.
   C. a temperature that fluctuates frequently between hot and cold.
   D. the difference between temperature extremes.

3. The Biosphere II project has been criticized by some scientists because:

   A. its human subjects were treated as objects of science with little regard for their human qualities and personal rights.
   B. additional resources were reportedly provided to the project after it had started.
   C. it was located in a desert environment.
   D. resources had to be recycled to provide water for plants, animals, and people.

4. Each of the following factors is usually managed in a controlled living environment *except*:

   A. populations of microorganisms.
   B. temperature.
   C. atmospheric gases.
   D. humidity.

5. Control of atmospheric gases in a controlled living environment is accomplished by installing:

   A. a thermostat.
   B. a dehumidifier.
   C. artificial lights.
   D. a ventilation system.

6. Which of the following light wavelengths or colors is reflected by the chlorophyll in plant stems and leaves?

   A. green
   B. red
   C. blue
   D. violet

7. The light wavelengths that are converted to plant energy consist of:

   A. absorbed light.
   B. reflected light.
   C. all colors of the visible spectrum.
   D. the absorption spectrum.

8. The contaminants least likely to be removed from a controlled living environment by a filter system are:

   A. dust particles.
   B. smoke particulates.
   C. ammonia gas.
   D. plant pollens.

9. The process by which an element passes from living tissues to nonliving compounds and back to living tissues is known as:

   A. an elemental cycle.
   B. nutrient biogenesis.
   C. denitrification.
   D. decomposition.

10. The most commonly controlled conditions that are designed in well-planned crop storage facilities include each of the following conditions *except*:

    A. temperature.
    B. light.
    C. pressure.
    D. humidity.

LEARNING ACTIVITIES

1. Assign class members to work in teams to research the ethical issues associated with confinement of large numbers of animals in confined spaces. Conduct a debate competition. Teams should defend both sides of the issue. Invite the speech class to a discussion of these issues in a joint class forum.

2. Plan a field trip to visit a facility that uses a controlled environment for agricultural purposes. Require the students to turn in a written report addressing each of these issues:

   a. control systems
   b. management of particulate and chemical contaminants
   c. special adaptations of the controlled living environment that fit the facility to its specific use
   d. safety features of the facility

CHAPTER 17

Protection of Natural Resources

The advance of technology has had a tremendous impact on the natural resources of the earth. Technology has affected the environment in both positive and negative ways. It has enabled humans to be both friends and foes to the environment. The contents of this chapter will concentrate on technologies that are used to improve the environment and conserve our natural resources.

OBJECTIVES

After completing this chapter, you should be able to:

- describe the causes and effects of soil erosion.
- identify soil conservation practices that have been demonstrated to be effective.
- assess the value of minimum tillage as a soil conservation practice.
- propose management practices that might be used to reduce the effects of destructive forces on forest and range lands.
- contrast the beneficial and destructive effects of fire on range and forest lands.
- suggest solutions that could be used to reduce or eliminate chemical contamination of surface and ground water.
- define the relationship between air pollution and acid rain.
- explain how technology has both added to and solved pollution problems associated with natural resources.
- describe ways that modern technology is used to develop and protect critical habitat for wildlife.

TERMS FOR UNDERSTANDING

The following vocabulary terms should be studied carefully as you read:

soil conservation
erosion
dike
conservation reserve program (CRP)

multiple-use
watershed
slash
smoke jumper
surface water
phosphate
algae
ground water
acid rain
ecosystem
smog
habitat
extinct
alien species
nonadaptive behavior
low biotic potential
threatened species
endangered species
ecologist

SOIL CONSERVATION

The soil formation process has required thousands of years to produce the thin layer of topsoil that covers the surface of the earth, Figure 17–1. **Soil conservation** includes all of the soil management practices that prevent soil losses. **Erosion** is a force that depletes topsoil and reduces fertility. It is caused by exposing unprotected soils to excessive flows of water or wind, Figures 17–2 and 17–3. Most of the topsoil that is eroded from farmland is lost when the forces of wind or water dislodge soil particles and remove them from the land as dust or silt.

Water erosion has resulted in deep scars on the land in the form of gullies and washes from which soil is sometimes carried completely away leaving exposed bedrock, Figures 17–4 and 17–5. Wind erosion tends to remove surface soil from large areas. Many attempts have been made to reduce erosion, but once soil has been lost, it cannot be replaced, Figure 17–6.

FIGURE 17–1 Man is totally dependent upon topsoil. It is his link to a supply of food and clothing. We must conserve it for there will be no more if it is lost. *(Photo courtesy of Utah Agricultural Experiment Station)*

FIGURE 17–2 The dust bowl was a name given to the massive soil erosion problem caused by the wind in the 1930s. It occurred in part because the soil was not properly managed. *(Photo courtesy of Utah Agricultural Experiment Station)*

FIGURE 17–3 Examples of wind erosion can be seen today in areas where wind-blown soil has piled up in drifts covering vegetation, farm lands, and even ancient cities. *(Photo courtesy of Utah Agricultural Experiment Station)*

FIGURE 17–4 The unrestricted movement of water over the surface of soil often results in lost topsoil and polluted water.

FIGURE 17–5 Water erosion cuts deep into fertile soil, leaving gullies and deep scars in the landscape, and carries tons of topsoil into rivers, lakes, and streams. *(Photo courtesy of Utah Agricultural Experiment Station)*

FIGURE 17–6 Scientists have studied soil samples from all over the world to learn how erosion of our soil resources can be controlled. Soil maps are useful tools for managing land as man attempts to reduce soil erosion. *(Photo courtesy of USDA/ARS #K–4914–2)*

Soil can be protected against erosion from wind and water by adopting management practices that hold the soil particles together. Plant roots and plant residues on the soil surface are both effective materials for conserving soils. Cultural practices that leave plant residues in place are recommended on erodible land, Figure 17–7. Practices such as contour farming, no-till farming, or minimum till farming are effective practices for reducing soil erosion.

In our zeal to develop new farm lands, some land has been developed that should be returned to other uses. Soils located on steep slopes that are

FIGURE 17–7 Rows of grain sorghum that are left standing in a field provide a windbreak that captures snow to replenish soil moisture, and they protect soil from wind erosion. *(Photo courtesy of Colorado State University)*

FIGURE 17–8 Local soil conservation districts have the responsibility to find solutions to soil erosion problems within local districts.

exposed to heavy surface runoff should be planted to permanent cover crops to prevent erosion.

Soil conservation districts are organized in most areas to identify soil conservation problems and to develop plans for solving them, Figure 17–8. Professional soil conservation experts, working with soil conservation districts, are available to provide advice and direction for soil conservation efforts.

Contour Farming

Contour farming is the practice of farming around slopes instead of up and down them, Figure 17–9. This practice tends to trap and absorb rain in the soil instead of allowing it to flow down slopes along the cultivation furrows. Farming on the contour reduces erosion by preventing water runoff. It reduces

FIGURE 17–9 Contour farming is the practice of tilling and planting while maintaining a constant elevation as equipment moves across the field. This practice reduces soil erosion. *(Photo courtesy of USDA Soil Conservation Service)*

or eliminates surface damage to soils due to the action of turbulent water on soil particles. This practice also makes more moisture available to plants.

A modified form of contour farming has been applied to mountains near populated areas to prevent flash floods. Deep contour furrows have been constructed in these areas to trap and hold water from heavy rainstorms and melting snow.

A **dike** is a bank of soil that is built up to prevent the flow of water. Construction of dikes on the contour of a slope is a proven practice for prevention of soil erosion on farm and pasture lands. In combination with contour farming practices, soil erosion due to water runoff is minimized. Excess water is often trapped in ponds where it provides water for domestic livestock and wild animals. In such cases, wild animal habitat is also improved.

Minimum Tillage

Minimum tillage is a farming practice in which the soil is tilled only enough to ensure that it will pack firmly around the seed. This improves the transfer of soil moisture to the seed, enhancing the ability of the seed to sprout and grow. Plowing is not practiced in this type of farming because it is considered to be a harsh soil tillage practice. It also leaves the soil surface unprotected from wind and rain.

Many farmers who practice minimum tillage perform several operations in each trip over the field. They may distribute fertilizer, till the soil, and plant the seed in a single operation. Advantages of minimum tillage include lower costs in performing field work, reduced loss of soil moisture, and reduced erosion and compaction of soils.

No-Till Farming

No-till farming is a practice in which crop seeds are planted without tilling the residues from the previous crop into the soil. A planter is used that cuts through the residue and plants the seeds in the soil beneath the surface cover, Figure 17–10. The soil is not disturbed by tillage, and the residue on the soil surface conserves moisture and reduces erosion.

AGRI-PROFILE

Conservation Reserve Program (CRP)

The **conservation reserve program** is a government program that provides payments to farmers who voluntarily take highly erodible land out of production. A cover crop must be established on the land, and harvesting is not approved except in special instances such as severe drought or natural disaster. The dense ground cover provides shelter and feed for many species of wildlife, and erosion problems are substantially reduced.

FIGURE 17–10 No-till planters are constructed of heavy materials to enable them to penetrate untilled soils. Seed is placed behind the heavy discs where they have cut through the soil surface. *(Photo courtesy of Utah Agricultural Experiment Station)*

Despite the advantages associated with no-till farming, it has not been widely adopted. This is partly due to the high cost of no-till planters, which are constructed of heavy materials and engineered to function in hard soils. Problems with seed germination during dry years have also interfered with the implementation of the no-till farming practice.

PROTECTION OF FOREST AND RANGE LANDS

Among our most valuable natural resources are the lands on which our forests and other vegetation grow, Figure 17–11. Our forests are important in many ways. They provide habitat for many species of wildlife. They are the basis for our wood industry. They provide fiber for paper production and lumber for construction of homes and buildings.

Multiple-use management is an important concept in the management of forest lands. It is a form of management in which the forest resource is used by many user groups, Figure 17–12. Wood products are harvested, mines are

FIGURE 17–11 The United States Department of Agriculture is responsible for the management of forest lands that are owned by the federal government.

operating, livestock grazing occurs, and recreation is encouraged, Figures 17–13, 17–14, and 17–15. All of these activities are appropriate uses of forest resources.

A **watershed** is the area from which water drains as it emerges from springs and moves into the streams, rivers, ponds, and lakes, Figure 17–16. It acts as a storage system by absorbing excess water and releasing it slowly throughout the year. Forests are important to watersheds because they con-

FIGURE 17–12 The multiple-use concept is an appropriate management strategy for public lands. It takes into account the needs of wildlife, domestic livestock, sportsmen, and the mining and timber industries in making management decisions. *(Photo courtesy of Utah Agricultural Experiment Station)*

FIGURE 17–13 Many range lands have been improved by "chaining" them to remove undesirable species of shrubs, trees, and brush. *(Photo courtesy of Utah Agricultural Experiment Station)*

FIGURE 17–14 Chaining is accomplished by dragging a huge chain across the surface of the land. Each end of the chain is hooked to a large tractor. *(Photo courtesy of Utah Agricultural Experiment Station)*

FIGURE 17–15 With undesirable plant species removed from the range, productive grasses are seeded. Range lands that are improved in this manner produce increased yields of forage for both domestic and wild animals. *(Photo courtesy of USDA)*

FIGURE 17–16 A watershed is an area in which rainwater and melting snow is absorbed to emerge as springs of water or artisan wells at lower elevations. *(Photo courtesy of Utah Agricultural Experiment Station)*

FIGURE 17–17 As the waters from several springs merge together, streams and rivers are formed. *(Photo courtesy of Utah Agricultural Experiment Station)*

trol erosion and slow the movement of water over the soil surface causing it to be absorbed into the soil. Later, this water is discharged from springs and provides a continuous flow in our streams and rivers, Figures 17–17 and 17–18. Forests also provide oxygen and moisture to the atmosphere. They help to cleanse the air and restore balance to nature.

Healthy forests require good management, Figures 17–19, 17–20, and 17–21. The trees in a forest are sometimes damaged or killed by insects or diseases. When these conditions affect large tracts of forest land, they may require chemical treatments to prevent destruction of a forest. Insect control in future forests will probably rely on integrated pest management and insect resistant varieties of plants and trees. Forestry specialists should continue to evaluate these conditions, and manage government and privately owned forest lands in an appropriate manner to preserve these important natural resources, Figure 17–22.

FIGURE 17–18 Lakes and man-made reservoirs capture the water from streams and rivers. Water from springs, marshes, streams, rivers, lakes, and oceans is called surface water. *(Photo courtesy of Utah Agricultural Experiment Station)*

FIGURE 17–19 Mature trees are valuable natural resources because they improve air quality and provide raw materials for the wood industry.

FIGURE 17–20 Mature trees provide renewable sources for wood products. Young trees of most species begin to grow in response to sunlight on the forest floor after mature trees are harvested.

Fire is sometimes used as a management tool to clear areas for seeding and to remove **slash** consisting of branches and waste materials from areas that have just been harvested. It is also used to control diseases that afflict forests. When fire is uncontrolled, however, it creates serious problems on private and public forests and range lands.

When drought conditions exist, fires destroy large tracts of trees, brush, grass, and other vegetation that provides the food supply and shelter for wild animals and domestic livestock, Figure 17–23. Fires also injure or kill many of the animals that live in the burned area.

FIGURE 17–21 Once a forest is mature, it should be harvested to make way for new growth.

FIGURE 17–22 Insects and diseases often kill trees. Failure to wisely manage our forests by controlling diseases and insects, and by harvesting mature timber in a timely manner is a mistake that wastes a valuable natural resource.

FIGURE 17–23 A forest fire leaves the land without its protective cover and the potential for serious erosion problems is high.

Many fires are accidentally started by people. For this reason, fire prevention is promoted every year by forest and range managers in efforts to create awareness of the huge cost of wildfires. The educational effort has paid off. The number of fires has decreased in the years since educational programs were adopted to prevent fires.

Once a fire starts, fire control becomes an important priority. Hundreds of workers are often needed to control a single fire. Fire crews work around the clock with only a few hours of rest when they are engaged in fighting a major

AGRI-PROFILE

Smokey Bear

Smokey Bear became the mascot for fire prevention after he survived a forest fire in the 1940s. After he recovered, he lived at the National Zoo in Washington, D.C. until his death. An educational program for prevention of forest fires was developed with Smokey as the star. Nearly everybody knew Smokey Bear. His one line message was always the same: "Remember, only you can prevent forest fires."

fire. Modern fire control technologies include the use of aircraft to drop fire retardants on the fire, and to carry airborne fire fighters called **smoke jumpers** to the fire zones.

Forest fires are sometimes allowed to burn. This practice is sometimes due to a shortage of firefighters. Other times it occurs because some forest managers believe that fire is a part of nature and should be allowed to burn.

PROTECTION OF SURFACE WATER

Surface water includes the water in our rivers, streams, lakes, and reservoirs, Figure 17–24. It flows across the surface of the land as it moves toward the oceans. These waters are used by towns and cities for human consumption. They are used by farms for irrigation of crops and for consumption by domestic animals.

Surface water is used by people for swimming and other forms of recreation. The water is also used by many species of wild animals. Some of them live in it, and nearly all of them drink it. Pollution of surface water is a serious problem to humans and all other life forms that inhabit the planet earth.

One source of water pollution is sewage from cities and industrial sites. Many new technologies have been developed to treat sewage, and legislation has mandated more effective treatment of sewage before it is discharged into rivers, streams, and lakes. Large amounts of pesticides enter surface and ground water in runoff water from lawns, golf courses, gardens, and parks. Other pollutants come from chemicals that have been dumped on city streets by residents who are trying to dispose of them. Examples of such materials include used oil, solvents, paint, and antifreeze.

Another source of surface water pollution is agricultural chemicals. These materials are applied to fields as fertilizers and pesticides. When irrigation water or water from excessive rainfall moves over the surface of a field, agricultural chemicals become dissolved in it. If this water is allowed to

FIGURE 17–24 Surface water is an important resource that must be protected. It is used by most living organisms, including humans.

drain off fields and enter rivers and streams, it becomes a source of pollution and a threat to the humans, fish, and wildlife that depend upon the water.

Two classes of chemical compounds that have become major concerns in water pollution are the nitrates and the **phosphates**. Nitrates are composed of nitrogen compounds in forms that are easily used by plants. Phosphates are compounds of phosphorous that provide nutrients to plants. When nitrates and phosphates are present in surface water, they promote the growth of tiny plants called **algae**. These plants are suspended in the water and they change the balance of dissolved gases in the water. Algae can reduce the oxygen level in water by using it up. This is harmful to some species of plants and animals that live in or near the water, and can destroy the living environments in streams and lakes.

Surface water pollution by residential and agricultural chemicals can be reduced by controlling the amount and type of chemicals used around homes, parks, gardens, and golf courses as well as on forest and crop lands. Careful attention must be given to following the directions on chemical containers. Excess chemicals should be properly disposed of instead of dumping leftover materials on the ground or in streets and drains.

One important practice in protection of surface water is preventing runoff. Irrigation water that is applied through sprinkler systems is entirely used in the fields, and no surface runoff occurs. When fields are irrigated by flooding, the water should be carefully controlled to prevent its flow back into rivers and streams.

Chemigation is the practice of applying agricultural chemicals to crops by mixing them with irrigation water, Figure 17–25. When this is done, great care must be taken to ensure that no chemical materials are spilled into irrigation canals and drainage ditches. High pressure sprinkler irrigation systems should have special one-way valves installed at the water source to prevent chemicals from being injected under pressure back into wells or canals. Agricultural chemicals can be used safely when they are applied as the manufacturer intended.

FIGURE 17–25 Some fertilizers that are used in crop production are delivered to plants through irrigation water. It is important that water containing agricultural chemicals is used in the fields, and that it does not find its way back into rivers and streams.

Runoff can be reduced on forested lands by using logging and cultural practices that trap water until it is absorbed in the soil. Logging roads must be carefully constructed to prevent erosion. A common practice on forest ground that has been harvested is to create an uneven surface on the forest floor. Water from rain or melting snow is trapped until it can penetrate the soil.

PROTECTION OF AQUIFERS

An aquifer is an underground layer of porous rock or sand that contains water, Figure 17–26. Large aquifers exist in many areas. Water obtained from an aquifer is called **ground water**. It is pumped from underground sources through wells for consumption by people and animals, and for irrigating crops.

Water from underground wells was once considered to be pure and free from contamination, but recent tests of water samples have demonstrated that water from this source is sometimes polluted. Major sources of pollution to aquifers are farm and industrial chemicals. They enter groundwater by moving down through the soil, or by seeping into the aquifer through porous rocks and sand as dissolved materials in surface water.

A partial solution to the problem of contaminated ground water is to avoid chemical spills, and to promptly clean areas where accidental spills occur. Excess chemical materials must be degraded or detoxified before they are disposed of. Fertilizers and pesticides should be applied wisely to avoid contamination of surface water and to prevent downward leaching of chemicals into the ground water.

Many agricultural chemicals are degraded by soil microorganisms. They are capable of destroying many chemicals and other pollutants by feeding upon them. The digestion process degrades many harmful materials by breaking the

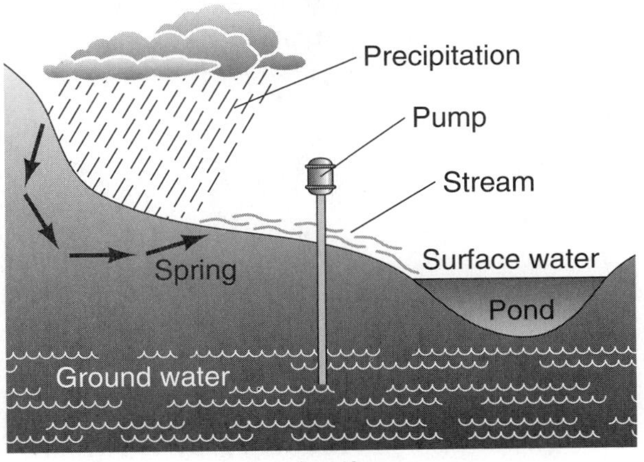

FIGURE 17–26 Water that is used to produce agricultural products is obtained from several sources. Some of it comes from precipitation which falls directly on the fields. Some areas depend upon surface water and ground water for use in irrigation.

chemical bonds that bind elements together as compound materials. Once the bonds are broken, many toxic chemicals can be broken down into products that are compatible with the environment.

Genetic engineering has made it possible to develop new strains of bacteria and other microorganisms. Some of these can be used to degrade or break down chemical wastes. They can even be used to clean up polluted water and chemical spills.

PROTECTION OF THE ATMOSPHERE

Major sources of atmospheric pollution include the smoke and gases released into the atmosphere by cities and factories. Included in these emissions are sulfur compounds that are released into the atmosphere when coal is burned. Sulfur and nitrogen compounds combine with moisture in the air to form **acid rain**, Figure 17–27. This damaging pollutant has been responsible for destruction of forests and other plant and animal life, and it inflicts severe damage on surface water.

An **ecosystem** is made up of all of the forms of life that inhabit a particular area. Acid rain is capable of serious damage to an ecosystem because it changes the pH of surface water and destroys many of the life forms that exist there. It also corrodes structures and damages trees and other plants. Because it is a destructive force, acid rain is a threat to agriculture in areas where it exists. The key to solving the acid rain problem is to reduce emissions of harmful pollutants such as nitrogen and sulfur compounds into the air.

Acid Rain

$2\ SO_2$	+	O_2	+	$2\ H_2O$	$\xrightarrow{NO,\ NO_2}$	$2\ H_2SO_4$
Sulfure dioxide		Atmospheric oxygen		Water	Nitrogen oxides (catalysts)	Sulfuric acid

or

$2\ NO$	+	O_2	$\longrightarrow$	$2\ NO_2$		
Nitric oxide		Atmospheric oxygen		Nitrogen dioxide		
$3\ NO_2$	+	H_2O	$\rightleftharpoons$	$2\ HNO_3$	+	NO
Nitrogen dioxide		Water		Nitric acid		Nitric oxide

FIGURE 17–27 Acid rain occurs when acids reduce the pH of the rain until it is at or below pH 5.

High levels of chemical pollutants and particulate matter can remain suspended in the atmosphere for long periods of time. When the pollution level is extreme and the polluted air becomes trapped in an area, it can block out the sunshine. This atmospheric pollution is called **smog**. It sometimes causes health problems for people who suffer from respiratory diseases.

Local government agencies have passed ordinances in recent years that are designed to reduce air pollution. In many areas it is illegal to operate wood burning stoves when pollution reaches critical levels. Exhaust emissions from cars and trucks are tested annually in many areas to ensure that pollutants are minimal. Factories and processing plants are required by law to meet smokestack emission standards.

Agriculture bears some of the responsibility for atmospheric pollution because some of the pollution comes from the factories that build agricultural machinery or process agricultural products. Agriculture also suffers from some of the consequences of atmospheric pollution. Crops and animals suffer from the effects of acid rain and other pollutants on their food supplies and living environments.

Technology is a double-edged sword. The use of new technologies in manufacturing and processing has led to much of the pollution that is suffered by natural resources and the environment. Technology must also be responsible for solutions to the pollution problems. New methods for reducing chemical and particulate pollutants are being developed as the people of the world become more conscious of environmental issues.

PROTECTION OF WILDLIFE

Management of our wildlife resources is an important activity. As the human population increases, the living environment or **habitat** of living creatures is changed to provide for the needs of people. In many cases, the movement of people into areas that once provided habitat for animals, birds, and other forms of life results in reduced populations of these species.

Many species of birds and animals have not survived. Populations of creatures that have all died are said to be **extinct**, Figure 17–28. Extinction of a species of organisms is not something new. It has been happening for as long as life has existed on the planet. Even before humans became competitive with other organisms, many species of organisms became extinct. Man has become a dominant species because humans are able to adapt to nearly any environment. They also act as predators toward many of the animals and birds with which they share the environment.

Some organisms are unable to adapt quickly to changes in environmental conditions or to heavy losses from predators. They decline in numbers as their environments are modified, or as predators increase. A single negative factor can result in the extinction of some species of animals. It is more likely, how-

FIGURE 17–28 Many species of living creatures have become extinct or endangered when humans or natural disasters have destroyed their habitats. The Trumpeter Swan is threatened by loss of wetlands. *(Photo courtesy of Utah Agricultural Experiment Station)*

ever, for a species to become extinct due to a combination of factors that impact the organisms in negative ways, Figure 17–29.

Destruction of habitat is the greatest single cause of extinction. When organisms lose their food supply, they soon starve to death. When their shelter is damaged or destroyed, they can be easily destroyed by natural enemies or unfavorable weather conditions. Even modest changes in weather conditions are dangerous to an organism that has lost the shelter to which it is accustomed. Natural disasters such as severe storms or extreme temperatures are among the greatest threats to endangered species of organisms.

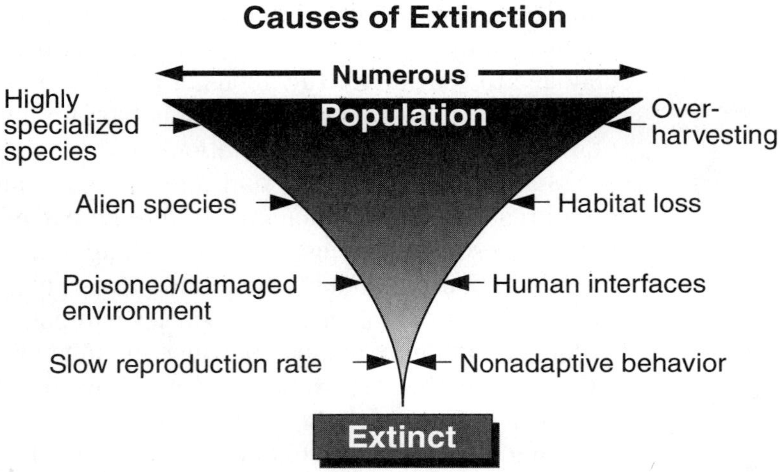

Causes of Extinction

Numerous
Population
Highly specialized species
Over-harvesting
Alien species
Habitat loss
Poisoned/damaged environment
Human interfaces
Slow reproduction rate
Nonadaptive behavior
Extinct

FIGURE 17–29 Extinction of an entire population of organisms can occur due to any of the causes cited here. In most instances, a combination of causes is responsible for extinction.

An important factor that sometimes leads to extinction of an established species is the introduction of an **alien species** into the ecosystem. The new species may compete with the native species for food and shelter or prey upon it as a source of food. When this happens, the fragile balance in the ecosystem is upset. The weaker species tends to decline, while the more competitive species increases in numbers.

Still another major factor that contributes to extinction is over-harvesting of a species by humans. In some instances this has been done commercially, but in other instances sport hunting of a vulnerable species has contributed to its extinction.

The degree of specialization in a species affects how vulnerable the species is to extinction. This is because a highly specialized animal or bird may depend on a single source of food or shelter. Such an animal may not even be able to eat or digest any other food. When the source of specialized food or shelter is gone, a highly specialized bird or animal faces extinction.

Most surviving species in the world today are able to adapt to modest changes in their environments. Abrupt changes in the environment allows no time for living organisms to adjust, and in some cases adjustment is not possible. Species that cannot adapt their behaviors or their diets to accommodate changes in their environments are highly vulnerable to extinction. Failure of a species to adapt to a changing environment is called **nonadaptive behavior**.

A slow rate of reproduction contributes to extinction by reducing the recovery rate of an endangered species. Biologists call this problem **low biotic potential**. Examples of animals and birds that fall into this class include the California condor that lays only one egg every two years or the whooping crane that takes several years to reach breeding age and then lays only two eggs per year. More successful species of birds often lay 8–10 eggs or nest several times per season, Figure 17–30.

FIGURE 17–30 A slow rate of reproduction reduces the recovery rate of a species and can contribute to extinction. Most successful species of birds lay 8–10 eggs per nesting attempt, or they nest several times each year.

The passenger pigeon is an example of one species of wild bird that was completely eliminated by man. These birds once numbered in the billions and caused serious damage to crops when flocks were feeding. They were killed in large numbers by farmers and others who considered them to be pests. They did not gain protection until it was too late to preserve them.

Some species of living things are classified as **threatened species**. This means that they require special protection to maintain their populations. Other declining populations are known as **endangered species**. This means that they are in danger of becoming extinct. The Endangered Species Act is a law that has been enacted to protect animals, birds, and other forms of life designated as endangered or threatened species. Some of these include bald eagles, grizzly bears, wolves, whooping cranes, several species of fish, and others, Figures 17–31, 17–32, 17–33, and 17–34.

One of the steps taken to protect endangered species is the development of wildlife preserves in which wild animals are protected from hunting and destruction by humans. One of the best known of these preserves is Yellowstone National Park located in Wyoming. Within the boundaries of the park, animals are protected. The checks and balances of nature control the numbers of the different animal species.

A management strategy that has been applied to waterfowl such as ducks and geese is the restoration of marshlands along major migration routes. Sportsmen have banded together to purchase large tracts of land on which wetlands have been restored for the benefit of many species of migratory birds,

FIGURE 17–31 The bald eagle is best known as the national bird of the United States. It is also protected under the Endangered Species Act. *(Photo by F.D. Schmidt © Zoological Society of San Diego)*

FIGURE 17–32 The Grizzly Bear is one of the animals the Endangered Species Act has been designed to protect. Under the act the animals are protected, and the checks and balances of nature control their numbers. *(Photo courtesy of Utah Agricultural Experiment Station)*

FIGURE 17–33 The wolf is a large predatory animal that is protected in some areas of the world. Conservationists are establishing populations of the species in wilderness areas and in Yellowstone National Park. Many sportsmen and livestock ranchers who use these areas are opposed to this plan. *(Photo courtesy of Utah Agricultural Experiment Station)*

FIGURE 17–34 The Whooping Crane is an endangered species. Conservation officers have attempted to establish new breeding colonies of Whooping Cranes by placing their eggs in the nests of Sandhill Cranes. *(Photo courtesy of Utah Agricultural Experiment Station)*

Figures 17–35 and 17–36. A similar strategy that has been used on government owned lands is the development of bird refuges.

Fish hatcheries have been established in many areas to provide environments in which large numbers of fish can be produced for recreational use. Some eggs are collected and fertilized by trapping fish that have returned to

FIGURE 17–35 Wetlands have been created along the migration routes of migratory waterfowl to replace natural wetlands that have been drained as the land was developed for farms and human population centers.

FIGURE 17–36 Canada goose populations have benefited from man-made nesting platforms. Nesting geese are protected from predators and flood water when they use the platforms.

FIGURE 17–37 Fish ladders are constructed at sites where dams or other obstructions prevent fish from migrating up rivers. Fish use the ladders to move above obstructed areas.

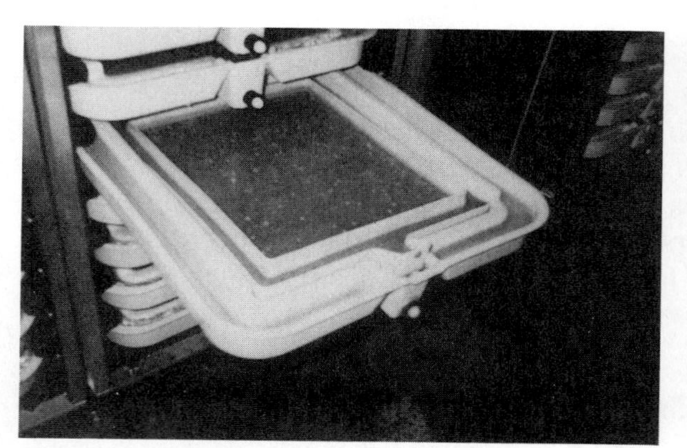

FIGURE 17–38 Fish eggs that have been collected from female fish are fertilized and placed in trays for hatching. Fresh water flows continually through the trays providing oxygen and maintaining a constant temperature.

FIGURE 17–39 Hatchery fish are usually raised in special rearing ponds until they are big enough to be released into streams and lakes for sport fishing.

their spawning areas, Figures 17–37, 17–38, and 17–39. Other eggs are obtained from mature fish that are maintained in captivity as breeding stock. After hatching, the immature fish are raised in large numbers to acceptable sizes before they are transported to rivers, streams, and lakes.

Wildlife managers have been able to develop new herds of endangered species of big game animals by trapping some of the animals in established herds, and releasing them in other areas where appropriate habitat is available to support them. The bighorn sheep is one animal species that has been introduced into new areas by this method, Figure 17–40. The establishment of many small herds is an important management strategy because it iso-

FIGURE 17–40 Bighorn sheep have been successfully transplanted back into some areas of their original range from which they had been eliminated. *(Photo courtesy of United States Fish and Wildlife Service)*

CAREER OPTION

Animal Ecologist

An animal **ecologist** studies the effects of the environment on animal life, Figure 17–41. He or she must have a strong background in the biological sciences and a good understanding of environmental issues. A college degree is required with graduate study recommended. An animal ecologist spends a lot of time outdoors, and will often be involved in solving problems in environments that have been damaged.

Oil and chemical spills create difficult situations for animals because they poison food supplies and radically change the environments in which wild creatures live. Animal ecologists work to restore damaged environments to make them habitable for wild animals, birds, and other living things.

FIGURE 17–41 An animal ecologist is a specialist on the interactions between animals and their environments. *(Photo courtesy of USDA)*

lates herds from diseases that could destroy all of the animals if they were managed as a single population.

A management strategy that has been used with success in protecting endangered species is the practice of creating habitat favorable to a particular species that is declining. This gives the endangered species a competitive edge over other animals that also inhabit the area, and it increases their chances of survival.

CHAPTER SUMMARY

Modern technologies have affected outdoor environments in both positive and negative ways. Many of our soils have been damaged by erosion. Much of the erosion that has occurred on cultivated land is due to lack of protective cover on plowed fields. New technologies, such as minimum tillage practices, continue to be developed that reduce erosion and protect the soil.

Forest and range lands are renewable resources that produce valuable products year after year. Many of these lands are managed using the multiple-use management concept. Sometimes these lands require protection from disease, insects, or fire. Tree varieties that are resistant to diseases and insects are being developed to reduce the need for chemicals in forest management.

Our fresh water supply requires protection from chemicals, sewage, petroleum products, and other pollutants. Biotechnology is developing new organisms that are capable of cleaning polluted water by digesting the pollutants. Wiser use of agricultural and industrial chemicals is also expected to protect surface and ground water.

Pollution of the atmosphere is a serious problem. New technologies are becoming available for reducing air pollution, acid rain, and smog.

Wildlife management involves management of the habitat that provides for the needs of living creatures. By creating habitat favorable to a particular animal, the survival rate of that animal is improved. Much work has been done to restore habitat for endangered species. Some endangered animals have also been introduced into areas where habitat is favorable to them.

CHAPTER REVIEW

Discussion and Essay Questions

1. Describe how soil erosion occurs and discuss how it affects the environment.
2. What are some soil conservation practices that have been demonstrated to be effective?
3. How does the practice of minimum tillage improve soil conservation?

4. What management practices might be used to reduce the harmful effects of water, wind, and fire on the environment?
5. Distinguish between the beneficial and destructive effects of fire on range and forest lands.
6. Suggest some agricultural and industrial practices that could be used to reduce or eliminate contamination of surface water and ground water.
7. Explain how a reduction in air pollution results in less damage to the environment from acid rain.
8. What are some examples of technologies that have had damaging and/or good effects on the environment?
9. What are some ways that endangered species of living organisms can be protected from extinction?
10. What are some strategies that might engage more citizens in providing protection to endangered species of living organisms?

Multiple Choice Questions

1. Which of the following forces is responsible for most of the soil erosion that occurs in North America?

   A. loss of the ozone layer
   B. wind and flowing water
   C. diversion of land for urban development
   D. unauthorized use of all-terrain vehicles

2. The use of proven soil management practices that reduce soil losses due to erosion is called:

   A. chemigation.
   B. protectionism.
   C. restoration.
   D. soil conservation.

3. A farming practice that reduces soil erosion by farming around slopes instead of up and down the slope is:

   A. contour farming.
   B. minimum tillage.
   C. no-till farming.
   D. the conservation reserve program.

4. A farming practice in which crops are planted without tilling the residues from the previous crop into the soil is:

   A. contour farming.
   B. minimum tillage.
   C. no-till farming.
   D. the conservation reserve program.

5. A farming practice in which the soil is tilled, fertilizer is incorporated, and seeds are planted in the same field operation is:
   A. contour farming.
   B. minimum tillage.
   C. no-till farming.
   D. the conservation reserve program.

6. The withdrawal of highly erodible land from crop production, and the planting of permanent cover crops and grasses for which a government incentive payment is provided is:
   A. contour farming.
   B. minimum tillage.
   C. no-till farming.
   D. the conservation reserve program.

7. Which of the following activities *is not* in harmony with the multiple-use concept of management for public lands?
   A. livestock grazing and wildlife habitat
   B. irrigation water storage and water sports
   C. logging activities and hunting
   D. restrictions for all uses except wildlife habitat

8. A natural drainage area where precipitation is absorbed into the soil to be released slowly from natural springs into streams is called a(n):
   A. watershed.
   B. forest.
   C. aquatic zone.
   D. biome.

9. Which of the following is a beneficial effect that is sometimes derived from fires on forested land?
   A. destruction of mature trees
   B. thinning of young trees
   C. loss of protective soil cover
   D. damage to wildlife habitat

10. Which of the following pollutants of surface water is most often attributed to agricultural practices?
    A. nitrates and phosphates
    B. acid rain
    C. petroleum products
    D. caustic chemicals

11. Water that is stored in natural underground basins or aquifers is called:
    A. artesian water.
    B. surface water.
    C. ground water.
    D. mineral water.

12. Smog is a form of pollution that is observed in:
    A. aquifers.
    B. surface water.
    C. sparsely populated areas.
    D. the atmosphere of densely populated areas.

13. Atmospheric pollutants that contain nitrogen and sulfur compounds combine with precipitation to form:
    A. acid rain or precipitation.
    B. recycled plant nutrients.
    C. mineral water.
    D. organic fertilizer.

14. A living organism whose population has been depleted to the extent that the entire population is likely to die is classified under the Endangered Species Act as a species that is:
    A. rare.
    B. threatened.
    C. endangered.
    D. extinct.

15. The greatest single cause leading to extinction of a population of organisms is:
    A. low biotic potential.
    B. nonadaptive behavior.
    C. a high degree of specialization.
    D. destruction of habitat.

LEARNING ACTIVITIES

1. Invite a local conservation officer to visit the class to discuss the effects of pollution on the environment and the creatures that live in it.

2. Survey your community to identify existing or potential environmental problems. Plan and carry out a conservation or restoration project that will solve or reduce one of the environmental problems identified in your survey.

Glossary

Abscisic acid—A natural plant hormone that prevents seeds from germinating until they are mature

Absorption spectrum—The visible range of colors of light that is reflected from an object instead of being absorbed

Acid rain—A condition that occurs when moisture in the air combines with sulphur or nitrogen compounds to form acid. It is a serious threat to plant and animal life.

Aerial mapping—A procedure that uses photographs taken from high flying planes or satellites to develop maps of fields and farms

Aerial seeding—The practice of applying seed to rough terrain by broadcasting it from aircraft

Aeroponics—A plant production system in which no soil is used, and a nutrient solution is sprayed at frequent intervals on the roots of plants that are suspended in the air

Agar—A nutrient gel, made from seaweed, that is used as a base for tiny plants generated using tissue culture techniques

Agricultural extension educator—University faculty members who transfer new technologies from the research labs to the field. They provide educational programs and expert consulting services to agricultural clients in their communities.

Agricultural research—The application of scientific principles to agricultural problems. It includes all activities that lead to greater understanding of plants and animals.

Agricultural technology—The tools and scientific processes used to produce agricultural products and prepare them for use by consumers

Agrobacterium tumefaciens—A naturally occurring strain of bacteria that is capable of inserting a portion of its genetic material into the chromosome of a plant cell

Algae—Tiny plants that multiply in water polluted by nitrates and phosphates that reduce the oxygen level in water causing threats to fish and other forms of life

Alien species—A species that is not native to an environment, and that competes with native species for available food and shelter

Alternative agriculture—A system of farming that attempts to maintain soil fertility using organic fertilizers and crop rotations

Alternating current (AC)—Electricity that changes its direction of flow many times per second

Alternator—A device that is used to produce electricity; often used to provide electric power for engines

Ambient temperature—The amount of heat in the environment that surrounds an organism

Amperage—The rate at which electricity flows through a circuit

AMS—Agricultural Marketing Service of the United States Department of Agriculture

Anaerobic respiration—A process also known as fermentation that occurs in the absence of oxygen and converts sugars obtained from plant materials into ethanol or lactic acid

Anaphase—A stage in cell division during which chromosome pairs pull apart and migrate to opposite sides of the cell

Anestrus—A time period, often initiated by changing seasons, when the ovaries of some female animals cease to produce ova

Anther—The plant organ in which pollen grains are produced

Antibiotic—A medicine that kills bacteria and other microorganisms that cause infections in animals

Aquaculture—The use of a water environment in the production of fish or aquatic animals and plants

Aquifer—A large body of water located underground

Arc welding—A process that applies heat to metals causing them to fuse or melt together

Armature—A magnet that rotates across lines of force in a motor or generator

Artificial insemination—The use of specialized equipment to place sperm in the reproductive tract of a female animal at a time when pregnancy can occur

Artificial vagina—A soft rubber cylinder containing a warm water jacket that is used to collect semen from a male animal

Asset—Anything that is owned and can be exchanged for something of value

ATP—Adenosine triphosphate, a high energy molecule that is produced during photosynthesis; source of energy for muscle movement in animals

Autosome—Any chromosome that is not a sex chromosome

Auxin—A natural plant hormone that causes shoots and stems to grow in length

Axis—An imaginary line around which an object rotates

Azolla—A small fern that is capable of nitrogen fixation and that is used as a source of nitrogen for rice plants

Bacteria—Microscopic single-celled organisms. Some forms cause diseases, but other types aid in digestion or fix nitrogen in a form that can be used by plants.

Balanced ration—A daily food supply that provides an animal with nutrients of the correct amount and proper quality to supply its nutritional needs

Basis—The difference in price between the local cash price of a commodity and the futures price on a given day

Bessemer process—A process that is used to make steel by blasting a jet of air through molten iron. A large amount of heat is generated, causing impurities in the iron to be burned.

Biodegradable—The tendency of materials to be decomposed by bacterial action

Biodiesel—A fuel similar to diesel that is obtained from vegetable oils

Biological control—A method of controlling undesirable insects and weeds by introducing their natural enemies into the environment

Biosphere II—An experimental prototype of a controlled living environment in which people, plants, and animals lived together in a sealed environment

Biotechnology—The use of engineering techniques to solve problems associated with living organisms

Birth rate—A measurement of the average number of births per year for females of breeding age

Blanch—To briefly submerge a product in a scalding hot water bath before freezing or packaging it

Bovine somatotropin—A growth hormone sometimes known as BST that is found in cattle. Injections of extra BST are used to increase the production of meat and milk.

Breed association—An organization that maintains records of animal pedigrees and issues certificates of registry to certify the pedigrees of animals

Breeding record—A record of the date of a mating and the identities of the dam and the sire

Budding—A form of grafting in which a live bud is attached to the upper shoot of an existing plant

Bulb—A short, starchy underground stem covered with layers or scales

Bulblet—Small bulbs that develop from the base of the bulb scales of the parent or original bulbs

Bulbourethral gland—One of the organs that functions to produce fluids that are added to the sperm in the male reproductive tract

Callus—Plant tissue that is not differentiated into leaf, root, stem, or other specialized tissue

Calvin cycle—A series of chemical reactions that convert carbon dioxide to simple sugars during the process of photosynthesis

Cambium—A porous layer beneath the bark of a woody plant through which water and nutrients are transported from the plant roots to the leaves

Carbon cycle—The recurring circular flow of carbon from living plant and animal tissues to nonliving materials

CD-ROM—Compact disk read only memory; a computer information storage system

Cell—A small structure that contains cytoplasm and a nucleus; the basic unit of life

Cell membrane—The outer flexible membrane that surrounds the contents of a cell

Cell wall—The outer layer or covering of a plant cell

Center pivot—An irrigation system that is self-propelled, moving continuously around a water source located in the center of the field

Centriole—A cell structure found at either end of a cell that anchors the fibers drawing chromosomes apart during cell division

Centromere—The point of attachment for a pair of chromatids

Cervix—A muscular female organ that seals off the passage from the uterus to the vagina during pregnancy

Chemical contaminant—A source of air pollution that is of a chemical nature such as ammonia or methane gas

Chemical control—The use of chemicals to control weeds and insects in crops

Chemical fertilizer—A plant nutrient produced chemically and derived from sources other than plants and animals

Chemigation—The practice of applying agricultural chemicals to crops by mixing them with irrigation water

Chlorophyll—A green substance found in plant cells that aids the plant in capturing energy from sunlight and converting it to sugars and starches through the process of photosynthesis

Chloroplast—A plant cell structure containing chlorophyll that converts raw materials to sugars and starches using energy obtained from sunlight

Chromatid—Half of a replicated chromosome

Chromosome—A genetic structure upon which genes are located. It is duplicated during the reproductive process.

Clone—An organism that is an exact genetic copy of another organism. It is derived from genetic material that can be traced to a single organism.

Cloning—A process through which genetically identical organisms are produced

Comfort zone—The range of environmental conditions to which an organism is best adapted

Commodity—Any item or basic agricultural product that is bought or sold

Commodity group—An organization of producers of a basic commodity who organize to promote the merits of their product

Compression ratio—A measurement of the volume inside the compression chamber of an engine at the beginning of the compression stroke in comparison with the volume inside the chamber at the end of the compression stroke

Computer—An electronic machine that is used to assemble, analyze and store data, and to perform complex calculations. It is also used to control machines as they perform tasks.

Computer network—A system in which information is shared between two or more computers through a direct cable connection or through the use of modems, utilizing telephone lines

Computerized records—Records that have been entered and stored in a computer

Conceive—To become pregnant

Conduction electrons—Electrons that flow through the network of atoms that makes up the wires and cables of electric power transmission lines

Conductor—A material through which electrons are capable of flowing

Confinement housing—Housing for livestock in which animals are managed in a controlled environment

Conservation reserve program (CRP)—A government program that provides payments to farmers who voluntarily remove highly erodible land from production of crops

Consumer goods—Processed products that have been converted to a form ready for use by consumers

Contour farming—A farming practice used on hilly fields in which field boundaries are laid out along elevation lines

Cooperative—A business organization that is owned by the people who use its services to collectively buy or sell commodities

Copulation—The act of transferring semen from the male reproductive tract to the reproductive tract of the female; natural breeding

Corm—A plant structure similar to a bulb except that it contains a longer stem and fewer scale leaves

Cotton gin—A machine that separates cotton fibers from cotton seeds

Cotyledon—The seed leaf or first leaf of a young plant

Cross-pollination—Transfer of pollen from the anthers to the stigmas of different plants of the same species

Crude oil—The form in which oil is found before it is processed or refined

Cryogenic liquid—A liquified gas such as liquid nitrogen or carbon dioxide that is used to freeze food products immediately on contact

Cryoprotectant—A product that is used to protect against cold temperatures

Cull—An inferior animal that is a candidate for removal from the herd

Cultivar—A particular variety of cultivated plant

Cutting—A plant part obtained from leaves, stems, roots, or buds that is treated with a hormone to stimulate rooting and growth of a new plant

Cycle—The production of electricity through one full revolution of a generator

Cytokinin—A plant hormone that acts to promote plant growth causing specialized plant tissues such as roots and shoots to develop

Cytoplasm—All of the structures and substances within a cell except for the nucleus

Database—The information contained in a set of computer files

Deficiency—The absence of an element that is essential to the health and production of an organism

Dehumidifier—A device that removes moisture from the air

Dehydration—Loss of water from plant or animal tissues; a method of preserving foods

Denitrification—A natural process by which nitrates are broken down by bacteria, resulting in nitrogen gas being released into the atmosphere

Depreciation—A reduction in the declared value of business assets as they diminish in value or as tax laws allow it

Dicot—A seed with two cotyledons

Diesel—Fuel oil that is a product obtained when crude oil is distilled. It is used by diesel engines as a fuel.

Diesel engine—An internal combustion engine in which fuel is ignited by injecting it into hot compressed air inside a sealed combustion chamber

Dike—A bank of soil that is built up to prevent water from flowing down a slope

Diploid—A cell that contains both homologues from each chromosome

Dipstick diagnosis—A method in which a plastic stick, coated with a specific monoclonal antibody, is used to diagnose the presence of disease organisms in body fluids

Direct current (DC)—Electricity that flows only in one direction

Direct sales—A marketing method in which the producer of a product sells it to the final consumer of the product

Distillation—A process by which liquids having different boiling temperatures are separated

Division—A form of plant propagation in which underground stems and roots are cut into smaller pieces and planted to produce new plants

Documentation—Written instructions describing how to use computer software

Domestic market—A market located within the borders of a country

Donor—A female animal from which a living embryo is obtained for the purpose of transferring it to a less valuable female

Dormant—A live but inactive state in which life processes slow down

Draft animal—An animal that provides power for performing work such as pulling farm implements

Drip irrigation—A form of irrigation that uses small flexible tubes to apply water to the root zone of each plant

Duty rating—A description of the ability of a motor to operate under a load

Dystocia—A condition in which a female animal experiences difficulty giving birth

Ecologist—A person who studies the relationships between the environment and living organisms

Economic threshold—The point at which the value of a loss equals the cost of controlling the factors that cause the loss

Ecosystem—All of the forms of life that inhabit a particular environment

Egg—A female gamete

Electricity—A form of energy that can be converted to light, heat, motion, and other useful purposes

Electromagnet—A magnet that consists of a conducting material coiled around an iron core through which electricity is passed to produce magnetism

Electronic mail—A computer network service that allows users to communicate by sending letters over the network

Electronic sensor—A device that is used to activate electric switches when changes or specific events occur in the environment where the switch is located

Electroporation—The use of electricity to make holes in cell membranes, allowing genes from other sources to be inserted in the cell nucleus

Elemental cycle—The recurring circular flow of elements from living organisms to nonliving materials and back again

Embryo—A tiny immature plant or animal

Embryo sac—A female gamete in plants consisting of the cell mass located in the ovule that develops into the embryo and the endosperm of a seed after fertilization

Embryo splitting—A form of cloning that is accomplished by dividing a growing embryo into equal parts using a surgical procedure performed with the aid of a microscope

Embryo transfer—This process is also referred to as ova transfer. It is a procedure for placing living embryos obtained from a donor animal in the reproductive tract of a recipient female animal.

Endangered species—Populations of organisms that are in danger of becoming extinct

Endosperm—The food supply for the tiny plant embryo that emerges from a seed when a plant germinates

Enterprise analysis—An accounting procedure that examines the profitability of a distinct part of a business

Entomology—The branch of science that studies insects

Entomologist—A scientist who studies insects

Environmental conditions—Factors in the area surrounding living organisms such as temperature, humidity, ventilation, and pollution that have an effect on the ability of an organism to survive

Environmental management—A system used to control some or all of the factors that maintain a stable life support system

Enzyme—A natural substance that acts as a catalyst to stimulate life processes when it is present in cells

Erosion—A force that removes topsoil and reduces soil fertility

Essential element—A nutrient that is required to promote the health, growth, and reproduction of an organism

Estrus—Heat period when a female animal becomes receptive to a male animal for the purpose of mating

Ethanol—An alcohol used for fuel that is produced by fermenting grain or other carbohydrates followed by distillation of the liquid

Ethylene—A natural plant hormone that causes fruits to soften and become ripe

Extender—A nutrient solution added to semen to dilute it and increase its volume

Extinct—Death of all members of a species

Eye—A bud on a potato tuber

Farmer's market—A form of marketing in which fresh farm produce is sold directly to the final consumers of the product at roadside stands or from market stalls located in urban population centers

FAX machine—A machine that transmits copies of documents to distant sites by sending signals between machines over a telephone line

Fermentation—A form of food preservation in which bacteria, yeasts, or enzymes are used to modify the chemical makeup of products

Fertigation—The distribution of liquid fertilizers through irrigation water

Fertility—The capacity of an animal to produce offspring

Fertilization—The fusion of male and female gametes; adding plant nutrients to soil

Fetus—An immature offspring of an animal that is developing inside the female uterus

Fiber—Materials supplied by plants and animals that can be used to produce cloth and other similar products. It may include but is not limited to cotton, linen, silk, wool, and mohair.

Field drain—A large porous pipe installed underground into which excess water seeps, allowing it to be removed from the root zone

Filament—The thin metal wire in a light bulb that gives off light when it is heated; the stalk of a flower stamen

Flood irrigation—A form of irrigation in which controlled flooding is used to apply water to a field

Flow cytometry—A process that utilizes fluorescent dyes and stains to separate sperm cells containing the sex chromosome from those that lack it

Fluorescent lamp—A specially constructed glass tube used to convert electricity to light

Follicle stimulating hormone (FSH)—A reproductive hormone that causes the female ovary to increase its production of ova or eggs

Food additives—Materials added to processed foods to reduce spoilage, improve quality, or add color

Force converter—A device that changes energy to motion or that modifies one kind of motion to another kind of motion

Forward contract—A contract to deliver a commodity of a certain quality grade at a later date for a specific price

Fossil fuel—A fuel obtained from coal or petroleum

Four-cycle engine—An engine in which each piston operates through four strokes during each complete cycle

Freeze-drying—A dehydration process in which products are frozen in a vacuum to draw off moisture
Freezing point liquid—A liquid that contains a substance such as sugar, salt, or glycerol that lowers its freezing temperature below 32 degrees for the purpose of quickly removing heat from food products during processing
Fruit—A ripened plant ovary surrounding a seed
Fulcrum—The point of support for a lever used to pry against an object
Futures contract—A legal obligation to deliver or accept delivery of a certain amount and quality of a commodity at a specified location and price during a particular month in the future
Futures trading—A marketing strategy in which contracts for agricultural commodities are bought and sold for delivery at a future date

Gamete—A haploid reproductive cell
Gametophyte—Same as gamete
Gasohol—A fuel that contains a mixture of ninety percent gasoline and ten percent ethanol
Gasoline—An important engine fuel obtained by distilling crude oil
Gasoline engine—An internal combustion engine that uses an electric spark to ignite a mixture of compressed air and gasoline
Gender selection—Managing the reproductive process to produce animals of the desired sex
Gene—A genetic structure that controls the traits and characteristics expressed in an organism
Gene mapping—The process of finding and recording the locations of genes on a chromosome
Gene pool—All of the genes that are present in a population of genetically similar organisms
Gene splicing—The process of removing a gene from its location on a chromosome and replacing it with another gene
Generative nucleus—One of two nuclei found in a pollen grain from which two sperm are produced
Generator—A device that converts the energy obtained from fuel, flowing water, or other sources into electricity
Genetic code—A sequence of genes on a chromosome that determines the inherited characteristics of an organism
Genetic engineering—The practice of modifying the heredity of an organism by inserting new genes from other organisms into the chromosome structure
Genetics—The study of inheritance or the manner in which traits found in parent organisms are passed to the offspring
Geographic information system (GIS)—The use of satellite and computer technology to map crop production and other field data for multiple sites in a field
Geothermal energy—Energy that is obtained from water that has been heated by the hot molten interior of the earth
Germination—The process by which a seed sprouts and begins to grow
Gibberellin—A natural plant hormone that causes the stored food in seeds to be converted to plant nutrients for use during germination and seedling growth
Global positioning system (GPS)—The use of satellite technology to accurately and consistently identify exact field locations
Glycolysis—The first stage of anaerobic fermentation during which sugar molecules are converted to pyruvic acid
Golgi apparatus—A network of fibers, rods, and granules found in cell cytoplasm
Grafting—A plant propagation procedure in which parts of two or more plants are united in a manner that allows their tissues to grow together forming a modified plant
Greenhouse—A structure designed to create an indoor growing environment for plants
Ground water—Water obtained from an aquifer

Habitat—The living environment of an organism
Handline—Sprinkler irrigation pipes that are constructed to be light in weight allowing them to be moved using hand labor
Haploid—Cells that contain a single chromosome from each homologous pair
Hard drive—An internal computer information storage device
Hardware—Mechanical and electronic devices and structures that are used to construct a computer
Hardwood—Plant materials obtained from mature plant parts; wood from a broad-leaved tree
Headgate—A mechanical device that controls the flow of water

Heat element—A special conductor that converts electricity to heat

Hedging—Trading in futures contracts

Herbicide—A chemical that is used to kill plants

Heredity—The passing of genetic traits and characteristics from parents to their offspring

Heritability factor—A numerical measurement of the tendency for an organism to pass characteristics on to its offspring

Heterosis—The tendency of plants and animals to produce more efficiently when they are the offspring of unrelated parents. This phenomenon is also known as hybrid vigor.

Home page—A computer access point on the internet system that provides information and access to computer files

Homologous chromosome—Each chromosome of an identical pair of chromosomes

Homologue—Each chromosome of an identical pair of chromosomes

Hormone—A substance produced in the body and carried by body fluids to tissues where it causes specific body functions to occur

Humidifier—A device that adds moisture to the air

Humidistat—A sensing device that detects changes in the moisture content of the air

Humidity—The amount of moisture that is present in the air

Hybrid—A cross between two similar organisms having different genetic makeups

Hybrid seed—Plant seed that is obtained by crossbreeding two pure but different strains of similar plants

Hybrid vigor—The tendency of plants and animals to produce more efficiently when they are the offspring of unrelated parents. This phenomenon is also known as heterosis.

Hybridoma—A cell that has been created by fusing a cancer cell to a cell that produces a useful antibody. It is used to produce large quantities of known antibodies that are used to control diseases.

Hydraulic power—The use of a fluid to transfer force from one location to another location

Hydraulic ram—A device that converts fluid force to mechanical force

Hydroelectric power—Electricity generated from the energy of moving water

Hydrogen ion—A hydrogen particle with a positive charge that is produced during photosynthesis as water molecules are split

Hydrogenation—A processing method that is used to preserve fats and oils by adding hydrogen to unsaturated chemical bonds

Hydrologist—A scientist who studies characteristics of water, including water quality

Hydroponics—A plant production system in which plant nutrients are provided in a water solution, and plants are grown without soil

Ice-minus bacteria—A genetically engineered bacteria that makes it possible for plants to tolerate temperatures several degrees below freezing

Immersion freezing—A process by which foods are dipped in liquids at temperatures below the freezing point of water

Implant—A device that is placed under the skin or in the female reproductive tract to slowly release hormones into the bloodstream of the animal

Implement—A device or machine used in the production of crops

Incandescent lamp—A glass bulb containing argon or nitrogen in which a thin electrical conductor or filament converts electricity to light

Index—A number that expresses the value of an animal in comparison with its herdmates

Induction—A process by which electrical or magnetic lines of force cause electricity to be generated or magnetism to occur in a conductor

Industrial revolution—The growth of manufacturing industries that resulted from the development of new technologies that improved production of industrial products

Information Age—The era in which we now live when nearly any information that is wanted or needed can be obtained by anyone who knows how to access the internet or other sources of computerized information

Information highway—Electronic services to which customers can subscribe to gain access to the internet system

Infrared photography—A method of photography that measures and records the amount of heat that is reflected off the surfaces of an object or organism

Infrared radiation—A form of energy that is detected as heat by special photographic film, and that can be used to map soil types and identify the locations of stressed plants

Infrared thermometry—A system that uses a special thermometer to scan a field of growing crops to identify problem areas

Inheritance—The manner in which traits found in parent organisms are passed to the offspring

Integrated pest management (IPM)—The use of natural insect enemies and limited chemical applications to control harmful insects while providing protection for useful insects

Internal combustion engine—An engine that burns fuel in a closed chamber and generates power from the expansion of heated gases

International market—A market located in another country

Internet—An international computer network on which information can be communicated to and accessed from the institutions, companies, and individuals who subscribe to the system

Inventory management—A system for managing business assets

Interphase—A resting or nonreproductive stage in the life span of a cell

Irradiation—A food processing method that uses radiant energy to kill microorganisms that cause food to spoil

Irrigation—The practice of applying water to land in areas where natural precipitation is not sufficient to produce a crop

Irrigation furrow—A water control structure resembling a small ditch that is created beside a row of plants

Land grant university—A system of education that started when Congress provided land and money to the states to establish agricultural colleges and universities

Laptop—A small portable computer that is equipped with a battery as a remote source of power

Laser—A narrow and intense beam of light that travels in a straight line from its source

Laser technology—The use of lasers to perform tasks

Lateral move sprinkler—An irrigation system that moves continuously across a square or rectangular field

Least cost ration—A ration composed of feed ingredients that satisfy the nutritional needs of an animal at the lowest cost

Ligase—An enzyme that aids in attaching a new gene to a chromosome in a location from which the original gene has been removed

Light reaction—A chemical reaction that occurs during photosynthesis in which energy is captured

Lines of force—Curving pathways of magnetic energy between the north and south poles of a magnet

LISA—Low Input Sustainable Agriculture is a farming strategy in which an attempt is made to reduce, but not eliminate, the use of agricultural chemicals to maintain soil fertility and control weeds and insects.

Locus—A specific site on a chromosome where the gene for a trait is located

Low pressure sprinkler—A form of irrigation that operates using lower water pressures than standard sprinkler irrigation systems, and that delivers water to the crop as a fine spray

Low biotic potential—A slow rate of reproduction due to few offspring, late reproductive maturity, and/or long time intervals between reproductive events

Lyophilization—The process of freeze-drying

Macronutrient—Any of six elements that is required by plants in moderate amounts

Magnet—An iron or steel object occupied by lines of force between the poles; sometimes placed in the stomach of a cow to trap and hold metal objects swallowed by the cow as she eats

Magnetic field—The area that is occupied by lines of force between the poles of a magnet

Magnetism—A force that attracts iron or steel

Main line—A large pipe through which water is delivered from its source to a sprinkler irrigation system

Marketing—All of the business activities involved in moving products from the producer to the consumer

Mature equivalent—A production index based on known production trends that takes into account the expected production of an animal at maturity

Mechanical pest control—Management of pests using machines to destroy or remove weeds and harmful insects from crops

Mechanical power—Force that is generated by using a tool or device to transfer or apply energy to a task, or to transfer a force from its source to the place where it is used

Megaspore—A haploid plant cell that occurs during the formation of female gametes

Megaspore mother cell—A diploid plant cell from which a female gamete is formed

Meiosis—A cell division process through which the number of chromosomes is reduced by half during the formation of male and female gametes

Metaphase—An intermediate step in cell reproduction during which the chromosomes become aligned at the center of the cell

Meteorologist—A person who collects and analyzes weather data, and who reports and predicts the weather

Methane gas—A component of natural gas that can also be generated from decomposing plant materials

Micronutrient—Any of several elements that is used by plants in small amounts

Micropyle—A small opening through which pollen enters the ovule of a flower during fertilization

Microspore—A haploid plant cell that divides to form a pollen grain

Microspore mother cell—A diploid plant cell from which male gametes are formed

Minimum tillage—A form of seedbed preparation in which only limited preparation of the soil precedes the planting of the crop

Mitochondrion—Cell structures that control enzyme activities

Mitosis—A type of cell division that occurs in an animal or plant, resulting in growth

Modem—An instrument that makes it possible for computers to exchange information using a telephone line

Monoclonal antibody—Specialized proteins that bind to specific disease organisms causing them to become incapable of causing disease

Monocot—A seed with a single cotyledon

Multiple-use—A form of management in which public lands are used by several user groups

Multi-spectral imagery—A system that uses computers to enhance special aerial photographs from which data describing soil types and plant vigor are recorded on field maps

Mutant gene—A gene that is different from the parent gene from which it was derived

NADPH—A high energy molecule that is produced in plant cells during photosynthesis

Nameplate—A metal plate fastened to a motor containing information about the motor

Natural selection—The tendency for important survival traits to appear in an increasing number of organisms in the population

Nitrates—Nitrogen compounds that are in a form that is easily used by plants; these compounds are known to cause serious pollution problems in water

Nitrogen cycle—The recurring circular flow of nitrogen from living plant and animal tissues to nonliving atmospheric nitrogen

Nitrogen fixation—A process by which certain strains of bacteria convert nitrogen gas from the atmosphere to nitrogen compounds useful to plants

Nodule—A site on the root of a legume plant in which colonies of nitrogen fixing bacteria live

Nonadaptive behavior—Failure of a species to adjust their living habits to a changing environment

Nonrenewable resource—A resource that is formed slowly and cannot be replaced once it is used up

Northern hemisphere—The part of the world located north of the equator

Notebook—A small portable computer that is equipped with a battery as a remote power source

No-till farming—A farming system in which land is not tilled prior to planting. The new crop is planted in the stubble of the old crop.

Nuclear energy—Heat that is released from an atom during a nuclear reaction

Nuclear waste—Radioactive waste materials including spent fuel rods that are by-products of controlled nuclear reactions

Nucleoplasm—The material that makes up a cell nucleus

Nucleus—A structure in a cell that contains hereditary materials

Offal—Animal body parts and other waste materials obtained from animal processing plants

Ohms—A measurement of the resistance within a circuit to the flow of electricity

Options trading—A form of marketing in which a producer buys an option giving him/her the right to purchase a futures contract for a commodity at a specific price by a specific future date

Organic farming—Farming without the use of chemicals to control weeds and insects; using only organic fertilizers and pest controls in the production of crops

Organism—An individual plant, animal, or other form of living thing that is capable of performing the activities necessary to sustain life

Ova—More than one ovum

Ova transfer—A procedure (also called embryo transfer) for placing living embryos obtained from a donor animal in the reproductive tract of a recipient female animal

Ovary—The primary female organ in which female gametes are produced

Oviduct—A tubular female organ in animals that transports the ovum from the ovary to the uterus

Ovule—An immature female germ cell in a plant

Ovum—A single egg from the female ovary

Packaging—Materials that are used to wrap or seal a commodity or product in a container

Parasite—Organisms that live in the bodies of animals or in plants that harm the hosts by feeding on fluids or damaging tissues and organs

Particulate matter—Tiny particles of dust suspended in the air

Parturition—The birth process

Parturition record—A record of an animal's date of birth and the identity of parents and offspring

Pasteurization—A food processing method that is accomplished by heating a product and maintaining high temperature long enough to kill the bacteria in the food

Pedigree—A record of the ancestors of an animal

Pelvic measurement—A measurement of the inside pelvic area of a heifer or cow to predict her ability to give birth to a calf without requiring assistance

Perishable—The tendency of a product to become spoiled

Permeable—The capability of a membrane to allow fluids to pass through it

Pessary—A sponge or other material to which hormones have been added and that is implanted under the skin or inserted in the female reproductive tract

Petiole—A plant leaf

Petiole testing—A procedure, sometimes called tissue analysis, in which plant tissue is tested in the laboratory to determine nutrient deficiencies in crops

Petroleum—A substance such as gasoline, diesel, or oil that is derived from crude oil

pH—A measurement of how basic or acidic a solution is

Pheromone—A chemical substance that is used by animals and insects to attract mates through their sense of smell

Phosphate—Phosphorous compounds that are essential nutrients to plants; compounds that are serious pollutants when they are dissolved in water

Photoelectric cell—A device that is used to activate an electrical circuit either in the presence or the absence of light

Photon—A unit of light energy

Photoperiod—The length of time in a day that organisms are exposed to light

Photosynthesis—A process that uses chlorophyll to capture energy from the sun and combine it with carbon dioxide, water, and nutrients to form plant tissues

Pistil—Female reproductive structures found in a flower

Placenta—An organ that encloses the developing fetus during pregnancy inside the female uterus and that transports nutrients and fetal waste between the blood supplies of the mother and her unborn baby

Plant growth regulator—Chemical compounds found in plants that control the growth of plants and the development of fruits and seeds

Plasmid—A chromosome that is found in bacterial cells which forms a circular ring

Pneumatic power—The use of compressed air to deliver power to a location that is remote to the power source

Polar nuclei—Haploid nuclei that develop during the formation of female gametes

Pollen—Male sex cell in plants that fertilizes female flower parts to produce fruits and seeds

Pollination—A stage in the reproductive process in plants; the transfer of pollen from the anther to the stigma within a flower or between flowers

Pollinator—An insect that aids the fertilization process by carrying pollen from one flower to another as it gathers nectar

Potential difference—A difference in voltage that exists between two points in a conductor

Power—Force or energy that is controlled to accomplish work

Precision farming—A crop management system that adjusts applications of fertilizers and other crop inputs on the basis of production differences that exist within a field

Prescription farming—The use of global positioning systems, geographic information systems, and computer management to match fertilizer, seeding, and chemical rates to the production of each specific site in a field

Primary nutrient—Any of three elements, nitrogen, phosphorous or potassium, that is required for plant growth and reproduction

Progeny—The offspring of an animal

Progeny testing—An evaluation system that determines the value of a breeding animal based on the production and performance of its offspring

Proof—A record of productive performance by breeding animals

Propagation—A process by which an organism reproduces

Prophase—The first stage of active cell reproduction

Prostate—An organ that functions to produce fluids that are added to the sperm in the male reproductive tract

Protoplasm—All of the structures and substances located within the cell

Pyruvic acid—An organic acid that is formed from sugar molecules during the glycolysis phase of the fermentation process

Quality—A measurement of how well a product conforms to the standards of excellence

Quality control—Testing of both raw and consumer ready products to ensure that they meet uniform quality standards

Quantum theory—A scientific theory proposed by Albert Einstein that describes the way atoms absorb and radiate energy

Radiant heat—Heat that is projected from the surface of an object or from living organisms

Radiation—Emission of potentially harmful nuclear particles from atoms and molecules of certain elements such as plutonium, uranium, radium, and similar materials

Ration—The amount of feed consumed by an animal in a day

Raw products—Unprocessed agricultural products as they are marketed following harvest or storage

Reaper—A machine, invented by McCormick in 1831, that cut stalks of ripened grain and tied them in bundles

Receptacle—The base of a flower

Recipient—A female animal in which a living embryo from another female is placed resulting in pregnancy and the birth of offspring that are not her genetic sons or daughters

Rectal probe—A device that is used to electrically stimulate a male animal to ejaculate for the purpose of collecting semen

Recycled nutrients—Undigested materials that are obtained from animal wastes, and used as feed to provide nutrients in the ration of an animal

Recycled waste—Animal and plant wastes, such as manure or animal and plant parts obtained from processing plants, that are useful in making a variety of products

Refrigeration—A process for removing heat from the atmosphere in a closed environment

Regeneration—The ability of a plant to replace missing parts by growing new parts

Relay—An inductive switch that is used to turn power on or off from remote or distant locations

Renewable resource—A resource that is not used up, but continues to produce more of its kind at regular intervals

Resistance—A reduction in flow of energy or materials due to friction

Respiration—A process in which energy and carbon dioxide are released due to digestion or the breakdown of plant tissues during periods of darkness

Restriction enzyme—A specific enzyme capable of removing a particular gene from its location on a chromosome

Retortable pouch—A food package made of three layers that consist of polyester, aluminum, and plastic

Rhizobium—The most important of the nitrogen-fixing bacteria that is capable of converting atmospheric nitrogen to nitrogen compounds that can be used by plants

Rhizome—A horizontal stem that grows underground

Robot—A computerized machine programmed to perform a specific task

Robotics—A study of the design and use of robots

Rootstock—The lower stem and root system of a grafted plant

Saline soil—Soils in which high concentrations of salts exist

Scale—Equipment that determines weights of animals, feed, harvested crops, and other materials

Scion—A short piece of shoot on which buds are present and from which the upper growth of a grafted plant is obtained

Seed—A mature ovule consisting of a tiny plant embryo and a food supply surrounded by a protective outer coat

Seed coat—A protective covering for the embryo and endosperm of a seed

Seed plate—A device shaped like a gear that is used to control the seeding rate of a planter. Different sizes of seeds may be planted using the same planter by changing the seed plates.

Seeding rate—The amount of seed that is planted per acre

Seedstock—High quality animals or plants that are used to produce breeding animals or plant seeds for use in commercial production

Selective breeding—A system of reproduction in which plants and animals are selected for breeding because they possess important traits or because they do not possess a harmful trait

Self-pollination—The transfer of pollen from the anther to the stigma of the same flower

Semen—The combination of sperm cells and the fluids from the male reproductive glands

Semiconductor—A material that is a poor conductor until it is acted upon by heat, light, or electricity

Sensor—Electrical devices that are used to detect changes in the environment

Separation—A form of asexual plant propagation in which new plants are grown from new generation bulbs and corms that have been separated from the parent bulbs and corms

Service factor—A rating of the capacity of a motor to handle an overload

Setting—The length of time that an irrigation system is operated in a single location; programmed instructions that control electronic devices

Sex chromosome—The chromosome in the cell nucleus that determines the sex of an animal

Sexual reproduction—The production of male and female gametes and the process by which they join together to produce offspring

Shelf life—The length of time that a perishable product can be stocked in a store before it loses quality

Short circuit—A leak in an electrical line that allows electricity to flow out of the desired path

Sine wave—A wave of electricity consisting of one positive and one negative pulse that is generated during a cycle corresponding with one full revolution of a generator

Site-specific farming—A crop management system that adjusts applications of fertilizers and other crop inputs on the basis of production differences that exist within a field

Slash—Waste materials in a forest area that has just been harvested

Smog—High concentrations of chemical pollutants and particulate matter suspended in the air

Smoke jumper—A fire fighter who parachutes from an aircraft to a site near the location of a fire

Software—Computer programs, data, and routines stored on magnetic tapes, disks, or CD-ROMs

Softwood—Immature plant materials obtained from new growth; wood from a conifer or needle-bearing tree

Soil conservation—Soil management practices that prevent soil losses and maintain fertility

Soil fertility—The amount and availability of plant nutrients in the soil

Solar cell—A device that uses silicon to capture and convert light energy to electricity

Solar energy—Energy obtained directly from the sun

Solar panel—A device that traps energy from the sun in the form of heat or light

Solenoid—An inductive device that converts electrical energy to lateral motion

Solid set—An irrigation system consisting of small diameter irrigation pipes that are placed at permanent field locations at the beginning of the irrigation season

Somatic embryogenesis—A process that is used to generate embryonic plants coated with a synthetic polymer material. The product is referred to as artificial seeds.

Southern hemisphere—The part of the world located south of the equator

Sperm—Male reproductive cells

Spindle—A bundle of fibers that functions to separate chromosome pairs during cell division

Spore—A male reproductive cell in a plant

Sprinkler irrigation—An efficient method of delivering water to crops using pipes. Water pressure is developed by pumps or by gravity, and water is sprayed on growing crops from special nozzles or sprinkler heads.

Stamen—The male part of a flower that consists of the anther and the filament

Static environment—An environment in which conditions tend to remain unchanged

Steam engine—An engine that converts steam pressure to motion

Stigma—A female flower part that functions as a pollen receptor

Style—A female flower part that connects the stigma to the ovary

Sublimation—The process by which a solid is converted to a gas without passing through the liquid state

Surface water—Water flowing in streams, rivers, and lakes as it moves across the land surface to the ocean

Switch—A device that opens or closes an electrical circuit

Symbiosis—A phenomenon in which two life forms exist together and each organism benefits from the presence of the other

Symbiotic relationship—A phenomenon, often called symbiosis, that occurs when two life forms exist together and each organism benefits from the presence of the other

Synchronize—To treat donor and recipient females with hormones to control and coordinate their reproductive cycles

Synthetic fertilizer—A plant nutrient produced through a chemical process from inorganic materials

Synthetic hormone—A hormone produced or refined in a laboratory

Tax management—A system for managing business assets, income, and expenses for tax purposes

Telemarketing—A telephone auction sale of animals or other commodities that have been inspected and graded prior to the sale

Teletype—A form of communication that transmits information over telephone lines to an automated printer on which copies of the information are printed

Telophase—The last stage of cell division during which cell cytoplasm is divided as two new cells begin to form

Temperature rise—A rating of how hot a motor is capable of operating above the temperature of the surrounding environment

Terrace—A bank of soil built up along the contour of a hill for the purpose of preventing soil erosion by trapping excess water and holding it until it is absorbed into the soil

Tetrad—A cluster of four haploid cells

Textile industry—An industry that processes fiber into threads, yarns, cloth, and similar products

Thermostat—A sensing device that is used to detect changes in temperature

Threatened species—Species of living things that require special protection to maintain the population

Tillage—Preparation of the soil for planting using machines that reduce the sizes of soil particles

Tillage tools—Farm implements or machines that are used to destroy weeds or prepare the soil for planting

Tissue analysis—A procedure, sometimes called petiole testing, in which plant tissue is tested in the laboratory to determine nutrient deficiencies

Tissue culture—The development of roots, stems, and leaves from callus tissue using a solution containing nutrients and hormones

Torque—The turning power that is exerted on or by an object

Transgenic animal—An animal that has developed from a genetically modified cell to which a gene has been transferred from another living organism

Tuber—Specialized underground plant stems or shoots that are distinguished by the presence of buds called eyes

Tuberous root—A specialized plant structure or root that has root growth on one end and buds on the other end

Turbine—A machine that converts the energy of falling water to rotary motion for the purpose of turning electrical generators

Two-cycle engine—An engine in which each piston operates through two strokes during each complete cycle

Uniformity—A measurement of sameness

Urethra—A tubular tract that provides a passage for the elimination of urine in most animals and through which semen is discharged in the case of males

Uterus—The female organ in animals in which the fetus is nourished and where it grows from the time of conception until birth

Vaccine—A substance that produces immunity to diseases that are known to cause death or illnesses in animals

Vacuole—A cell structure that gathers excess water and wastes that are discharged through the cell wall

Vagina—The female organ in animals located between the external genitals and the uterus

Variable switch—A device that controls the rate at which electricity is allowed to flow through a circuit

Variety—Plants that have inherited the same genetic characteristics or traits and have descended from the same parent stock

Ventilation—The movement of fresh air into a closed environment and the movement of fouled air from the environment

Video-merchandising—A marketing technology that allows potential buyers in distant locations to inspect a sale offering by viewing a video tape prior to or during the sale

Visible spectrum—All of the colors of light that are visible to the human eye when the colors are separated by shining the light source through a prism

Voltage—The measurement of electrical pressure

Watershed—The area from which water drains as it emerges from springs and moves into the streams and rivers

Wheel line—An irrigation system using pipe similar to handlines except that each pipe is mounted in the center of a large lightweight wheel. The entire sprinkler line is rolled to a new setting using power from a hydraulic pump or a small gasoline engine.

Windmill—A machine that converts the energy of wind to rotary motion

Withdrawal period—The waiting period that must be observed following treatment of animals or crops with chemicals or medications before products can be safely marketed

Yield—The amount of a product that is obtained from a production unit

Zygote—A fertilized egg

Index

Italics indicate illustrations.